# CGEIT

## Review Manual

### 7th Edition

## About ISACA

Nearing its 50th year, ISACA® (isaca.org) is a global association helping individuals and enterprises achieve the positive potential of technology. Technology powers today's world and ISACA equips professionals with the knowledge, credentials, education and community to advance their careers and transform their organizations. ISACA leverages the expertise of its half-million engaged professionals in information and cybersecurity, governance, audit, assurance, risk and innovation, as well as its enterprise performance subsidiary, CMMI® Institute, to help advance innovation through technology. ISACA has a presence in over 185 countries, including more than 215 chapters and offices in both the United States and China.

In addition, ISACA advances and validates business-critical skills and knowledge through the globally respected Certified Information Systems Auditor® (CISA®), Certified Information Security Manager® (CISM®), Certified in the Governance of Enterprise IT® (CGEIT®) and Certified in Risk and Information Systems Control™ (CRISC™) credentials.

## Disclaimer

ISACA has designed and created *CGEIT® Review Manual 7th Edition* primarily as an educational resource to assist individuals preparing to take the CISA certification exam. It was produced independently from the CISA exam and the CISA Certification Committee, which has had no responsibility for its content. Copies of past exams are not released to the public and were not made available to ISACA for preparation of this publication. ISACA makes no representations or warranties whatsoever with regard to these or other ISACA publications assuring candidates' passage of the CISA exam.

## Reservation of Rights

© 2015 ISACA. All rights reserved. No part of this publication may be used, copied, reproduced, modified, distributed, displayed, stored in a retrieval system or transmitted in any form by any means (electronic, mechanical, photocopying, recording or otherwise) without the prior written authorization of ISACA.

## ISACA

1700 E. Golf Road, Suite 400
Schaumburg, IL 60173, USA
Phone: +1.847.660.5505
Fax: +1.847.253.1755
Contact us: https://support.isaca.org
Website: www.isaca.org

**Provide feedback:** https://support.isaca.org
**Participate in the ISACA Knowledge Center:** www.isaca.org/knowledge-center

**Twitter:** www.twitter.com/ISACANews
**LinkedIn:** www.linkd.in/ISACAOfficial
**Facebook:** www.facebook.com/ISACAHQ
**Instagram:** www.instagram.com/isacanews

ISBN 978-1-60420-373-8
*CGEIT® Review Manual 7th Edition*
Printed in the United States of America

CRISC is a trademark/service mark of ISACA. The mark has been applied for or registered in countries throughout the world.

This publication, *CGEIT® Review Manual, 7th Edition*, includes an excerpt from Professional Accountants in Business (PAIB) Committee's Enterprise Governance: Getting the Balance Right, and is used with permission of IFAC. Any views or opinions that may be included in this publication are solely those of the authors, and do not express the views and opinions of IFAC or any independent standard setting board supported by IFAC.

# CGEIT REVIEW MANUAL 7TH EDITION

ISACA is pleased to offer the *CGEIT® Review Manual 7th Edition*. The purpose of the manual is to provide CGEIT candidates with technical information and references to assist in the preparation and study for the Certified in the Governance of Enterprise IT (CGEIT) exam.

The *CGEIT® Review Manual* is the result of contributions of volunteers across the globe who are actively involved in the governance of enterprise IT and who have generously contributed their time and expertise. The *CGEIT® Review Manual* will be updated to keep pace with rapid changes in the field of IT governance. As such, your comments and suggestions regarding this manual are welcome. Once you have completed your exam, please take a moment to complete the online evaluation that corresponds to this publication (*www.isaca.org/studyaidsevaluation*). Your observations will be extremely valuable for the preparation of the 8th edition of the manual.

No representations or warranties are made by ISACA in regard to these or other ISACA/IT Governance Institute® (ITGI®) publications assuring candidates' passage of the CGEIT exam. This publication was produced independently of the CGEIT Certification Working Group, which has no responsibility for the content of this manual.

Copies of the CGEIT exam are not released to the public. The sample practice questions in this manual are designed to provide further clarity to the content presented in the manual and to depict the type of questions typically found on the CGEIT exam. The CGEIT exam is a practice-based exam. Simply reading the reference material in this manual will not properly prepare candidates for the exam. The sample questions are included for guidance only. Your scoring results do not indicate future individual exam success. You may also want to obtain a copy of the *CGEIT® Review Questions, Answers & Explanations Manual 4th Edition*, which consists of 250 multiple-choice study questions, answers and explanations.

Certification has resulted in a positive impact on many careers, including worldwide recognition for professional experience and enhanced knowledge and skills. CGEIT is designed for professionals who have a management, advisory or assurance role related to satisfying the IT governance needs of an enterprise. We wish you success with the CGEIT exam.

# ACKNOWLEDGMENTS

The *CGEIT® Review Manual 7th Edition* is the result of the collective efforts of many volunteers. ISACA members from throughout the global IT governance profession participated, generously offering their talent and expertise. This international team exhibited a spirit and selflessness that has become the hallmark of contributors to ISACA manuals. Their participation and insight are truly appreciated.

Special thanks go to Steven De Haes, Professor, University of Antwerp—Antwerp Management School, who worked on the 7th Edition of the *CGEIT® Review Manual*.

Special thanks go to the following reviewers of the 7th Edition of the *CGEIT® Review Manual*:
Miguel Garcia-Menendez, CISA, CISM, CGEIT, CRISC, Innovation & Technology Trends Institute (iTTi), Spain
W. Noel Haskins-Hafer, CISA, CISM, CGEIT, CRISC, CIA, USA
Ramaswami Karunanithi, CISA, CGEIT,CRISC, CA, CAMS, CBCI, CFE, CFSA, CGAP, CGMA, CIA, CMA, CPA,
　　CRMA, CSCA, FCS, Prince2 Practitioner, PMP, New South Wales Government, Australia
Sharanbir S. Khurana, CISA, CGEIT, CRISC, Alberta Gaming and Liquor Commission, Canada
Tapiwa Mvere, CISA, CGEIT, AgilePM, COBIT 5, ISO 27001, ITIL, Medscheme, South Africa
Manolo Palao, CISA, CISM, CGEIT, Accredited COBIT 5 Trainer, COBIT 5 Certified Assessor, P&T: S,
　　SLU | Innovation & Technology Trends Institute (iTTi), Spain
Upesh Parekh, CISA, India
Sylvia Tosar Piaggio, CGEIT, Uruguay
Ravikumar Ramachandran, CISA, CISM, CGEIT, CRISC, CAP, CEH, CFE, CHFI, CIA, CISSP-ISSAP, CRMA,
　　ECSA, FCMA, PMP, SSCP, Hewlett-Packard India Sales Pvt. Ltd, India
Nancy J. Thompson, CISA, CISM, CGEIT, PMP, Nancy Thompson Consulting, USA

ISACA has begun planning the 8th Edition edition of the *CGEIT® Review Manual*. Volunteer participation drives the success of the manual. If you are interested in becoming a member of the select group of professionals involved in this global project, we want to hear from you. Please email us at *studymaterials@isaca.org*.

# About This Manual .................................................................................................................................xiii
Overview .............................................................................................................................................................xiii
Organization of This Manual ..............................................................................................................................xiii
Format of This Manual ........................................................................................................................................xiv
About the CGEIT Review Questions, Answers and Explanations Manual ..........................................................xv

# Chapter 1: Framework for the Governance of Enterprise IT ..................................1

## Section One: Overview ..................................................................................................................................2
**Domain Definition** ................................................................................................................................................2
**Domain Objectives** ..............................................................................................................................................2
**Learning Objectives** ............................................................................................................................................2
**CGEIT Exam Reference** .....................................................................................................................................2
**Task and Knowledge Statements** .......................................................................................................................2
   Tasks ....................................................................................................................................................................2
   Knowledge Statements ........................................................................................................................................3
   Relationship of Task to Knowledge Statements ..................................................................................................5
**Self-assessment Questions** ................................................................................................................................6
**Answers to Self-assessment Questions** ............................................................................................................7
**Suggested Resources for Further Study** ..........................................................................................................8

## Section Two: Content ....................................................................................................................................9
**1.1 Components of a Framework for the Governance of Enterprise IT** .........................................................9
   Enterprise Governance .........................................................................................................................................9
   Conformance and Performance ...........................................................................................................................9
   Governance of Key Assets, Including IT ............................................................................................................10
**1.2 IT Governance Industry Practices, Standards and Frameworks** ..........................................................11
   Forest of Frameworks, Standards and Good Practices ....................................................................................11
      Governance of Enterprise IT ........................................................................................................................11
      Management of Enterprise IT .......................................................................................................................12
   COBIT 5 .............................................................................................................................................................12
**1.3 Business Drivers Related to IT Governance** ............................................................................................14
   Typical Pain Points .............................................................................................................................................14
   Trigger Events in the Internal and External Environments ................................................................................15
**1.4 IT Governance Enablers** ..............................................................................................................................16
   Holistic Approach of Structures, Processes and Relational Mechanisms ........................................................16
   COBIT 5 Enablers ..............................................................................................................................................17
**1.5 Techniques Used to Identify IT Strategy** ...................................................................................................19
   SWOT Analysis ..................................................................................................................................................20
   BCG's Growth Share Matrix ...............................................................................................................................20
   Other Models and Methods for IT Strategy .......................................................................................................21
**1.6 Components, Principles and Concepts Related to Enterprise Architecture** ........................................22
   Understanding Enterprise Architecture ..............................................................................................................22
   Layers of Enterprise Architecture ......................................................................................................................23
**1.7 Organizational Structures and Their Roles and Responsibilities** .........................................................24
   Specifics of Organizational Structures ..............................................................................................................25
   Structures as a Basis to Build RACI Charts ......................................................................................................25
   Governance of Enterprise IT Arrangements .....................................................................................................27
**1.8 Methods to Manage Organizational, Process and Cultural Change** .....................................................28
   Change Enablement ..........................................................................................................................................28
   Kotter's Implementation Life Cycle ....................................................................................................................29
   Lewin/Schein's Change Theory—Unfreeze–Change–Refreeze .......................................................................30

# Table of Contents

1.9 Models and Methods to Establish Accountability for Information Requirements,
Data and System Ownership, and IT Processes ........................................................................................... 30
    Distinction Between Governance and Management ........................................................................................ 31
    Governance and Management Roles, Activities and Relationships ................................................................ 31
    Accountability ................................................................................................................................................... 32
1.10 IT Governance Monitoring Processes/Mechanisms ..................................................................................... 32
    Importance of IT Performance Management .................................................................................................. 33
    Current IT Performance Management Governance Approaches ..................................................................... 33
    Good Practices for IT Performance Management ........................................................................................... 34
1.11 IT Governance Reporting Processes/Mechanisms ........................................................................................ 36
1.12 Communication and Promotion Techniques ................................................................................................ 39
    Importance of Communication and Marketing ............................................................................................... 39
    Communication Strategy and Plan ................................................................................................................... 39
    Content of Governance of Enterprise IT Communication Related to Risk ..................................................... 40
1.13 Assurance Methodologies and Techniques ................................................................................................... 40
    Components of Assurance Initiatives ............................................................................................................... 40
    IT Assurance Road Map ................................................................................................................................... 42
1.14 Continuous Improvement Techniques and Processes .................................................................................. 43
    Components of a Continuous Improvement Cycle .......................................................................................... 43
    Phases in an Implementation Life Cycle ......................................................................................................... 44
        Phase 1—What Are the Drivers? ............................................................................................................. 44
        Phase 2—Where Are We Now? ............................................................................................................... 44
        Phase 3—Where Do We Want to Be? ..................................................................................................... 44
        Phase 4—What Needs to Be Done? ........................................................................................................ 44
        Phase 5—How Do We Get There? .......................................................................................................... 44
        Phase 6—Did We Get There? .................................................................................................................. 44
        Phase 7—How Do We Keep the Momentum Going? ........................................................................... 44
**Endnotes** .................................................................................................................................................................. 45

## Chapter 2: Strategic Management ............................................................................................................... 47

### Section One: Overview ............................................................................................................................. 48
**Domain Definition** ................................................................................................................................................... 48
**Domain Objectives** ................................................................................................................................................... 48
**Learning Objectives** ................................................................................................................................................. 48
**CGEIT Exam Reference** ........................................................................................................................................... 48
**Task and Knowledge Statements** ............................................................................................................................. 48
    Tasks .................................................................................................................................................................. 48
    Knowledge Statements ..................................................................................................................................... 48
    Relationship of Task to Knowledge Statements ............................................................................................... 49
**Self-assessment Questions** ....................................................................................................................................... 51
**Answers to Self-assessment Questions** ................................................................................................................... 52
**Suggested Resources for Further Study** ................................................................................................................ 53

### Section Two: Content ................................................................................................................................. 54
**2.1 An Enterprise's Strategic Plan and How It Relates to Information Technology** ............................................ 54
    Strategic Alignment Model ............................................................................................................................... 54
    Strategic Fit ....................................................................................................................................................... 54
    Functional Integration ....................................................................................................................................... 55
    The Complexity of Strategic Alignment .......................................................................................................... 55

## Table of Contents

**2.2 Strategic Planning Processes and Techniques** .................................................................................................56
    The COBIT 5 Goals Cascade and Strategic Planning.........................................................................................56
        Step 1. Stakeholder Drivers Influence Stakeholder Needs..........................................................................57
        Step 2. Stakeholder Needs Cascade to Enterprise Goals............................................................................57
        Step 3. Enterprise Goals Cascade to IT-related Goals ...............................................................................58
        Step 4. IT-related Goals Cascade to Enabler Goals....................................................................................60
    Value of the COBIT 5 Cascade for Strategic Planning ......................................................................................64
**2.3 Impact of Changes in Business Strategy on IT Strategy** ..................................................................................65
    Agility ................................................................................................................................................................65
    Enterprise Agility...............................................................................................................................................66
    IT Agility............................................................................................................................................................66
    Agility Loops.....................................................................................................................................................66
**2.4 Barriers to the Achievement of Strategic Alignment**........................................................................................67
    Expression Barriers............................................................................................................................................67
    Specification Barriers ........................................................................................................................................67
    Implementation Barriers....................................................................................................................................67
**2.5 Policies and Procedures Necessary to Support IT and Business Strategic Alignment**...................................67
    Practices Supporting Strategic Alignment ........................................................................................................67
    Role of the IT Strategy Committee...................................................................................................................68
    Importance of Policies and Procedures.............................................................................................................68
        Policies........................................................................................................................................................68
        Procedures...................................................................................................................................................68
**2.6 Methods to Document and Communicate IT Strategic Planning Processes** ..................................................69
    Business Strategy and the Business Balanced Scorecard..................................................................................69
    IT Strategy and the IT Balanced Scorecard......................................................................................................70
**2.7 Components, Principles and Frameworks of Enterprise Architecture**..........................................................72
    Components of Enterprise Architecture ............................................................................................................72
    COBIT 5 View on Enterprise Architecture.......................................................................................................73
    Information Governance and Management ......................................................................................................74
**2.8 Current and Future Technologies**......................................................................................................................74
**2.9 Prioritization Processes Related to IT Initiatives**.............................................................................................77
    Investment Portfolio Categorizations................................................................................................................77
    IT-enabled Investment Programs.......................................................................................................................78
**2.10 Scope, Objectives and Benefits of IT Investment Programs** ..........................................................................79
    Current Practice in Business Case Development ..............................................................................................79
    Business Case Components ...............................................................................................................................79
    Business Cases as Operational Tools.................................................................................................................80
**2.11 IT Roles and Responsibilities and Methods to Cascade Business and IT Objectives to IT Personnel**............81
    Illustrating and Quantifying the IT Strategy.....................................................................................................81
    Continuous Communication ..............................................................................................................................81
    Focus on Explanation and Training ..................................................................................................................82
    Using a Participatory Style of Decision-making Process..................................................................................82
    Mastering the "Operational" Art.......................................................................................................................82
    Risk Considerations at the CIO Level ..............................................................................................................83
**Endnotes**.......................................................................................................................................................................83

*Table of Contents*

# Chapter 3: Benefits Realization ......85

## Section One: Overview ......86
**Domain Definition** ......86
**Domain Objectives** ......86
**Learning Objectives** ......86
**CGEIT Exam Reference** ......86
**Task and Knowledge Statements** ......86
   Tasks ......86
   Knowledge Statements ......87
   Relationship of Task to Knowledge Statements ......89
**Self-assessment Questions** ......90
**Answers to Self-assessment Questions** ......91
**Suggested Resources for Further Study** ......92

## Section Two: Content ......93
**3.1 IT Investment Management Processes, Including the Economic Life Cycle of Investments** ......93
   The Business Case ......93
   Program Management ......94
   Benefits Realization ......94
   Full Economic Life Cycle Management ......94
**3.2 Basic Principles of Portfolio Management** ......95
**3.3 Benefit Calculation Techniques** ......100
   The Investment Cycle and the Service Management Cycle ......100
   Financially Oriented Cost-benefit Techniques ......101
   Nonfinancially Oriented Cost-benefit Techniques ......101
**3.4 Process and Service Measurement Techniques** ......102
   Balanced Scorecard and Metrics ......103
   Process Capability Model ......104
**3.5 Processes and Practices for Planning, Development, Transition, Delivery, and Support of Solutions and Services** ......105
   ITIL and ISO 20000 as Reference Frameworks ......105
   Service Management Processes ......106
   ITIL 2011 ......107
   System Development Life Cycle and Agile Development ......108
   The Link Between Solution Delivery and Service Management ......108
**3.6 Continuous Improvement Concepts and Principles** ......109
   Six Sigma ......109
   Total Quality Management ......110
   Plan-Do-Check-Act ......110
**3.7 Outcome and Performance Measurement Techniques** ......110
   The Need for Metrics ......110
   Metrics Defined in COBIT ......111
   Metrics Defined in ITIL ......111
   SMART Metrics ......112
**3.8 Procedures to Manage and Report the Status of IT Investments** ......112
   Business Case Life Cycle ......113
**3.9 Cost Optimization Strategies** ......116

**3.10 Models and Methods to Establish Accountability Over IT Investments** ............................................................. 117
    Val IT as a Framework for GEIT and Value Management ............................................................................................ 118
    Case Study: Analyzing IT Value Management at the Dutch Airline Company KLM
    Through the Lens of Val IT ........................................................................................................................................... 119
        The Case Company: KLM ..................................................................................................................................... 120
        Value Governance at KLM .................................................................................................................................... 121
        Portfolio and Investment Management at KLM ................................................................................................... 122
        Reported Benefits, Lessons Learned and Future Challenges ................................................................................ 124
**3.11 Value Delivery Frameworks** ................................................................................................................................... 126
    Val IT Principles ............................................................................................................................................................ 126
    Val IT Domains .............................................................................................................................................................. 127
    Val IT Terminology ....................................................................................................................................................... 128
    Val IT Processes and Key Management Practices .......................................................................................................... 129
**3.12 Business Case Development and Evaluation Techniques** .................................................................................... 132
    How Business Cases Relate to Value Management ....................................................................................................... 132
    Business Case Development .......................................................................................................................................... 134
    Business Case Maintenance ........................................................................................................................................... 134
    Business Case Customization ........................................................................................................................................ 134
**Endnotes** ............................................................................................................................................................................ 136

## Chapter 4: Risk Optimization ............................................................................................................. 139

### Section One: Overview ............................................................................................................................ 140
**Domain Definition** ............................................................................................................................................................. 140
**Domain Objectives** ............................................................................................................................................................. 140
**Learning Objectives** ........................................................................................................................................................... 140
**CGEIT Exam Reference** .................................................................................................................................................... 140
**Task and Knowledge Statements** ..................................................................................................................................... 140
    Tasks ............................................................................................................................................................................... 140
    Knowledge Statements .................................................................................................................................................. 140
    Relationship of Task to Knowledge Statements ........................................................................................................... 143
**Self-assessment Questions** ............................................................................................................................................... 145
**Answers to Self-assessment Questions** .......................................................................................................................... 146
**Suggested Resources for Further Study** ........................................................................................................................... 147

### Section Two: Content ............................................................................................................................. 148
**4.1 The Application of Risk Management at the Strategic, Portfolio, Program,**
    **Project and Operations Levels** ............................................................................................................................ 148
    Risk and Risk Management ........................................................................................................................................... 148
    Risk Hierarchy ............................................................................................................................................................... 148
**4.2 Risk Management Frameworks and Standards** ................................................................................................. 150
    Risk IT Framework ........................................................................................................................................................ 151
    COBIT 5 for Risk ........................................................................................................................................................... 153
    COSO ERM Framework ................................................................................................................................................ 153
    ISO 31000:2009 Principles and Guidelines on Implementation of Risk Management ................................................ 154
    M_o_R Framework ....................................................................................................................................................... 155
    OCTAVE ........................................................................................................................................................................ 156
    Other Risk Management Standards and Framework .................................................................................................... 157
**4.3 The Relationship of the Risk Management Approach to Legal and Regulatory Compliance** ........................... 158
    Enterprise Goal Categories ............................................................................................................................................ 158
    Objective Setting ............................................................................................................................................................ 159
    Critical Success Factors ................................................................................................................................................. 159

*Table of Contents*

**4.4 Methods to Align IT and Enterprise Risk Management** ................................................................................................160
    IT Risk in the Risk Hierarchy..........................................................................................................................................160
    The Risk IT Framework.................................................................................................................................................161
**4.5 The Relationship of the Risk Management Approach to Business Resiliency** ................................................................161
    Resilience......................................................................................................................................................................162
    The Business Continuity Process...................................................................................................................................162
    ISO 22301:2012—Societal Security—Business Continuity Management Systems ....................................................164
    Other Business Continuity Standards............................................................................................................................164
**4.6 Risk, Threats, Vulnerabilities and Opportunities Inherent in the Use of IT** .....................................................................165
    Risk Categories .............................................................................................................................................................165
    Risk Scenarios...............................................................................................................................................................166
    Opportunities and Risk..................................................................................................................................................166
        Business Process Reengineering.............................................................................................................................166
        Cybersecurity..........................................................................................................................................................168
        Cloud Computing....................................................................................................................................................168
        Social Media ..........................................................................................................................................................169
        Big Data..................................................................................................................................................................169
        Consumerization of IT and Mobile Devices ..........................................................................................................170
**4.7 Types of Business Risk, Exposures and Threats That Can Be Addressed Using IT Resources**........................................172
    IT Risk Analytics, Monitoring and Reporting..............................................................................................................172
    Segregation of Duties....................................................................................................................................................172
    Risk Management Information System .........................................................................................................................173
    Locked-down Operations ..............................................................................................................................................173
    Decision Support, Risk Analytics and Reporting .........................................................................................................173
**4.8 Risk Appetite and Risk Tolerance** ......................................................................................................................................174
    Risk Appetite.................................................................................................................................................................174
    Risk Tolerance ..............................................................................................................................................................175
    Process to Determine Risk Appetite .............................................................................................................................175
**4.9 Quantitative and Qualitative Risk Assessment Methods** ..................................................................................................177
    Qualitative Risk Assessment.........................................................................................................................................178
    Quantitative Risk Assessment.......................................................................................................................................178
    Combining Qualitative and Quantitative Methods—Toward Probabilistic Risk Assessment......................................178
    Practical Guidance on Analyzing Risk .........................................................................................................................178
**4.10 Risk Mitigation Strategies Related to IT in the Enterprise** .............................................................................................179
    Risk Avoidance.............................................................................................................................................................179
    Risk Reduction/Mitigation............................................................................................................................................179
    Risk Sharing/Transfer ...................................................................................................................................................179
    Risk Acceptance ...........................................................................................................................................................179
    Risk Response Selection and Prioritization..................................................................................................................180
    Developing a Risk Action Plan .....................................................................................................................................181
**4.11 Methods to Monitor Effectiveness of Mitigation Strategies and/or Controls**................................................................181
    Six Sigma......................................................................................................................................................................182
        Core Principles......................................................................................................................................................182
        Impact ....................................................................................................................................................................183
    Service Level Management...........................................................................................................................................183
        Core Principles......................................................................................................................................................183
        Impact ....................................................................................................................................................................183
    IT Balanced Scorecard..................................................................................................................................................184
        Core Principles......................................................................................................................................................184
        Impact ....................................................................................................................................................................184

**Table of Contents**

| | |
|---|---:|
| **4.12 Stakeholder Analysis and Communication Techniques** | 184 |
| Risk Awareness—Risk Culture | 185 |
| Risk Communication—What to Communicate | 185 |
| Effective Risk Communication | 186 |
| Risk Communication—Stakeholders | 187 |
| **4.13 Methods to Establish Key Risk Indicators** | 188 |
| Risk Indicators | 189 |
| Risk Indicators as Communication Instruments | 189 |
| **4.14 Methods to Manage and Report the Status of Identified Risk** | 190 |
| Status Reports | 191 |
| Issue Logs | 192 |
| Evaluations | 192 |
| Risk Audits | 193 |
| **Endnotes** | 193 |

## Chapter 5: Resource Optimization .................................................................. 195

| | |
|---|---:|
| **Section One: Overview** | 196 |
| **Domain Definition** | 196 |
| **Domain Objectives** | 196 |
| **Learning Objectives** | 196 |
| **CGEIT Exam Reference** | 196 |
| **Task and Knowledge Statements** | 196 |
| Tasks | 196 |
| Knowledge Statements | 197 |
| Relationship of Task to Knowledge Statements | 198 |
| **Self-assessment Questions** | 199 |
| **Answers to Self-assessment Questions** | 200 |
| **Suggested Resources for Further Study** | 201 |
| **Section Two: Content** | 202 |
| **5.1 IT Resource Planning Methods** | 202 |
| Outsourcing | 202 |
| Multisourcing | 202 |
| Good Practices in IT Resource Planning | 203 |
| **5.2 Human Resource Procurement, Assessment, Training and Development Methodologies** | 204 |
| Human Capital | 204 |
| The Objective of Human Resource Management | 205 |
| Human Resource Management and IT Personnel | 205 |
| **5.3 Processes for Acquiring Application, Information and Infrastructure Resources** | 206 |
| IT Demand | 206 |
| IT Supply | 206 |
| Acquisition and Outsourcing | 206 |
| Information Services Procurement Library | 207 |
| **5.4 Outsourcing and Offshoring Approaches That May Be Employed to Meet the Investment Program and Operational Level Agreements and Service Level Agreements** | 207 |
| Benefits and Risk Considerations for Outsourcing | 207 |
| Business Process Outsourcing | 208 |
| Outsourcing Stakeholders | 209 |
| Outsourcing Responsibilities | 209 |

*Table of Contents*

**5.5 Methods Used to Record and Monitor IT Resource Utilization and Availability** ........................210
    Demand Management ........................210
    Availability and Capacity Management ........................210
        BAI04.01 Assess current availability, performance and capacity and create a baseline. ........................210
        BAI04.02 Assess business impact. ........................211
        BAI04.03 Plan for new or changed service requirements. ........................211
        BAI04.04 Monitor and review availability and capacity. ........................211
        BAI04.05 Investigate and address availability, performance and capacity issues. ........................212
**5.6 Methods Used to Evaluate and Report on IT Resource Performance** ........................212
    Capacity Management Information Systems ........................212
    Availability Management ........................213
**5.7 Interoperability, Standardization and Economies of Scale** ........................214
    Resource Optimization ........................214
    Economies of Scale ........................214
    Interoperability ........................215
    Standardization ........................215
**5.8 Data Management and Data Governance Concepts** ........................215
    The Information Cycle ........................216
    COBIT 5 Enabler: Information ........................216
    Further Considerations About Information ........................219
**5.9 Service Level Management Concepts** ........................221
    Service Level Management ........................221
    Service Level Agreement Types ........................221
    Service Level Management and COBIT ........................222
**Endnotes** ........................223

# General Information ........................225
Requirements for Certification ........................225
Description of the Exam ........................225
Registration for the CGEIT Exam ........................226
CGEIT Program Accreditation Renewed Under ISO/IEC 17024:2012 ........................226
Preparing for the CGEIT Exam ........................226
Types of Exam Questions ........................226
Administration of the Exam ........................227
Sitting for the Exam ........................227
Budgeting Time ........................227
Rules and Procedures ........................228
Grading the CGEIT Exam and Receiving Results ........................228
Confidentiality ........................229

# Glossary ........................231

# Index ........................243

# Evaluation ........................249

# About This Manual

## OVERVIEW

The *CGEIT® Review Manual 7th Edition* is intended to assist candidates in preparing for the CGEIT exam. **The manual is one source of preparation for the exam, but should not be thought of as the only source nor should it be viewed as a comprehensive collection of all the information and experience that is required to pass the exam.** No single publication offers such coverage and detail.

As candidates read through the manual and encounter topics that are new to them or ones in which they feel their knowledge and experience are limited, additional references should be sought. The exam will be composed of questions testing candidates' technical and practical knowledge, and their ability to apply the knowledge (based on experience) in given situations.

## ORGANIZATION OF THIS MANUAL

The *CGEIT® Review Manual 7th Edition* provides coverage of the knowledge and tasks related to the various responsibilities associated with the domains as detailed in the CGEIT job practice, which can be found on the ISACA web site (*www.isaca.org/cgeitjobpractice*). A job practice serves as the basis for the exam and the experience requirements to earn the CGEIT certification. This job practice consists of task and knowledge statements, organized by domains.

The task and knowledge statements are intended to depict the tasks performed by individuals who have a management, advisory or assurance role related to the governance of enterprise IT and the knowledge required to perform these tasks. They are also intended to serve as a definition of the roles and responsibilities of the professionals performing governance of enterprise IT work.

For purposes of the task and knowledge statements, the terms "enterprise" and "organization" or "organizational" are considered synonymous. The *CGEIT® Review Manual 7th Edition* also uses the term "IT governance" synonymously with the term "governance of enterprise IT."

There are five domains in the CGEIT job practice and each is tested on the exam in the percentages listed below:

| Domain 1 | Framework for the Governance of Enterprise IT | 25 percent |
| --- | --- | --- |
| Domain 2 | Strategic Management | 20 percent |
| Domain 3 | Benefits Realization | 16 percent |
| Domain 4 | Risk Optimization | 24 percent |
| Domain 5 | Resource Optimization | 15 percent |

The objectives of each of these domains appear at the beginning of each chapter with the corresponding task and knowledge statements that are tested on the exam. Exam candidates should evaluate their strengths, based on knowledge and experience, in each of these domains.

**About This Manual**

## FORMAT OF THIS MANUAL

The manual has been developed and organized to assist in the study of these domains. Each of the five chapters deals with one of the domains and is divided into two sections:

Section one of each chapter includes:
- A definition for the domain
- Objectives for the domain as a practice area
- Learning objectives for the domain
- A listing of task and knowledge statements for the domain
- CGEIT exam reference
- A map of the relationship of each task to the knowledge statements for the domain
- Sample self-assessment questions and answers with explanations
- Suggested resources for further study of the domain

Section two of each chapter includes:
- A discussion of each knowledge statement in the domain
- Endnotes

Understanding the textual material is a barometer of one's knowledge, strengths and weaknesses, and an indication of domains in which one needs to seek reference sources over and above this manual. However, written material is not a substitute for experience. **Actual exam questions will test the candidate's practical application of this knowledge.** The self-assessment questions at the end of section one of each chapter should not be used independently as a source of knowledge nor should they be considered a measure of one's ability to answer questions correctly on the exam for that area. The sample self-assessment questions and answers (with explanations) are intended to familiarize candidates with question structure and general content, and may or may not be similar to questions that will appear on the actual exam. The reference material includes other publications that could be used to further acquire and better understand detailed information on the topics addressed in the manual.

Throughout the manual, "association" refers to ISACA, formerly known as Information Systems Audit and Control Association, and "institute" or "ITGI" refers to the IT Governance Institute. Please note that the manual has been written using standard American English, except where material has been imported from publications written in International English.

> **Note:** The *CGEIT® Review Manual* is a living document. As the IT governance job practice advances, the manual will be updated to reflect such advances. Further updates to this document before the date of the exam may be viewed at *www.isaca.org/studyaidupdates*.

Suggestions to enhance the review manual or suggested reference materials should be submitted to *studymaterials@isaca.org*.

## ABOUT THE CGEIT REVIEW QUESTIONS, ANSWERS AND EXPLANATIONS MANUAL

Candidates may also wish to enhance their study and preparation for the exam by using the *CGEIT® Review Questions, Answers & Explanations Manual 4th Edition*.

The *CGEIT® Review Questions, Answers & Explanations Manual 4th Edition* consists of 250 multiple-choice study questions, answers and explanations arranged in the domains of the current CGEIT job practice.

Questions in this publications are representative of the types of questions that could appear on the exam and include explanations of the correct and incorrect answers. Questions are sorted by the CGEIT domains and as a sample test. This publication is ideal for use in conjunction with the *CGEIT® Review Manual 7th Edition*. These manuals can be used as study sources throughout the study process or as part of a final review to determine where candidates may need additional study. It should be noted that these questions and suggested answers are provided as examples; they are not actual questions from the examination and may differ in content from those that actually appear on the exam.

> **Note:** When using the CGEIT review materials to prepare for the exam, it should be noted that they cover a broad spectrum of governance of enterprise IT issues. **Again, candidates should not assume that reading these manuals and answering review questions will fully prepare them for the examination.** Because actual exam questions often relate to practical experiences, candidates should refer to their own experiences and other reference sources, and draw upon the experiences of colleagues and others who have earned the CGEIT designation.

Page intentionally left blank

# Chapter 1: Framework for the Governance of Enterprise IT

## Section One: Overview

Domain Definition ............................................................................................................................................................ 2
Domain Objectives ........................................................................................................................................................... 2
Learning Objectives ......................................................................................................................................................... 2
CGEIT Exam Reference ................................................................................................................................................... 2
Task and Knowledge Statements ..................................................................................................................................... 2
Self-assessment Questions ............................................................................................................................................... 6
Answers to Self-assessment Questions ............................................................................................................................ 7
Suggested Resources for Further Study .......................................................................................................................... 8

## Section Two: Content

1.1 Components of a Framework for the Governance of Enterprise IT ....................................................................... 9
1.2 IT Governance Industry Practices, Standards and Frameworks ........................................................................... 11
1.3 Business Drivers Related to IT Governance ......................................................................................................... 14
1.4 IT Governance Enablers ........................................................................................................................................ 16
1.5 Techniques Used to Identify IT Strategy .............................................................................................................. 19
1.6 Components, Principles and Concepts Related to Enterprise Architecture ......................................................... 22
1.7 Organizational Structures and Their Roles and Responsibilities ......................................................................... 24
1.8 Methods to Manage Organizational, Process and Cultural Change ..................................................................... 28
1.9 Models and Methods to Establish Accountability for Information Requirements,
    Data and System Ownership, and IT Processes ................................................................................................... 30
1.10 IT Governance Monitoring Processes/Mechanisms ............................................................................................. 32
1.11 IT Governance Reporting Processes/Mechanisms ............................................................................................... 36
1.12 Communication and Promotion Techniques ........................................................................................................ 39
1.13 Assurance Methodologies and Techniques .......................................................................................................... 40
1.14 Continuous Improvement Techniques and Processes .......................................................................................... 43
Endnotes ......................................................................................................................................................................... 45

# Chapter 1—Framework for the Governance of Enterprise IT
## Section One: Overview

# Section One: Overview

## DOMAIN DEFINITION
Ensure the definition, establishment and management of a framework for the governance of enterprise IT in alignment with the mission, vision and values of the enterprise.

## DOMAIN OBJECTIVES
The objective of this domain is to define, establish and maintain an information technology (IT) governance framework, by having the requisite leadership, organizational structures and processes to:
- Ensure alignment with enterprise governance
- Control the business information and information technology environment through the implementation of good practices
- Assure compliance with external requirements

IT governance is concerned with ensuring that enterprise objectives are achieved by evaluating stakeholder needs, conditions and options; setting direction through prioritization and decision making; and monitoring performance, compliance and progress against agreed-on direction and objectives (evaluate, direct, monitor [EDM]). IT management is concerned with ensuring that planning, building, running and monitoring (PBRM) activities are in alignment with the direction set by the governance body to achieve the enterprise objectives.

The premise of this domain is that governance of IT has become a major concern of the board and executive management in enterprises globally. IT has evolved over the past decades to become fundamentally critical to sustaining growth, innovation and transformation, to reduce and contain costs, and to support the ongoing business operations of most enterprises. IT is ever-present, and has become an asset that belongs in boardroom meetings, in the same way other resources (i.e., human, financial, etc.) do.

## LEARNING OBJECTIVES
The purpose of this domain is to have boards of directors, professionals and executives understand their required involvement in governance of enterprise IT (GEIT). Personnel involved with GEIT need to know how to advise the board of directors so that they are able to evaluate, direct and monitor the use of IT in their organization. Personnel also need to understand the broad requirements for effective governance, the elements and actions required to develop a framework if one is not already in place, and a plan of action to implement it or improve it if there is one.

## CGEIT EXAM REFERENCE
This domain represents 25 percent of the CGEIT exam (approximately 38 questions).

## TASK AND KNOWLEDGE STATEMENTS

### TASKS
There are 11 tasks within this domain that a CGEIT candidate must know how to perform. These relate to having to develop, or be part of the development of, an IT governance framework.

T1.1　Ensure that a framework for the governance of enterprise IT is established and enables the achievement of enterprise goals and objectives to create stakeholder value, taking into account benefits realization, risk optimization and resource optimization.

T1.2　Identify the requirements and objectives for the framework for the governance of enterprise IT, incorporating input from enablers such as principles, policies and frameworks; processes; organizational structures; culture, ethics and behavior; information; services, infrastructure and applications; people, skills and competencies.

T1.3　Ensure that the framework for the governance of enterprise IT addresses applicable internal and external requirements (for example, principles, policies and standards, laws, regulations, service capabilities and contracts).

T1.4 Ensure that strategic planning processes are incorporated into the framework for the governance of enterprise IT.

T1.5 Ensure the incorporation of enterprise architecture (EA) into the framework for the governance of enterprise IT in order to optimize IT-enabled business solutions.

T1.6 Ensure that the framework for the governance of enterprise IT incorporates comprehensive and repeatable processes and activities.

T1.7 Ensure that the roles, responsibilities and accountabilities for information systems and IT processes are established.

T1.8 Ensure that issues related to the framework for the governance of enterprise IT are reviewed, monitored, reported and remediated.

T1.9 Ensure that organizational structures are in place to enable effective planning and implementation of IT-enabled business investments.

T1.10 Ensure the establishment of a communication channel to reinforce the value of the governance of enterprise IT and transparency of IT costs, benefits and risk throughout the enterprise.

T1.11 Ensure that the framework for the governance of enterprise IT is periodically assessed, including the identification of improvement opportunities.

## KNOWLEDGE STATEMENTS

The CGEIT candidate must have a good understanding of each of the 14 areas delineated by the following knowledge statements. These statements are the basis for the exam. Each statement is defined and its relevance and applicability to this job practice are briefly described as follows:

Knowledge of:

**KS1.1 components of a framework for the governance of enterprise IT**
GEIT does not occur in a vacuum. It is an integral part of the enterprise and the system by which organizations are directed and controlled. Because of business dependency on IT, enterprise governance issues (including business governance, corporate governance) can no longer be solved without considering IT. Enterprise governance should therefore drive and establish the framework for GEIT.

**KS1.2 IT governance industry practices, standards and frameworks**
COBIT is the ISACA business framework for the governance and management of enterprise IT. Understanding this framework and its underlying principles is fundamental to establishing and improving practices and processes for effective governance of enterprise IT. Many industry practices, standards and frameworks are complementary to COBIT. Therefore, knowledge of these industry practices, standards and frameworks—especially what, when and how they are used—becomes important for effective GEIT.

**KS1.3 business drivers related to IT governance**
In a constantly changing business and economic context, it is important to detect business drivers (pain points or trigger events in the internal or external environment) that constitute a need for a new or revised GEIT practice.

**KS1.4 IT governance enablers**
Efficient and effective governance and management of enterprise IT requires a holistic approach, taking into account several interacting components. These components are enablers supporting implementation of a comprehensive governance and management system for enterprise IT. Enablers are broadly defined as anything that can help to achieve the objectives of the enterprise.

**KS1.5 techniques used to identify IT strategy**
In setting direction and crafting approaches to implement enterprisewide improvements in GEIT, it is crucial to tie these initiatives to the organizational mission, vision and strategy. To be effective, GEIT needs to exist within these defining organizational statements. Therefore, it is important for practitioners to understand why the IT strategy is important and how it can be linked to the enterprise strategy by leveraging techniques such as the strengths, weaknesses, opportunities and threats (SWOT) analysis and the Boston Consulting Group's (BCG) growth share matrix.

**KS1.6 components, principles and concepts related to enterprise architecture (EA)**
GEIT and EA are related and intertwined concepts. Understanding EA concepts is important for analyzing and anticipating the interdependencies between GEIT and EA adoptions in the organization.

**KS1.7 organizational structures and their roles and responsibilities**
Effective GEIT includes the governance of organizational structures to ensure that IT-related decisions occur in a transparent environment and to enable effective contact and exchange between business and IT management. These structures can be seen as a blueprint for how the framework for GEIT will function conceptually.

**KS1.8 methods to manage organizational, process and cultural change**
Successful implementation or improvement depends on implementing the appropriate change in the correct way. In many enterprises, there is not enough emphasis on managing the human, behavioral and cultural aspects of the change and motivating stakeholders to buy in to the change. Change enablement[1] is one of the biggest challenges to GEIT implementation.

**KS1.9 models and methods to establish accountability for information requirements, data and system ownership, and IT processes**
Accountability is crucial for effective governance. For GEIT, accountability has a special context—for information requirements, data and system ownership, and IT-related processes. How accountability is established, both at management (plan, build, run, monitor) and governance (evaluate, direct, monitor) levels needs to be well understood for good GEIT to be in place.

**KS1.10 IT governance monitoring processes/mechanisms**
The concept of performance management and monitoring and its associated techniques are highly useful to improve and map GEIT to organizational imperatives. Use of the balanced scorecard (BSC) and its mechanisms to translate strategy into measurable action plays an important role in this context.

**KS1.11 IT governance reporting processes/mechanisms**
For effective communication with all stakeholders, it is important to ensure that the enterprise IT performance and conformance measurement and reporting are transparent.

**KS1.12 communication and promotion techniques**
For effective GEIT, stakeholder groups need a high level of buy-in to the GEIT process. Knowledge of marketing and communication methods and techniques (for example, asserting, persuading, bridging, attracting) plays an important role in helping to define and sell key messages to targeted audiences.

**KS1.13 assurance methodologies and techniques**
Knowledge and application of assurance methodologies and techniques are the basis for enforcement of the governance initiatives and practices. Without appropriate GEIT monitoring and controls, as prescribed by assurance methodologies and techniques (planning assurance, scoping assurance, executing assurance), the quality and sustainability of initiatives and practices may not meet internal and external requirements.

**KS1.14 continuous improvement techniques and processes**
Applying a continuous improvement life cycle approach provides a method for enterprises to address the complexity and challenges typically encountered during GEIT implementation. There are three interrelated components to the life cycle:
- the core continuous improvement cycle
- the enablement of change
- the management of the program

## RELATIONSHIP OF TASK TO KNOWLEDGE STATEMENTS

The task statements are what the CGEIT candidate is expected to know how to do. The knowledge statements delineate what the CGEIT candidate is expected to know to perform the tasks. The task and knowledge statements are mapped in **figure 1.1** insofar as it is possible to do so. Note that although there often is overlap, each task statement will generally map to several knowledge statements.

| Figure 1.1—Task and Knowledge Statement Mapping—Framework for the Governance of Enterprise IT Domain | |
|---|---|
| **Task Statement** | **Knowledge Statements** |
| T1.1 Ensure that a framework for the governance of enterprise IT is established and enables the achievement of enterprise goals and objectives to create stakeholder value, taking into account benefits realization, risk optimization and resource optimization. | KS1.1 components of a framework for the governance of enterprise IT<br>KS1.2 IT governance industry practices, standards and frameworks<br>KS1.8 methods to manage organizational, process and cultural change |
| T1.2 Identify the requirements and objectives for the framework for the governance of enterprise IT, incorporating input from enablers such as principles, policies and frameworks; processes; organizational structures; culture, ethics and behavior; information; services, infrastructure and applications; people, skills and competencies. | KS1.1 components of a framework for the governance of enterprise IT<br>KS1.3 business drivers related to IT governance<br>KS1.4 IT governance enablers |
| T1.3 Ensure that the framework for the governance of enterprise IT addresses applicable internal and external requirements. | KS1.1 components of a framework for the governance of enterprise IT<br>KS1.3 business drivers related to IT governance<br>KS1.5 techniques used to identify IT strategy |
| T1.4 Ensure that strategic planning processes are incorporated into the framework for the governance of enterprise IT. | KS1.5 techniques used to identify IT strategy |
| T1.5 Ensure the incorporation of enterprise architecture (EA) into the framework for the governance of enterprise IT in order to optimize IT-enabled business solutions. | KS1.6 components, principles and concepts related to enterprise architecture (EA) |
| T1.6 Ensure that the framework for the governance of enterprise IT incorporates comprehensive and repeatable processes and activities. | KS1.8 methods to manage organizational, process and cultural change |
| T1.7 Ensure that the roles, responsibilities and accountabilities for information systems and IT processes are established. | KS1.8 methods to manage organizational, process and cultural change<br>KS1.9 models and methods to establish accountability for information requirements, data and system ownership, and IT processes |
| T1.8 Ensure that issues related to the framework for the governance of enterprise IT are reviewed, monitored, reported and remediated. | KS1.10 IT governance monitoring processes/mechanisms<br>KS1.11 IT governance reporting processes/mechanisms |
| T1.9 Ensure that organizational structures are in place to enable effective planning and implementation of IT-enabled business investments. | KS1.7 organizational structures and their roles and responsibilities<br>KS1.8 methods to manage organizational, process and cultural change<br>KS1.9 models and methods to establish accountability for information requirements, data and system ownership, and IT processes<br>KS1.11 IT governance reporting processes/mechanisms<br>KS1.12 communication and promotion techniques |
| T1.10 Ensure the establishment of a communication channel to reinforce the value of the governance of enterprise IT and transparency of IT costs, benefits and risk throughout the enterprise. | KS1.11 IT governance reporting processes/mechanisms<br>KS1.12 communication and promotion techniques |
| T1.11 Ensure that the framework for the governance of enterprise IT is periodically assessed, including the identification of improvement opportunities. | KS1.13 assurance methodologies and techniques<br>KS1.14 continuous improvement techniques and processes |

*Chapter 1—Framework for the Governance of Enterprise IT*
*Section One: Overview*

## SELF-ASSESSMENT QUESTIONS

CGEIT self-assessment questions support the content in this manual and provide an understanding of the type and structure of questions that have typically appeared on the exam. Questions are written in a multiple-choice format and designed for one best answer. Each question has a stem (question) and four options (answer choices). The stem may be written in the form of a question or an incomplete statement. In some instances, a scenario or a description problem may be included. These questions normally include a description of a situation and require the candidate to answer two or more questions based on the information provided. Many times a question will require the candidate to choose the **MOST** likely or **BEST** answer among the options provided.

In each case, the candidate must read the question carefully, eliminate known incorrect answers and then make the best choice possible. Knowing the format in which questions are asked, and how to study and gain knowledge of what will be tested, will help the candidate correctly answer the questions.

1-1    In addition to corporate governance, which of the following is a key component of an enterprise governance framework?

     A. Value governance
     B. Key asset governance
     C. Business governance
     D. Financial governance

1-2    The **MOST** effective way to implement governance of enterprise IT in an enterprise is through the use of a:

     A. business case.
     B. IT balanced scorecard (BSC).
     C. phased life cycle.
     D. set of IT performance metrics.

1-3    Which of the following should be achieved **FIRST** to enable implementation of a framework for the governance of enterprise IT?

     A. Establishing the desire to change
     B. Forming an implementation team
     C. Empowering role players
     D. Embedding new approaches

# ANSWERS TO SELF-ASSESSMENT QUESTIONS
Correct answers are shown in **bold**.

1-1   A.  Value governance only deals with the limited perspective of value, which is not one of the specified key components of enterprise governance.

B.  Key asset governance represents only the limited perspective of key assets, which is not one of the specified key components of enterprise governance.

**C.  Business governance and corporate governance lie at the core of enterprise governance. Business governance relates to the performance element of enterprise governance and corporate governance relates to the conformance element of enterprise governance.**

D.  Financial governance represents only the limited perspective of finance, which is not one of the specified key components of enterprise governance.

1-2   A.  A business case is used to justify and monitor the progress of an investment program.

B.  An IT BSC is used to define the strategy for IT and to measure the performance of key IT initiatives in four specific perspectives or dimensions.

**C.  The phased life cycle approach to governance of enterprise IT implementation is a best practice because it addresses the complexity and challenges typically encountered in governance of enterprise IT implementation.**

D.  A set of IT performance metrics is used to monitor IT performance.

1-3   **A.  Any plan to significantly modify existing processes and behaviors should start with establishing a common desire to change or a "call to action," which can often be linked to current pain points or trigger events.**

B.  Getting the involvement and participation of the optimal implementation team is often dependent on the common vision or desire to change.

C.  Role players cannot be empowered before a desire to change is established.

D.  New approaches can only be embedded once the desire to change and call to action is established.

---

**Note:** For more self-assessment questions, you may also want to obtain a copy of the *CGEIT® Review Questions, Answers & Explanations Manual 4th Edition*, which consists of 250 multiple-choice study questions, answers and explanations.

*Chapter 1—Framework for the Governance of Enterprise IT*
*Section One: Overview*

## SUGGESTED RESOURCES FOR FURTHER STUDY

In addition to the resources cited throughout this manual, the following resources are suggested for further study in this domain (publications in **bold** are stocked in the ISACA Bookstore):

Calder, Alan; *IT Governance Today: A Practitioner's Handbook*, IT Governance Ltd., UK, 2005

De Haes, Steven; Anant Joshi; Wim Van Grembergen; "State and Impact of Governance of Enterprise IT in organizations: Findings of an International Study," *ISACA Journal*, Vol. 4, 2015

**ISACA, *Benchmarking and Business Value Assessment of COBIT® 5*, USA, 2015**

**ISACA, *COBIT 5*, USA, 2012,** *www.isaca.org/cobit*

**ISACA, *COBIT 5—Implementation*, USA, 2012,** *www.isaca.org/cobit*

**ISACA, *COBIT 5 for Assurance*, USA, 2013,** *www.isaca.org/cobit*

**ISACA, *COBIT 5—Enabling Information*, USA, 2013,** *www.isaca.org/cobit*

Selig, Gad J.; *Implementing IT Governance: A Practical Guide to Global Best Practices in IT Management*, Van Haren Publishing, The Netherlands, 2008

Van Grembergen, Wim; *Strategies for Information Technology Governance*, Idea Group Inc., USA, 2004

Van Grembergen, Wim; Steven De Haes; *Business Strategy and Applications in Enterprise IT Governance*, IGI Global, USA, 2012

Van Grembergen, Wim; Steven De Haes; *Enterprise Governance of IT: Achieving Alignment and Value*, Second Edition, Springer, USA, 2015

Van Grembergen, Wim; Steven De Haes; *Implementing Information Technology Governance: Models, Practices and Cases*, IGI Global, USA, 2008

Weill, Peter; Jeanne Ross; *IT Governance: How Top Performers Manage IT Decision Rights for Superior Results*, Harvard Business School Press, USA, 2004

# Section Two: Content

## 1.1 COMPONENTS OF A FRAMEWORK FOR THE GOVERNANCE OF ENTERPRISE IT

Governance of enterprise IT (GEIT) does not occur in a vacuum. GEIT is an integral part of the enterprise and the system by which organizations are directed and controlled. Because of business dependency on IT, enterprise governance issues (including business and corporate governance) can no longer be solved without considering IT. Therefore, enterprise governance should drive and establish the framework for GEIT.

GEIT is an integral part of enterprise governance. Therefore, it is important to understand the need for and components of enterprise governance before discussing governance of enterprise IT.

### Enterprise Governance

Recent high-profile cases of corporate failure and fraud have brought enterprise governance to the top of business and political agendas. Increasingly, educated and assertive stakeholders are concerned about the sound management of their interests. This has led to the emergence of principles and standards for enterprise governance. In recent years, several codes of conduct have been published internationally and legislation has been enacted to engender the practice of good enterprise governance. These regulations establish board responsibilities and require the board of directors to exercise due diligence. The McKinsey Global Survey 2009 results indicate that governance programs do create shareholder value and the current economic turmoil has increased the importance of governance programs.

Kotter defines enterprise governance as the following:[2]

> *Enterprise governance is a set of responsibilities and practices exercised by the board and executive management with the goal of providing strategic direction, ensuring that objectives are achieved, ascertaining that risks are managed appropriately and verifying that the enterprise's resources are used responsibly.*

### Conformance and Performance

In October 2003, the International Federation of Accountants (IFAC) published a significant survey report exploring the concept of enterprise governance, featuring a series of case studies covering 10 countries and 10 market sectors.[3] The report states that "enterprise governance constitutes the entire accountability framework of the organization" and it identifies two key dimensions of enterprise governance—conformance and performance—and indicates that these two dimensions need to be in balance. **Figure 1.2** illustrates the components.

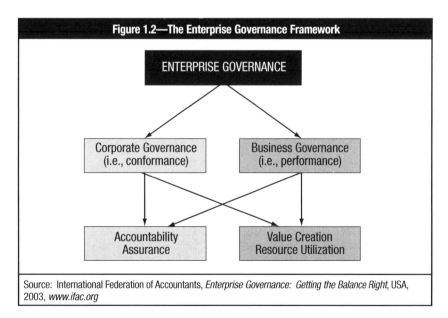

Conformance is also called corporate governance, and covers issues such as board structure, roles and executive remuneration. Codes and/or standards can generally address this dimension with compliance being subject to assurance/audit. There are also well-established oversight mechanisms for the board to use to ensure that good corporate governance processes are effective (e.g., audit committees).

The performance dimension focuses on strategy and value creation, and on helping the board make strategic decisions, understand its appetite for risk and its key drivers of performance, and identify its key points of decision making. This dimension does not lend itself easily to a regimen of standards and audit. Instead, it is usually desirable to develop a range of good practice tools and techniques that can be applied intelligently within different types of enterprises. Unlike the conformance dimension, there are typically no dedicated oversight mechanisms such as audit committees. In their survey project on which the report is based, the IFAC project team explored and confirmed that this oversight gap was a significant issue and proposed an approach (a strategic scorecard) whereby enterprises can better govern the performance dimension.

The conformance dimension takes a historical view while the performance view looks forward. It is clear that good corporate governance is only part of the story; strategy is also important. The report states that while the heavy emphasis on corporate governance issues has been necessary in the light of recent corporate scandals, it is important to remember that good governance on its own cannot make an enterprise successful. It is important to complement and balance good corporate governance with the creation of sustainable value.

Through the case studies undertaken, the IFAC report has identified that in the case of successful enterprises there was generally a "virtuous circle" that emerged based on a conscious decision to take good corporate governance seriously because it was good for the enterprise rather than because it was required by law or formal codes of best practice. In those successful cases where good governance did not feature strongly as a key factor for success, it was found that good corporate governance is a necessary, but not sufficient, foundation for success. In other words, while bad corporate governance can ruin an enterprise, good corporate governance cannot, on its own, ensure success. Enterprise governance, with its focus on both the conformance and performance aspects of business, ensures that enterprises do not lose sight of this perspective.

In the previous discussion, enterprise governance is decomposed into Corporate Governance (conformance) and Business Governance (performance). Other reports and frameworks use different terminology and refer to Corporate Governance for what ISACA refers to as Enterprise Governance. As of this printing, ISO/IEC uses the term "governance of IT for the organization" to refer to the GEIT knowledge area.

## Governance of Key Assets, Including IT

In the enterprise governance discussion, Weill and Ross identify six key assets that need to be governed, through which an organization can accomplish its strategies and generate business value: human assets, financial assets, physical assets, intellectual property (IP) assets, information and IT assets, and relationship assets (**figure 1.3**). According to Weill and Ross, "Senior executive teams create mechanisms to govern the management and use of each of these assets both independently and together. […] Governance of the key assets occurs via a large number of organizational mechanisms, for example structures, processes, procedures and audits."[4] In the current digital environment, the governance of IT assets becomes increasingly important to ensure value creation out of IT while managing its business risk.

For more information on the components of a GEIT framework, in terms of required enablers, see section 1.5 Techniques Used to Identify IT Strategy.

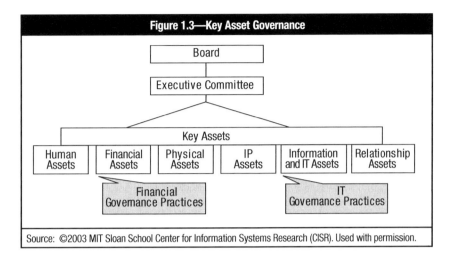

Figure 1.3—Key Asset Governance

Source: ©2003 MIT Sloan School Center for Information Systems Research (CISR). Used with permission.

## 1.2 IT GOVERNANCE INDUSTRY PRACTICES, STANDARDS AND FRAMEWORKS

COBIT is the ISACA business framework for the governance and management of enterprise IT. Understanding this framework and its underlying principles is fundamental to establishing and improving practices and processes for effective governance of enterprise IT. Many industry practices, standards and frameworks are complementary to COBIT. Therefore, knowledge of these industry practices, standards and frameworks—especially what, when and how they are used—becomes important for effective GEIT.

Successful enterprises have realized that the board of directors and executives need to embrace IT like any other significant part of doing business. The board of directors often handles departments such as human resources, finance, etc., but often due to a lack of knowledge, this is not the case with IT. Boards and management—both in the business and IT functions—must collaborate and work together, so that IT is included within the governance and management approach. In addition, legislation is being passed increasingly and regulations implemented to address this need.

### Forest of Frameworks, Standards and Good Practices

It has been said that there is a "forest" of frameworks, standards and good practices when it comes to governance and management of enterprise IT:
- A **framework** is a generally accepted, business-process-oriented structure that establishes a common language and enables repeatable business processes.
- A **standard** is a mandatory requirement, code of practice or specification approved by a recognized external standards organization, such as ISO.
- A **practice** is a frequent or usual action performed (willingly or by imperative).

This forest is an indication of both the need for, and response to, the demands of today's enterprises in terms of governance and management of IT. This situation has stimulated the development, import from other disciplines and the acceptance of managerial frameworks, methods and methodologies that support the proper functioning of today's IT function. The result has been the proliferation of standards and frameworks, of which the following are representative examples.

### Governance of Enterprise IT
- **AS/NZS 8016:2013**—Governance of IT Enabled Projects (issued in 2013, based on ISO/IEC 38500:2008)
- **COBIT 5**—A business framework for the governance and management of enterprise IT
- **ISO/IEC 38502:2014**—Governance of IT—Framework and Model
- **ISO/IEC TS 38501:2015**—Governance of IT—Implementation Guide

### Management of Enterprise IT
- **ASL**—Application Services Library
- **BiSL**—Business Information Services Library
- **CMMI**—Capability Maturity Model Integration (staged and continuous)
- **COBIT 5**—A business framework for the governance and management of enterprise IT
- **EFQM**—European Foundation for Quality Management model
- **eSCM**—eSourcing Capability Model of Carnegie Mellon University (CMU) Software Engineering Institute (SEI)
- **eTOM**—Enhanced Telecom Operations Map
- **ISO/IEC 27000 series**—Family of ISO/IEC Information Security Management Systems standard (ISMS)
- **ISO 9000 series**—Set of international standard for quality management and quality assurance
- **ISO/IEC 20000-1:2011 (formerly BS15000)**—Information technology—Service management—Part 1: Service management system requirements
- **ISO 31000 series**—Family of standards relating risk management
- **ISO 22301:2012**—Societal security—Business continuity management systems—Requirements
- **ISO/IEC 33000 series**—Family of ISO Process Assessment Standards
- **ISO/IEC/IEEE 42010**—Systems and software engineering—Architecture description
- **ISPL**—Information Services Procurement Library
- **ITIL**—IT Infrastructure Library
- **M_o_R**—Management of Risk
- **MSP**—Managing Successful Programs
- **PMBOK**—Project Management Body of Knowledge
- **PRINCE2**—PRojects IN Controlled Environments
- **Six Sigma**—Six Sigma model for quality management
- **TickIT**—Quality management for IT
- **TOGAF**—The Open Group Architecture Framework
- **TQM**—Total quality management

Most of these standards focus on specific aspects of IT governance and management. COBIT 5 aligns with these standards and frameworks at a high level and, thus, can serve as the overarching framework for governance and management of enterprise IT. The CGEIT candidate is not expected to learn the detailed standards thoroughly for the exam. The relevance of the standards from a GEIT perspective and COBIT 5 as an integrating framework should be the learning focus.

> **Note:** COBIT 5 is provided as an example of a CGEIT framework throughout this manual in order to illustrate the concepts discussed. The CGEIT candidate will not be tested on the specific aspects of COBIT 5 or any standard or framework.

## COBIT 5

In the field, many good practice frameworks are developed and promoted to guide managers in implementing GEIT.[5] One of these frameworks is COBIT; ISACA issued COBIT 5 in April 2012. COBIT is a freely available framework that describes a set of best practices for governance, management, control and assurance of IT, and organizes them around a logical framework of IT-related processes.

COBIT 5 provides a comprehensive framework that assists enterprises in achieving their objectives for the governance and management of enterprise IT. Simply stated, it helps enterprises create optimal value from IT by maintaining a balance between benefits and risk levels and resource use. COBIT 5 enables IT to be governed and managed in a holistic manner for the entire enterprise, taking in the full end-to-end business and IT functional areas of responsibility, considering the IT-related interests of internal and external stakeholders. COBIT 5 is generic and useful for enterprises of all sizes, whether commercial, not-for-profit or in the public sector. COBIT 5 identifies 37 governance and management processes for IT, as shown in **figure 1.4**.

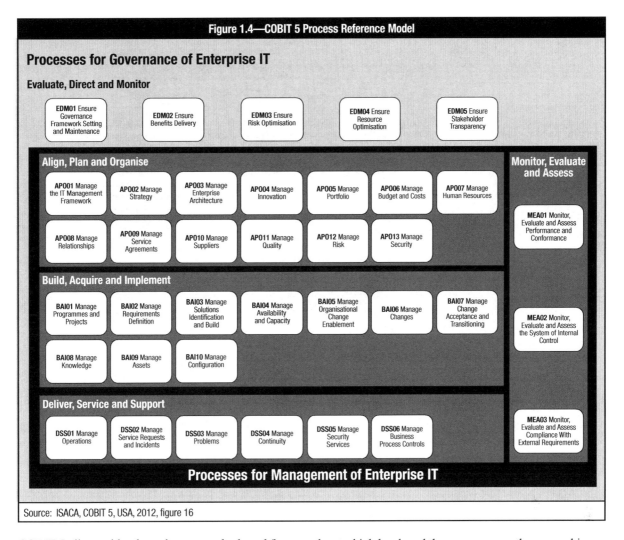

Figure 1.4—COBIT 5 Process Reference Model

Source: ISACA, COBIT 5, USA, 2012, figure 16

COBIT 5 aligns with other relevant standards and frameworks at a high level, and thus can serve as the overarching framework for governance and management of enterprise IT. Some important complementary frameworks are depicted in **figure 1.5** and positioned toward the COBIT 5 processes and domains as shown in the previous figure.

COBIT 5 is a single and integrated framework because:
- It aligns with other recent, relevant standards and frameworks, and thus allows the enterprise to use COBIT 5 as the overarching governance and management framework integrator.
- It is complete in enterprise coverage, providing a basis to effectively integrate other frameworks, standards and practices used.
- It provides a simple architecture for structuring guidance materials and producing a consistent product set.
- It integrates all knowledge previously dispersed over different ISACA frameworks. ISACA has researched the key area of GEIT for many years and has developed frameworks such as COBIT, Val IT, Risk IT, Business Model for Information Security (BMIS), *Board Briefing on IT Governance* and ITAF.

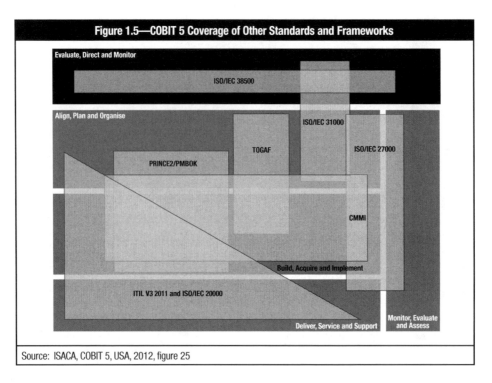

Figure 1.5—COBIT 5 Coverage of Other Standards and Frameworks

Source: ISACA, COBIT 5, USA, 2012, figure 25

## 1.3 BUSINESS DRIVERS RELATED TO IT GOVERNANCE

In a constantly changing business, economic and technological (cloud computing, BYOD, big data, Internet of Things, etc.) context, it is important to detect business drivers (pain points or trigger events in the internal or external environment) that constitute a need for new or revised GEIT practices.

Many factors may indicate a need for new or revised GEIT practices. It is, however, important to note that these symptoms may not only point to underlying issues that need to be addressed, but could also be indicative of other issues (or a combination of factors). For example, if business has the perception that IT costs are unacceptably high, this may be due to governance and/or management issues (such as inappropriate criteria being used in the IT investment management process), but it could also be due to a legacy of underinvestment in IT that now manifests in significant investments being required.

By using pain points or trigger events as the launching point for GEIT initiatives, the business case for improvement will be related to issues being experienced, which will improve buy-in. A sense of urgency can be created within the enterprise that is necessary to kick off the implementation. In addition, quick wins can be identified and value-add can be demonstrated in areas that are the most visible in the enterprise. This provides a platform for introducing further changes and can assist in gaining widespread senior management commitment and support for more pervasive changes.

### Typical Pain Points

New or revised GEIT practices can typically solve or be part of a solution to the following symptoms:
- **Business frustration with failed initiatives, rising IT costs and a perception of low business value**—While many enterprises continue to increase their investments in IT, the value of these investments and overall performance of IT are often questioned or not fully realized. This can be indicative of a GEIT issue where communication between IT and the business needs to be improved and a common view on the role and value of IT needs to be established. It can also be a consequence of suboptimal portfolio and project formulation, proposal and approval mechanisms.
- **Significant incidents related to IT-related business risk, such as data loss or project failure**—These significant incidents are often the tip of the iceberg and the impacts can be exacerbated if they receive public and/or media attention. Further investigation often leads to the identification of deeper and structural misalignments or even a complete lack of an IT risk-aware culture within the enterprise. Stronger GEIT practices are then required to get a complete view and a solid understanding of IT-related risk and how it should be managed.

- **Outsourcing service delivery problems, such as agreed-on service levels, not being consistently met**—Issues with service delivery by external service providers may be due to governance issues such as a lack of defined or inadequate tailoring of third-party service management processes (including control and monitoring) with associated responsibilities and accountabilities to fulfill business IT service requirements.
- **Failure to meet regulatory or contractual requirements**—In many enterprises, ineffective or inefficient governance mechanisms prevent complete integration of relevant laws, regulations and contractual terms into organizational systems or lack an approach for managing them. Regulations and compliance requirements are generally increasing globally, often with an impact on IT-enabled activities.
- **IT's limiting of enterprise's innovation capabilities and business agility**—A common complaint is that IT's role is that of a support function, whereas there is a requirement for innovation capabilities to provide a competitive edge. These are symptoms that may point to a lack of true bidirectional alignment between business and IT, which could be due to communication issues or suboptimal business involvement in IT decision making. It could also be due to business involving IT at too late a stage during strategic planning and business-driven initiatives. This issue typically can be highlighted when economic conditions require rapid enterprise responses such as the introduction of new products or services.
- **Regular audit findings about poor IT performance or reported IT quality of service problems**—This may be indicative of service levels not being in place or not functioning well, or inadequate business involvement in IT decision making.
- **Hidden and rogue IT spending**—A sufficiently transparent and comprehensive view of IT expenditures and investments is often lacking. IT spending can often be "hidden" in business unit budgets or not classified as IT spending in the accounts, creating an overall biased view of IT costs.
- **Duplication or overlap between initiatives or wasting resources**—This is often due to a lack of a portfolio/holistic view of all IT initiatives and indicates that process and decision structure capabilities around portfolio and performance management are not in place.
- **Insufficient IT resources, staff with inadequate skills or staff burnout/dissatisfaction**—These are significant IT human resource management issues that require effective oversight and good governance to ensure that people management and skills development are addressed effectively. These issues could also be indicative of other factors, such as underlying weaknesses in IT demand management and internal service delivery practices.
- **IT-enabled changes frequently failing to meet business needs and delivered late or over budget**—These pain points could be related to problems with business-IT alignment, definition of business requirements, lack of a benefit realization process, or suboptimal implementation and project/program management processes.
- **Multiple and complex IT assurance efforts**—This could be indicative of poor coordination between the business and IT regarding the need for and execution of IT-related assurance reviews. An underlying cause could be a low level of business trust in IT, causing the business to initiate its own reviews, or a lack of adequate business accountability for IT assurance reviews, resulting in the business being unaware when they take place.
- **Board members, executives or senior managers who are reluctant to engage with IT, or a lack of committed and satisfied business sponsors for IT**—These pain points often relate to a lack of business understanding and insight into IT and a lack of IT visibility at the appropriate levels. Root causes can be found in issues with board mandates, poor communication between the business and IT, lack of governance arrangements and management structures, etc.
- **Complex IT operating models**—The complexity inherent in, for example, decentralized or federated IT models that often have different structures, practices and policies requires a strong focus on GEIT to ensure optimal IT decision making and effective and efficient operations. This pain point often becomes more significant in international contexts because each territory or region may have specific and potentially unique internal and external environmental factors to be addressed. Too often this problem is increased by a lack of a well-thought-out and developed EA.

## Trigger Events in the Internal and External Environments

In addition to the symptoms described previously, other events in the enterprise's internal and external environments, such as the following, can signal or trigger a focus on GEIT and drive it high on the enterprise agenda:
- **Merger, acquisition or divestiture**—The strategic and operational consequences relating to IT may be significant following a merger, acquisition or divestiture. During due diligence reviews there will be a need to gain an understanding of IT issues in the environment(s). Also, among all of the other integration or restructuring requirements, there will be a need to design the appropriate GEIT mechanisms for the new environment.

- **A shift in the market, economy or competitive position**—For example, an economic downturn could lead enterprises to revise GEIT mechanisms to enable large-scale cost reductions or performance improvement.
- **Change in business operating model or sourcing arrangements**—For example, a move from a decentralized or federated model toward a more centralized operating model will require changes to GEIT practices to enable more central IT decision making. Another example could be the implementation of shared service centers for areas such as finance, human resources (HR) or procurement. This may have IT impacts such as the consolidation of fragmented IT application or infrastructure domains with associated changes to the IT decision-making structures or processes that govern them. The outsourcing of some IT functions and business processes may similarly lead to a focus on GEIT.
- **New regulatory or compliance requirements**—As an example, expanded corporate governance reporting requirements and financial regulations trigger a need for better GEIT as well as the focus on information privacy caused by the pervasiveness of IT.
- **An enterprisewide governance focus or project**—Such projects will likely trigger initiatives in the GEIT area.
- **A new chief information officer (CIO), chief financial officer (CFO), chief executive officer (CEO) or board member**—The appointment of new C-level representatives can often trigger an assessment of current GEIT mechanisms and initiatives to address any weak areas found.
- **External audit or consultant assessments**—An assessment by an independent third party against appropriate practices can typically be the starting point of a GEIT improvement initiative.
- **A new business strategy or priority**—Pursuing a new business strategy will have GEIT implications. For example, a business strategy of being close to customers (i.e., knowing who they are, their requirements, and responding to these requirements in the best possible manner) may require more freedom of IT decision making for a business unit/country as opposed to central decision making at the corporate or holding level.
- **Desire to significantly improve the value to be gained from IT**—A need to improve competitive advantage, be innovative, or create new business opportunities can call attention to GEIT.

The need to act should be clear and widely solicited and communicated. This communication can be either in the form of a "wake-up call" (where pain points are being experienced) or an expression of the improvement opportunity to be pursued and benefits that will be realized. Current GEIT pain points or trigger events provide a starting point—the identification of these can typically be done through high-level health checks, diagnostics or capability assessments. These techniques have the added benefit of creating consensus on the issues to be addressed. It can be beneficial to ask a third party to perform a review to obtain an independent and objective high-level view on the current situation, which may increase buy-in to take action.

## 1.4 IT GOVERNANCE ENABLERS

Efficient and effective governance and management of enterprise IT requires a holistic approach, taking into account several interacting components. These components are enablers supporting implementation of a comprehensive governance and management system for enterprise IT. Enablers are broadly defined as anything that can help to achieve the objectives of the enterprise.

Governance enablers are the organizational resources for governance, such as frameworks, principles, structures, processes and practices, through or toward which action is directed and objectives can be attained. Enablers also include the enterprise's resources (e.g., service capabilities [IT infrastructure, applications, etc.], people and information).

### Holistic Approach of Structures, Processes and Relational Mechanisms

Efficient and effective governance and management of enterprise IT requires a holistic approach, taking into account several interacting components. These components are a set of enablers to support the implementation of a comprehensive governance and management system for enterprise IT.

This implementation challenge is related to what is described in strategic management literature as the need for an organizational system (i.e., "the way a firm gets its people to work together to carry out the business"[6]). Such an organizational system requires the definition and application, in a holistic manner, of structures (e.g., organizational units and functions) and processes (to ensure that tasks are coordinated and integrated), and attention to people and relational aspects (e.g., culture, values, joint beliefs).

De Haes and Van Grembergen[7] have applied this organizational system theory to the discussion of GEIT. These authors conclude that organizations can and are deploying GEIT by using a holistic mixture of various structures, processes and relational mechanisms. GEIT structures include organizational units and roles responsible for making IT decisions and for enabling contacts between business and IT management decision-making functions (e.g., IT steering committee). This can be seen as a blueprint for how the governance framework will be organized structurally. GEIT processes refer to the formalization and institutionalization of strategic IT decision making and IT monitoring procedures to ensure that daily behaviors are consistent with policies and provide input back to decisions (e.g., IT BSC). The relational mechanisms are ultimately about the active participation of, and collaborative relationship among, corporate executives, IT management and business management, and include mechanisms such as announcements, advocates and education efforts.

## COBIT 5 Enablers

COBIT 5 builds on these insights and talks about "enablers" in its framework. Enablers are defined as factors that, individually and collectively, influence whether something will work—in this case, governance and management over enterprise IT. The COBIT 5 framework describes seven categories of enablers, of which the processes enabler, the organizational structures enabler, and the culture, behavior and ethics enabler are closely related to the organizational systems concept (**figure 1.6**).

- **Principles, policies and frameworks** are the vehicles to translate the desired strategy into practical guidance for day-to-day management.
- **Processes** describe a set of practices and activities to achieve certain objectives and produce a set of outputs in support of achieving overall IT-related goals.
- **Organizational structures** are the key decision-making entities in an enterprise.
- **Culture, ethics and behavior** refer to the set of individual and collective behaviors within an enterprise. This enabler is key to effective performance along with other enablers such as process, organization structure and principles and policies.
- **Information** is pervasive throughout any organization and includes all information produced and used by the enterprise. Information is required for keeping the organization running and well governed, but at the operational level, information is very often the key product of the enterprise itself.
- **Services, infrastructure and applications** include the infrastructure, technology and applications that provide the enterprise with IT processing and services.
- **People, skills and competencies** are linked to people and are required for successful completion of all activities and for making correct decisions and taking corrective actions.

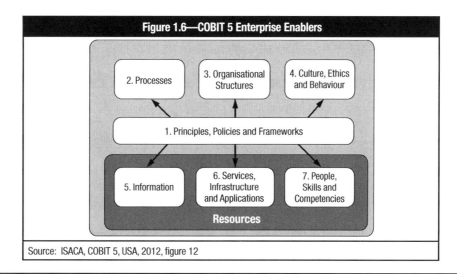

Source: ISACA, COBIT 5, USA, 2012, figure 12

Any enterprise must always consider an interconnected set of enablers. That is, each enabler:
- Needs the input of other enablers to be fully effective (e.g., processes need information, organizational structures need skills and behavior).
- Delivers output to the benefit of other enablers (e.g., processes deliver information, skills and behavior make processes efficient).

When dealing with governance and management of enterprise IT, good decisions can be made only when the systemic nature of governance and management enablers is taken into account. This means that all interrelated enablers must be analyzed for relevance and modified if necessary to address stakeholder needs. To be effective, senior management and the board of directors must drive the process.

Recent research performed by the University of Antwerp—Antwerp Management School and ISACA suggests that all of the enablers are considered to be very important by professionals, all having scored averages higher than 4 on a scale of 1 to 5 (**figure 1.7**).[8]

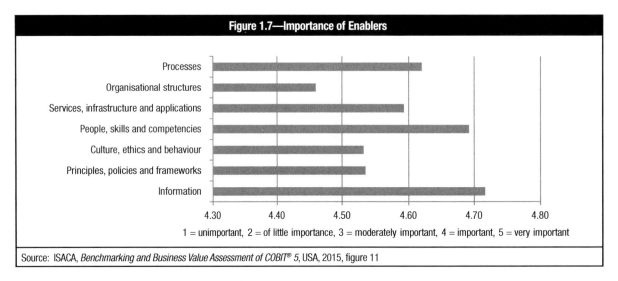

Source: ISACA, *Benchmarking and Business Value Assessment of COBIT® 5*, USA, 2015, figure 11

This would suggest that the seven GEIT enablers, as proposed by COBIT 5, are seen by the market as highly relevant as well as holistic and related to each other. Comparing the seven enablers relative to each other, it appears that the Information enabler and the People, Skills and Competencies enabler are perceived as the most important, closely followed by the Processes enabler.[9] Detailed guidance is published by ISACA regarding the Processes and Information enablers. The Skills Framework for the Information Age (SFIA), currently in version 5,[10] or the European e-Competence Framework[11] provide sound guidance in support of the People, Skills and Competencies enabler.

In the same research, it was analyzed whether GEIT enablers contribute to the achievement of IT-related goals and, by extension, enterprise goals. The study found a strong positive association between overall average score of implementation or management of seven enablers and the overall average score of IT-related goals (**figure 1.8**). The straight line in the graph implies a positive correlation between the enabler implementation or management to the IT-related goals. The study also confirmed a strong positive association between IT-related goals and overall enterprise goals, which suggests that IT-related outcomes positively link to the goals of the enterprise (**figure 1.9**).

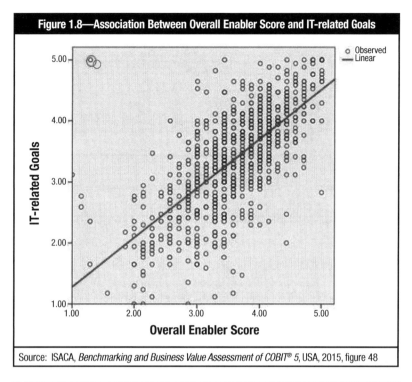

Source: ISACA, *Benchmarking and Business Value Assessment of COBIT® 5*, USA, 2015, figure 48

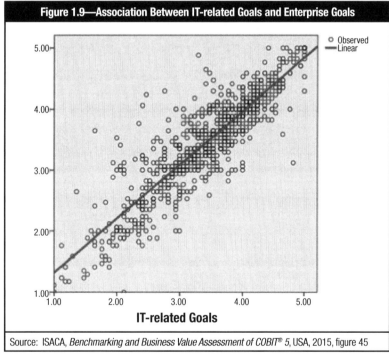

Source: ISACA, *Benchmarking and Business Value Assessment of COBIT® 5*, USA, 2015, figure 45

## 1.5 TECHNIQUES USED TO IDENTIFY IT STRATEGY

In setting direction and crafting approaches to implement enterprisewide improvements in GEIT, it is crucial to tie these initiatives to the organizational mission, vision and strategy. To be effective, GEIT needs to exist within these defining organizational statements. Therefore, it is important for practitioners to understand why the IT strategy is important and how the IT strategy can be linked to the enterprise strategy by leveraging techniques such as strengths, weaknesses, opportunities and threats (SWOT) analysis and the Boston Consulting Group's (BCG) growth share matrix.

Strategy setting is a fundamental part of governance. From the board and executive management viewpoint, the overall objective in this matter is to understand the issues and the strategic importance of IT so that the enterprise can sustain its operations and implement the strategies required to extend its activities into the future.[12]

Important strategic planning techniques that can help in understanding and defining the strategic importance of IT are SWOT analysis and the BCG's growth share matrix.

## SWOT Analysis

SWOT is a strategic planning method used to evaluate the strengths, weaknesses, opportunities and threats involved in a project or in a business venture, as shown in **figure 1.10**. Sometimes SWOT is referred to as SLOT (strengths, limitations, opportunities and threats). It involves specifying the objective of the business venture or project and identifying the internal and external factors that are favorable and unfavorable to achieve that objective. The objective should be set again after the SWOT analysis has been performed. This ensures achievable goals or objectives are set for the organization.

In more detail, the SWOT domains stand for:
- **Strengths**—Characteristics of the business or project team that give it an advantage over others
- **Weaknesses (or Limitations)**—Characteristics that place the team at a disadvantage relative to others
- **Opportunities**—*External* chances to improve performance (e.g., make greater profits) in the environment
- **Threats**—*External* elements in the environment that could cause trouble for the business or project

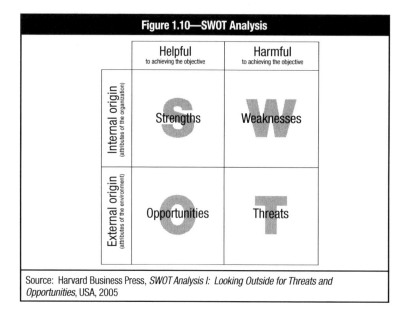

SWOT analyses can be applied to IT strategic discussions by evaluating current and future strengths, weaknesses, opportunities and threats of IT for the organization. Results of such a SWOT analysis can be used to understand the strategic role and direction of IT for the organization.

## BCG's Growth Share Matrix

The BCG's growth share matrix is a framework created by Bruce Henderson for the BCG in 1968 to help corporations with analyzing their business units or product lines. To use the chart, analysts plot a scatter graph to rank the business units (or products) on the basis of their relative market shares and growth rates.

BCG terms are:
- **Cash cows**—Units with high market share in a slow-growing industry. These units typically generate cash in excess of the amount of cash needed to maintain the business. They are regarded as staid and boring, in a "mature" market, and every corporation would be thrilled to own as many as possible. They are to be "milked" continuously with as little investment as possible, since such investment would be wasted in an industry with low growth.

- **Dogs** (also known as pets)—Units with low market share in a mature, slow-growing industry. These units typically "break even," generating barely enough cash to maintain the business's market share. Although owning a break-even unit provides the social benefit of providing jobs and possible synergies that assist other business units, from an accounting point of view such a unit is worthless, not generating cash for the company. They depress a profitable company's return on assets ratio, used by many investors to judge how well a company is being managed. Dogs, it is thought, should be sold off.
- **Question marks** (also known as problem children)—Are growing rapidly and thus consume large amounts of cash, but because they have low market shares they do not generate much cash. The result is a large net cash consumption. A question mark has the potential to gain market share and become a star, and eventually a cash cow when the market growth slows. If the question mark does not succeed in becoming the market leader, then after perhaps years of cash consumption it will degenerate into a dog when the market growth declines. Question marks must be analyzed carefully to determine whether they are worth the investment required to grow market share.
- **Stars**—Units with a high market share in a fast-growing industry. The hope is that stars become the next cash cows. Sustaining the business unit's market leadership may require extra cash, but this is worthwhile if that is what it takes for the unit to remain a leader. When growth slows, stars become cash cows if they have been able to maintain their category leadership, or they move from brief stardom to dogdom.

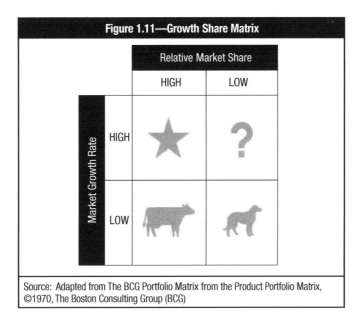

Source: Adapted from The BCG Portfolio Matrix from the Product Portfolio Matrix, ©1970, The Boston Consulting Group (BCG)

Growth share matrix analyses, as shown in **figure 1.11**, can be applied to the IT strategic discussion by evaluating the role of IT in each of the quadrants, and assessing how technology can help in moving from one quadrant to another. These insights can be used to understand the strategic role and direction of IT for the organization.

For information on the BSC, as a model to clarify the IT strategy, objectives and related metrics, see section 1.10 IT Governance Monitoring Processes/Mechanisms and section 1.11 IT Governance Reporting Processes/Mechanisms.

## Other Models and Methods for IT Strategy

Many other models are being used by organizations to think about their IT strategy and/or the role of IT in business strategy and operations of the enterprise. Well-known examples include:
- The IT balanced scorecard: see detailed explanation in section 1.10 IT Governance Monitoring Processes/Mechanisms, and section 1.11 IT Governance Reporting Processes/Mechanisms.
- Porter's five forces analysis: a framework for industry analysis and business strategy development addressing market forces such as rivalry among competitors, threats of new entrants in the market and substitute products and bargaining power of customers and clients.

- Porter's value chain model: a model that describes a chain of activities that a firm operating in a specific industry performs in order to deliver a valuable product or service to the market, composed of primary activities such as production and sales and secondary (supporting) activities such as procurement and human resources.
- The McKinsey 7S Framework: a tool often used to assess and monitor changes in the internal situation of an enterprise, addressing seven factors including strategy, skills, structure, system, staff, style and shared values.
- The McFarlan matrix on the strategic importance of IT: a framework that analyzes the role of IT for the enterprise in terms of the dependency on technology for innovation (value creation) and the dependency on reliable technology for running the enterprise.

COBIT 5 also proposes a very specific way to identify and link enterprise goals and IT-related goals, as a way to build IT strategic plans to translate these plans into action. Information regarding this cascade of enterprise goals and IT related goals can be found in section 2.2 Strategic Planning Processes and Techniques.

## 1.6 COMPONENTS, PRINCIPLES AND CONCEPTS RELATED TO ENTERPRISE ARCHITECTURE

GEIT and enterprise architecture (EA) are related and intertwined concepts. Understanding EA concepts is important for analyzing and anticipating the interdependencies between GEIT and EA adoptions in the organization.

Architecture can be defined as a representation of a conceptual framework of components and their relationships at a point in time.[13] Architecture discussions have traditionally focused on technology issues. It is, however, important to analyze architecture from the enterprise point of view, beginning at the top level, business architecture, and then drilling down.

EA is a key process in the context of IT strategy and of GEIT in general. By depicting the interrelationship between business processes, applications, underlying data and infrastructure, EA helps in strategizing IT and GEIT initiatives.

### Understanding Enterprise Architecture

EA takes a broader view of business, matching it with the associated information. It provides the framework for ensuring that enterprisewide goals, objectives and policies are properly and accurately reflected in decision making related to building, implementing or changing information systems and to provide reasonable assurance that standards for interprocess communication, data naming, data representation, data structures and information systems will be consistently and appropriately applied across the enterprise. As such, EA discussions in many organizations are a crucial part in GEIT programs.

The widely accepted US National Institute of Standards and Technology (NIST) model for EA[14] is shown in **figure 1.12**.

An enterprise is composed of one or more business units that are responsible for a specific business area. As the US National Association of State CIOs (NASCIO) EA Development Tool-Kit[15] states:

> *Adopting EA increases the utility of an enterprise's data by facilitating information sharing between data stores. Committing to an ongoing renewable EA process fosters a technology-adaptive enterprise. EA becomes a road map, guiding all future technology investments and identifying and aiding in the resolution of gaps in the entity's business and IT infrastructures.*

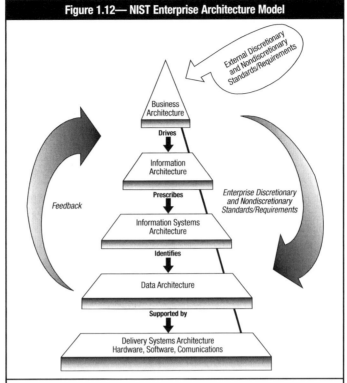

Figure 1.12— NIST Enterprise Architecture Model

Source: National Institute of Standards and Technology; *NIST Special Publication 500-167: Information Management Decisions: The Integration Challenge*, USA, 1989. Reprinted courtesy of the National Institute of Standards and Technology, US Department of Commerce. Not copyrightable in the United States.

## Layers of Enterprise Architecture

Typical EA is the inclusive term used to describe the five layers of architecture:
- Business unit architecture (or business architecture)
- Information architecture
- Information systems architecture (sometimes called solution architecture)
- Data architecture
- Delivery system architecture (sometimes called technology architecture)

These five levels of EA were first introduced in 1989 by NIST and remain valid and can be readily applied in today's extended enterprise environment. A number of popular EA frameworks such as Zachman, TOGAF, Federal Enterprise Architecture Framework (FEAF), ISO 15704, ISO/IEC 15288 and others are broadly based on this paradigm. The five-layered model is able to allow for organizing, planning and building an integrated set of information and IT architecture. Each of these layers is described as follows:
- **Business unit architecture (business processes)**—This component describes the core business processes that support the enterprise's missions. Components for the business unit architecture generally focus on external and internal reporting requirements and functional areas. From the discretionary standards perspective that an enterprise may select as part of its architecture, standards could be based on policies used by like industries, both nationally and internationally—standards that would provide reusability of assets and migration from the current environment to a proposed environment, as well as standards for information sharing. From the mandatory standards perspective, an enterprise must adhere to best business practices and legislation. The major component of this architecture is a high-level analysis of the work performed in support of the enterprise's mission, vision and goals. Business processes can be described by decomposing the processes into derivative business activities. Analysis of the business processes determines the information needed by the enterprise. Each business process should incorporate performance management structure in accordance with the Plan-Do-Check-Act (PDCA or Shewhart) cycle.

- **Information architecture (information flows and relationships)**—This component analyzes the information used by the enterprise in its business processes—identifying the information used and the movement of the information within the enterprise. Components of this architecture include: original documents, data, revisions, classification, and responsible organizational units. Relationships among the various flows of information also need to be described in this component—to indicate where the information is needed and how the information is shared to support mission functions. This architecture level represents technical and management information flow as well as the impact of time on information integrity and meaning.
- **Information systems architecture (applications)**—This component identifies, defines and organizes the activities that capture, manipulate and manage the business information to support mission operations as well as the logical dependencies and relationships among business activities. It establishes a framework to meet the specific information requirements required by the information architecture. It uses its components to acquire and process data, shows the automated and procedure-oriented information system that supports the information flow, and produces and distributes information according to the architecture requirements and standards. Components for the information systems architecture refer to specifications, requirements, applications, modules, databases and procedures.
- **Data architecture (data descriptions)**—This component identifies how data are maintained, accessed and utilized. At a high level, it defines the data and describes the relationship among data elements in the enterprise's information systems. It also interfaces with the application system component—to store or locate information required for processing or for subsequent storage by application systems. Components for this architecture layer can include data models that describe the nature of the data underlying the business and information needs, such as physical database design, database and file structures, data definitions, data dictionaries and data elements underlying the information systems of the enterprise. In formulating these components, it is important that redundancy be minimized and that new applications are supported.
- **Delivery system architecture (technology infrastructure)**—The delivery system architecture (technology and communication infrastructure) describes and identifies the information service layer, network service layer, and components, including the functional characteristics, capabilities and interconnections of the hardware, software and communications (networks, protocols and nodes). It represents the "wiring diagram" of the physical IT infrastructure and facility support requirements so that these assets can be properly accommodated and connected in an integrated manner.

The architecture layers outlined previously are mutually interdependent and interrelated. For example, the first four layers are logically connected and related in a top-down dependency. The delivery system is the foundation of the architecture and is dependent on the definition of the business goals and objectives. An architecture may be a description of one of these layers at a particular point in time and may represent a view of a current situation with islands of automation, redundant processes and data inconsistencies. It can also be a representation of a future integrated automation structure or end state that is in the enterprise's migration plan and gives context and guidance for future activities.

More information on EA can be found in section 2.7 Components, Principles and Frameworks of Enterprise Architecture and section 2.8 Current and Future Technologies.

## 1.7 ORGANIZATIONAL STRUCTURES AND THEIR ROLES AND RESPONSIBILITIES

Effective governance of enterprise IT includes the governance of organizational structures to ensure that IT-related decisions occur in a transparent environment and to enable effective contact and exchange between business and IT management. These structures can be seen as a blueprint that shows how the framework for GEIT will function conceptually.

As discussed in section 1.4 IT Governance Enablers, efficient and effective governance and management of enterprise IT requires a holistic approach, taking into account several interacting components. These components are a set of enablers to support the implementation of a comprehensive governance and management system for enterprise IT.

The COBIT 5 framework describes seven categories of enablers, including organizational structures, which are defined as the key decision-making entities in an enterprise.

## Specifics of Organizational Structures

The specifics for the organizational structures enablers, in terms of stakeholders, goals, life cycle and good practices are discussed below:

- **Stakeholders**—Organizational structures stakeholders can be internal and external to the enterprise, and they include the individual members of the structure, other structures, organizational entities, clients, suppliers and regulators. Their roles vary, and include decision making, influencing and advising. The interests of each stakeholder vary relative to the decisions made by the structure.
- **Goals**—The goals for the organizational structures enabler itself would include having a proper mandate, well-defined operating principles and application of other good practices. The outcome of the organizational structures enabler should include a number of good activities and decisions.
- **Life cycle**—An organizational structure has a life cycle. It is created, exists and is adjusted, and finally it can be disbanded. During its inception, a mandate—a reason and purpose for its existence—has to be defined.
- **Good practices**—A number of good practices for organizational structures can be distinguished such as:
  - Operating principles—These are the practical arrangements regarding how the structure will operate, such as frequency of meetings, documentation and housekeeping rules.
  - Composition—Structures have members, who are internal or external stakeholders.
  - Span of control—This includes the boundaries of the organizational structure's decision rights.
  - Level of authority/decision rights—These are the decisions that the structure is authorized to take.
  - Delegation of authority—The structure can delegate (a subset of) its decision rights to other structures reporting to it.
  - Escalation procedures—The escalation path for a structure describes the required actions in case of problems in making decisions.

## Structures as a Basis to Build RACI Charts

The COBIT 5 process reference model also provides a number of illustrative organizational structures as a basis to build RACI charts (Responsible, Accountable, Consulted, Informed); see **figure 1.13**. These illustrations do not necessarily need to correspond with actual functions that enterprises have implemented, but nonetheless they provide value in the sense that the described purpose of the structure or the role remains valid for most enterprises.

| Figure 1.13—COBIT 5 Roles and Organizational Structures | |
|---|---|
| **Role/Structure** | **Definition/Description** |
| Board | The group of the most senior executives and/or non-executive directors of the enterprise who are accountable for the governance of the enterprise and have overall control of its resources |
| CEO | The highest-ranking officer who is in charge of the total management of the enterprise |
| CFO | The most senior official of the enterprise who is accountable for all aspects of financial management, including financial risk and controls and reliable and accurate accounts |
| Chief Operating Officer (COO) | The most senior official of the enterprise who is accountable for the operation of the enterprise |
| CRO | The most senior official of the enterprise who is accountable for all aspects of risk management across the enterprise. An IT risk officer function may be established to oversee IT-related risk. |
| CIO | The most senior official of the enterprise who is responsible for aligning IT and business strategies and accountable for planning, resourcing and managing the delivery of IT services and solutions to support enterprise objectives |
| Chief Information Security Officer (CISO) | The most senior official of the enterprise who is accountable for the security of enterprise information in all its forms |
| Business Executive | A senior management individual accountable for the operation of a specific business unit or subsidiary |
| Business Process Owner | An individual accountable for the performance of a process in realising its objectives, driving process improvement and approving process changes |
| Strategy (IT Executive) Committee | A group of senior executives appointed by the board to ensure that the board is involved in, and kept informed of, major IT-related matters and decisions. The committee is accountable for managing the portfolios of IT-enabled investments, IT services and IT assets, ensuring that value is delivered and risk is managed. The committee is normally chaired by a board member, not by the CIO. |

*Chapter 1—Framework for the Governance of Enterprise IT*
*Section Two: Content*

| Figure 1.13—COBIT 5 Roles and Organizational Structures *(cont.)* ||
|---|---|
| **Role/Structure** | **Definition/Description** |
| (Project and Programme) Steering Committees | A group of stakeholders and experts who are accountable for guidance of programmes and projects, including management and monitoring of plans, allocation of resources, delivery of benefits and value, and management of programme and project risk |
| Architecture Board | A group of stakeholders and experts who are accountable for guidance on enterprise architecture-related matters and decisions, and for setting architectural policies and standards |
| Enterprise Risk Committee | The group of executives of the enterprise who are accountable for the enterprise-level collaboration and consensus required to support enterprise risk management (ERM) activities and decisions. An IT risk council may be established to consider IT risk in more detail and advise the enterprise risk committee. |
| Head of HR | The most senior official of an enterprise who is accountable for planning and policies with respect to all human resources in that enterprise |
| Compliance | The function in the enterprise responsible for guidance on legal, regulatory and contractual compliance |
| Audit | The function in the enterprise responsible for provision of internal audits |
| Head of Architecture | A senior individual accountable for the enterprise architecture process |
| Head of Development | A senior individual accountable for IT-related solution development processes |
| Head of IT Operations | A senior individual accountable for the IT operational environments and infrastructure |
| Head of IT Administration | A senior individual accountable for IT-related records and responsible for supporting IT-related administrative matters |
| Programme and Project Management Office (PMO) | The function responsible for supporting programme and project managers, and gathering, assessing and reporting information about the conduct of their programmes and constituent projects |
| Value Management Office (VMO) | The function that acts as the secretariat for managing investment and service portfolios, including assessing and advising on investment opportunities and business cases, recommending value governance/management methods and controls, and reporting on progress on sustaining and creating value from investments and services |
| Service Manager | An individual who manages the development, implementation, evaluation and ongoing management of new and existing products and services for a specific customer (user) or group of customers (users) |
| Information Security Manager | An individual who manages, designs, oversees and/or assesses an enterprise's information security |
| Business Continuity Manager | An individual who manages, designs, oversees and/or assesses an enterprise's business continuity capability, to ensure that the enterprise's critical functions continue to operate following disruptive events |
| Privacy Officer | An individual who is responsible for monitoring the risk and business impacts of privacy laws and for guiding and co-ordinating the implementation of policies and activities that will ensure that the privacy directives are met. Also called data protection officer. |
| Source: ISACA, COBIT 5, USA, 2012, figure 33 ||

The roles mentioned in **figure 1.13** are the basis to build a RACI chart:
- **Responsible**—Refers to the person who must ensure that activities are completed successfully
- **Accountable**—Refers to the person or group who has the authority to approve or accept the execution of an activity
- **Consulted**—Refers to those people whose opinions are sought on an activity (two-way communication)
- **Informed**—Refers to those people who are kept up to date on the progress of an activity (one-way communication)

Such a RACI chart typically decomposes a process into different subprocesses and then indicates roles for each of the structures identified. **Figure 1.14** shows the example of a RACI chart for the COBIT process APO02 *Manage Strategy*, which clearly defines responsibilities and accountabilities at both the business and IT sides for this process.

### Figure 1.14—COBIT 5 APO02—Manage Strategy RACI Chart

| Key Management Practice | Board | Chief Executive Officer | Chief Financial Officer | Chief Operating Officer | Business Executives | Business Process Owners | Strategy Executive Committee | Steering (Programmes/Projects) Committee | Project Management Office | Value Management Office | Chief Risk Officer | Chief Information Security Officer | Architecture Board | Enterprise Risk Committee | Head Human Resources | Compliance | Audit | Chief Information Officer | Head Architect | Head Development | Head IT Operations | Head IT Administration | Service Manager | Information Security Manager | Business Continuity Manager | Privacy Officer |
|---|---|---|---|---|---|---|---|---|---|---|---|---|---|---|---|---|---|---|---|---|---|---|---|---|---|---|
| APO02.01 Understand enterprise direction. | | C | C | C | A | C | C | | | | C | C | | C | | | | R | C | R | R | | R | R | R | |
| APO02.02 Assess the current environment, capabilities and performance. | | C | C | C | R | C | C | | | | C | | | | | C | C | A | R | R | R | C | C | C | C | |
| APO02.03 Define the target IT capabilities. | | A | C | C | C | I | R | | I | | C | | C | | | C | C | R | C | C | C | C | C | C | C | |
| APO02.04 Conduct a gap analysis. | | | | | R | R | C | | | | C | | | | C | R | R | A | R | R | R | R | R | R | C | |
| APO02.05 Define the strategic plan and road map. | | C | I | C | C | | C | R | | | C | C | | | | C | C | A | C | C | C | C | C | C | C | |
| APO02.06 Communicate the IT strategy and direction. | I | R | I | I | R | I | A | I | I | I | I | I | I | I | I | I | I | R | I | I | I | I | I | I | I | I |

Source: ISACA, *COBIT 5: Enabling Processes*, USA, 2012, page 58

## Governance of Enterprise IT Arrangements

The importance of organization structure for GEIT has also been discussed by Weill and Ross.[16] According to the authors, IT governance is all about specifying the decision rights and accountability framework to encourage desirable behavior in the use of the IT. More specifically, they studied how firms make decisions in five key interrelated IT domains:
- IT principles
- IT infrastructure
- IT architecture
- Business applications needs
- IT investments and prioritization

In answer to the question "Who should make governance decisions?" they define six governance of enterprise IT archetypes, or styles, as shown in **figure 1.15**, which describes who within the enterprise has decision rights or provides input to IT decisions.

They found that the enterprise in the study often showed different governance archetypes for different IT decision domains, both for making decisions and for providing input to decisions. The federal governance model is typically used as input for the three more business-related IT decisions (IT principles, business applications and IT investments). Enterprises mainly rely on IT monarchies when choosing an IT architecture or making IT infrastructure strategy decisions because both are seen as more technical activities. In their research, Weill and Ross also concluded that the federal style is the most effective for input to all five key IT decisions. Indeed, "the federal model for input provides a broad-based vehicle for capturing the tradeoffs between the desires of the senior corporate managers and the managers in the business units." For making decisions, on the other hand, the federal model in

Chapter 1—Framework for the Governance of Enterprise IT
Section Two: Content

general is experienced as less effective mainly because too many people are involved, which can slow down the decision-making process and create too many compromises, which may block the real needs of the business. Top governance performers often used duopolies for both IT principles and investments, enabling joint decision making between the business leaders and IT professionals.

| Figure 1.15—Who Should Make Governance Decisions? | |
|---|---|
| **Governance of Enterprise IT Style** | **Who has decision or input rights?** |
| Business monarchy | A group of business executives or individual executives (CxOs); includes committees of senior business executives (may include CIO); excludes IT executives acting independently |
| IT monarchy | Individuals or groups of IT executives |
| Feudal | Business unit leaders, key process owners or their delegates |
| Federal | C-level executives and business groups (e.g., business units or processes); may also include IT executives as additional participants |
| Duopoly | IT executives and at least one other group (e.g., CxO or business unit or process leaders) |
| Anarchy | Each individual business process owner or end user |
| Source: ©2003 MIT Sloan School Center for Information Systems Research (CISR). Used with permission. | |

## 1.8 METHODS TO MANAGE ORGANIZATIONAL, PROCESS AND CULTURAL CHANGE

Successful implementation or improvement depends on implementing the appropriate change in the correct way. In many enterprises, there is not enough emphasis on managing the human, behavioral and cultural aspects of the change and motivating stakeholders to buy into the change. Change enablement[17] is one of the biggest challenges to GEIT implementation.

It is important for the appropriate environment and culture to exist when implementing GEIT improvements. This helps ensure that the initiative itself is governed and adequately guided and supported by management. Major IT initiatives often fail due to inadequate management direction, support and oversight and lack of appropriate cultural context. GEIT implementations are no different; they have more chance of success if they are well governed and well managed, with sufficient attention to the human behavior aspect.

Inadequate support and direction from key stakeholders can, for example, result in GEIT initiatives producing new policies and procedures that have no proper ownership. Process improvements are unlikely to become normal business practices without a management structure that assigns roles and responsibilities, commits to their continued operation and monitors conformance. An appropriate environment should be created and maintained to ensure that GEIT is implemented as an integral part of an overall governance approach within the enterprise.

The elements of change, culture and human behavior should also be addressed when writing a business case around a GEIT program. The importance of addressing these elements in the business case is addressed in section 3.1 IT Investment Management Processes, Including the Economic Life Cycle of Investments.

### Change Enablement

Change enablement is one of the biggest challenges to GEIT implementation. It should not be assumed that the various stakeholders involved in, or affected by, new or revised governance arrangements will readily accept and adopt the change. The possibility of ignorance and/or resistance to change needs to be addressed through a structured and proactive approach. Also, optimal awareness of the program should be achieved through a communication plan that defines what will be communicated, in what way and by whom throughout the various phases of the program.

COBIT 5 defines change enablement as:

> *A systematic process of ensuring that all stakeholders are prepared and committed to the changes involved in moving from a current state to a desired future state.*

All key stakeholders should be involved. At a high level, change enablement typically entails:
- Assessing the impact of the change on the enterprise, its people and other stakeholders
- Establishing the future state (vision) in human/behavioral terms and the associated measures to describe it
- Building "change response plans" to manage change impact proactively and maximize engagement throughout the process. These plans may include training, communication, organization design, process redesign and updated performance management systems.
- Continuously measuring the change progress toward the desired future state

In terms of a typical GEIT implementation, the objective of change enablement is having enterprise stakeholders from the business and IT leading by example and encouraging staff at all levels to work according to the desired new way. Examples of desired behavior include:
- Follow agreed-on processes.
- Participate in the defined GEIT structures such as a change approval or advisory board.
- Enforce guiding principles, policies, standards, processes or practices such as a policy regarding new investment or security.

This can be best achieved by gaining the commitment of the stakeholders (diligence and due care, leadership, and communicating and responding to the workforce) and selling the benefits. If necessary, it may be required to enforce compliance. In other words, human, behavioral and cultural barriers must be overcome so that there is a common interest to properly adopt, create a will to adopt and ensure the ability to adopt a new way of doing something. It may be useful to draw on change enablement skills within the enterprise or, if necessary, from external consultants to facilitate the change in the organization.

## Kotter's Implementation Life Cycle

Various approaches to enabling change have been defined over the years, and they provide valuable input that could be utilized during the implementation life cycle. One of the most widely accepted approaches to change enablement has been developed by Kotter:[18]

1. Establish a sense of urgency.
    If a change is backed by the whole enterprise, it is much more likely that the change will be implemented. A sense of urgency needs to be developed around the need for change in order to establish initial motivation. In terms of GEIT implementation, the identification and communication of pain points discussed in section 1.3 Business Drivers Related to IT Governance could help to establish a sense of urgency in the organization for GEIT implementation.
2. Form a powerful guiding coalition.
    It is important to convince people that change is necessary. This often takes strong leadership and visible support from key people within the enterprise. Managing change is not enough—someone needs to lead it.
3. Create a clear vision that is expressed simply.
    When change is first considered, there will probably be many great ideas and solutions floating around. These concepts should be linked to an overall vision that people can grasp easily and remember.
4. Communicate the vision.
    Communicating the vision is key to change success. The message of change will probably have strong competition from other day-to-day communications within the company, so it needs to be communicated frequently and powerfully.
5. Empower others to act on the vision.
    By following these steps, buy-in from all levels of the enterprise can be created. Hopefully, staff members will be enthusiastic about the benefits that have been promoted in the change message.
6. Plan for and create short-term wins.
    Nothing motivates more than success. A taste of victory early in the change process can be very powerful. Within a short time frame (this could be a month or a year, depending on the type of change), it is important to have results that staff can see. Without this, critics and negative thinkers might hurt the progress.
7. Consolidate improvements and produce more change.
    Kotter argues that many change projects fail because victory is declared too early. Real change runs deep. Quick wins are only the beginning of what needs to be done to achieve long-term change.
8. Institutionalize new approaches.
    Finally, to make any change stick, it should become part of the core of the enterprise. Corporate culture often determines what gets done, so the values behind the vision must show in day-to-day work.

The Kotter approach has been chosen as an example and adapted for the specific requirements of a GEIT implementation or improvement. This is illustrated by the implementation life cycle in **figure 1.16**, providing a high-level, but holistic, overview of each phase of the change enablement life cycle as applied to a typical GEIT implementation.

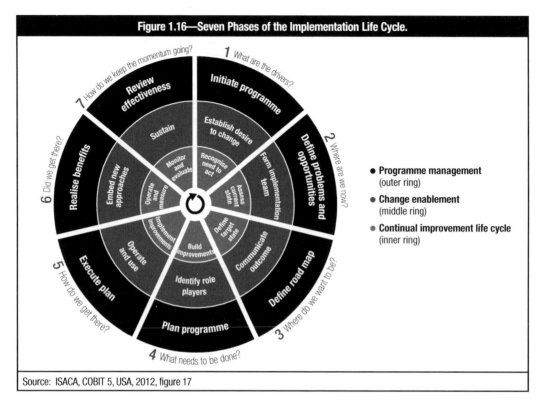

Source: ISACA, COBIT 5, USA, 2012, figure 17

### Lewin/Schein's Change Theory—Unfreeze–Change–Refreeze

One of the cornerstone models for understanding organizational change was developed in the 1940s by Lewin/Schein, and the model still holds true today. Their model is known as Unfreeze–Change–Refreeze, which refers to a three-stage process of change. By recognizing these three distinct stages of change, a plan can be put in place to implement the change required. The first stage is creating the motivation to change (unfreeze). The next step in the change process is promoting effective communications and empowering people to embrace new ways of working (change). The process ends with returning the enterprise to a sense of stability (refreeze), which is necessary for creating the confidence from which to embark on the next, inevitable change.

More information on continuous improvement techniques is discussed in section 1.14 Continuous Improvement Techniques and Processes.

## 1.9 MODELS AND METHODS TO ESTABLISH ACCOUNTABILITY FOR INFORMATION REQUIREMENTS, DATA AND SYSTEM OWNERSHIP, AND IT PROCESSES

Accountability is crucial for effective governance. For GEIT, accountability has a special context—for information requirements, data and system ownership and IT-related processes. How accountability is established, both at management (plan, build, run, monitor) and governance (evaluate, direct, monitor) levels, needs to be well understood for good GEIT to be in place.

One of the guiding principles in COBIT 5 is the distinction made between governance and management. In line with this principle, every enterprise would be expected to implement a number of governance processes and a number of management processes to provide comprehensive governance and management of enterprise IT.

## Distinction Between Governance and Management

In the discussion around accountability for governance, it is crucial to understand the difference between management and governance roles. The COBIT 5 framework makes a clear distinction between governance and management, as shown in **figure 1.17**. These two disciplines encompass different types of activities, require different prioritization structures and serve different purposes.

The COBIT 5 view on this key distinction between governance and management is:

- *Governance ensures that stakeholder needs, conditions and options are evaluated to determine balanced, agreed-on enterprise objectives to be achieved; setting direction through prioritization and decision making; and monitoring performance and compliance against agreed-on direction and objectives.*

    *In most enterprises, governance is the responsibility of the board of directors under the leadership of the chairperson.*

- *Management plans, builds, runs and monitors activities in alignment with the direction set by the governance body to achieve the enterprise objectives.*

    *In most enterprises, management is the responsibility of the executive management under the leadership of the CEO.*

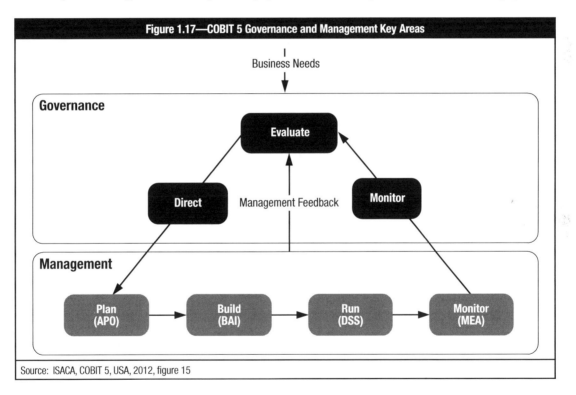

Source: ISACA, COBIT 5, USA, 2012, figure 15

## Governance and Management Roles, Activities and Relationships

In COBIT 5, clear differentiation is made between governance and management activities in the governance and management domains, as well as the interfacing between them and the role players that are involved. **Figure 1.18** details the interactions between the different roles.

# Chapter 1—Framework for the Governance of Enterprise IT
## Section Two: Content

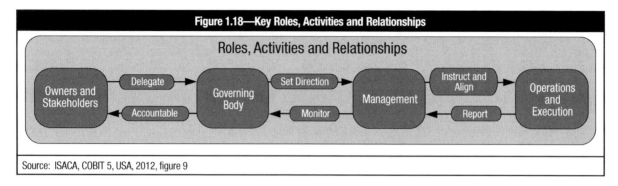

In general, responsible parties refer to the person who must ensure that activities are completed successfully. A RACI chart (as defined in section 1.7 Organizational Structures and Their Roles and Responsibilities) answers the question "Who is getting the task done?" Accountable parties are the individual, group or entities that are ultimately responsible for a subject matter, process or scope. A RACI chart also answers the question "Who accounts for the success of the task?"

### Accountability

Within the context of GEIT, the establishment of accountability is of particular importance in relation to information requirements, data and system ownership, and IT-related processes. Information requirements are rooted in business. The following information is borrowed from the ISACA publication *COBIT 5: Enabling Information*.[19]

Information governance ensures that:
- Stakeholder needs, conditions and options are evaluated to determine balanced, agreed-upon enterprise objectives, which are to be achieved through the acquisition and management of information resources.
- Direction is set for information management capabilities through prioritization and decision making.
- Performance and compliance of the information resource are monitored against agreed-upon direction and objectives.

Information management plans, builds, runs and monitors the practices, projects and capabilities that acquire, control, protect, deliver and enhance the value of data and information assets, in alignment with the direction set by the data and information governance body.

## 1.10 IT GOVERNANCE MONITORING PROCESSES/MECHANISMS

The concept of performance management and monitoring and its associated techniques are highly useful to improve and map GEIT to organizational imperatives. The use of the BSC and its mechanisms to translate strategy into measurable action plays an important role in this context.

Measuring IT performance should be a key concern of both business and IT executives because it demonstrates the effectiveness and added business value of IT. Many methods, tools and good practices exist to support these executives with performance management responsibilities. Traditional performance methods such as return on investment (ROI) capture the financial worth of IT projects and systems but reflect only a limited (tangible) part of the value that can be delivered by IT. The more sophisticated IT BSC is an evaluation method that incorporates both tangible and intangible values.

The IT BSC can be leveraged as a management system to enable fusion between IT and the business, and it can also be an effective means for IT management to communicate with and report to the board of directors and executive management about the business value of IT.

Combining these practices with good IT portfolio management, which helps in achieving an optimal mix of programs, creates a solid foundation for a balanced IT governance approach in the enterprise. This section addresses the importance of IT performance management and discusses some methods, tools and good practices to support this concept.

## Importance of IT Performance Management

Investments in IT are growing extensively, and business managers often worry that the benefits of IT investments might not be as high as expected. The same worry applies to the perceived ever-increasing total cost of the IT department, without clear evidence of the value derived from it. This phenomenon is called the "IT black hole"; large sums go in, but no returns (seem to) come out.

Getting business value from IT and measuring that value are, therefore, important governance objectives. They are responsibilities of both the business and IT and should take both tangible and intangible costs and benefits into account. In this way, good IT performance management should enable both the business and IT to fully understand how IT is contributing to the achievement of business goals, in the past and in the future. Measuring and managing IT performance should provide answers to questions such as:
- If I spend extra funds on IT, what do I get back?
- How does my IT benchmark against competitors?
- Do I get back from IT what was promised?
- How do I learn from past performance to optimize my organization?
- Is my IT implementing its strategy in line with the business strategy?

## Current IT Performance Management Governance Approaches

IT performance management is aimed at identifying and quantifying IT costs and IT benefits. There are different monitoring instruments available, depending on the features of the costs and benefits. When both costs and benefits can be easily quantified and assigned a monetary value, traditional performance measures such as ROI, net present value (NPV), internal rate of return (IRR) and payback method work well, as shown in **figure 1.19**.

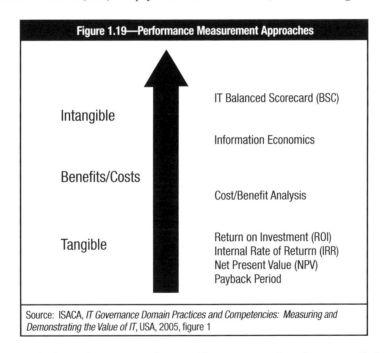

Source: ISACA, *IT Governance Domain Practices and Competencies: Measuring and Demonstrating the Value of IT*, USA, 2005, figure 1

Because the traditional methods need monetary values, problems emerge when they are applied to information systems, which often generate intangible benefits such as better customer service. Moreover, different levels of management and users perceive the value of IT differently. Weill and Broadbent refer in this context to the "business value hierarchy."[20] Very successful investments in IT have a positive impact on all levels of the business value hierarchy. Less successful investments are not strong enough to impact the higher levels and consequently influence only the lower levels. The higher one goes in the measurement hierarchy, the more dilution that occurs from factors such as pricing decisions and competitors' moves. This dilution means that measuring the impact of an IT investment is much easier at the bottom of the hierarchy than at the top.

Multicriteria measurement methods may solve this problem because they account for both tangible and intangible impacts, where the latter are more typical for the higher business value hierarchies. One of the best known multicriteria methods is information economics (IE), which in essence is a scoring technique whereby a mix of tangible benefits (typically ROI) and intangible benefits are scored.

The aforementioned performance measurement methods are measurement instruments for individual IT projects and investments. A broader performance measurement technique is the BSC, which can be applied to IT projects, investments and even entire IT departments. The BSC, initially developed on the enterprise level by Kaplan and Norton,[21] is a performance management system that enables businesses to drive strategies based on measurement and follow-up. The idea behind the BSC is that the evaluation of a firm should not be restricted to the traditional financial measures, but should be supplemented with a mission, objectives and measures regarding customer satisfaction, internal processes, and the ability to innovate and prepare for the future. Results achieved within the additional perspectives should assure financial results. The objectives and measures of a BSC can be used as a cornerstone of a management system that uncovers and communicates strategies, establishes long-term strategic targets, aligns initiatives, allocates long- and short-term resources and, finally, provides feedback and learning about the strategies.

## Good Practices for IT Performance Management

"Use of an IT BSC is one of the most effective means to aid the board and management to achieve IT and business alignment."[22] In these words, the IT Governance Institute promotes the IT BSC as a good practice for performance measurement and alignment. This is supported by testimonials of several executives, such as:

> "The major advantage of the IT BSC is that it provides a systematic translation of the strategy into critical success factors and metrics, which materialises the strategy." (CIO of a financial organization)[23]

> "The Balanced Scorecard gives a balanced view of the total value delivery of IT to the business. It provides a snapshot of where your IT organization is at a certain point in time. Most executives, like me, do not have the time to drill down into the large amount of information." (Vice president of an insurance organization)[24]

To apply this best practice to the IT function as an internal service provider, the four perspectives of the generic BSC (as described above) should be changed accordingly. In **figure 1.20**, a generic IT BSC for an IT department is shown.[25]

The user orientation perspective represents the user evaluation of IT. The operational excellence perspective represents the IT processes employed to develop and deliver the applications. The future orientation perspective represents the human and technology resources needed by IT to deliver its services over time. The business contribution perspective captures the business value created from the IT investments.

Each of these perspectives must be translated into corresponding metrics and measures that assess the current situation. As noted previously, the cause-and-effect relationships between measures are essential components of the IT BSC, and these relationships are articulated by two types of measures:
• Outcome measures
• Performance drivers

Outcome measures, such as programmers' productivity (e.g., number of function points per person per month), need performance drivers, such as IT staff education (e.g., number of education days per person per year), to communicate how the outcomes are to be achieved.

Performance drivers need outcome measures to ensure a way to determine whether the chosen strategy is effective, especially important in cases where a significant investment is made. These cause-and-effect relationships must be defined throughout the entire scorecard: more and better education of IT staff (future orientation) is an enabler (performance driver) for a better quality of developed systems (operational excellence perspective) that in turn is an enabler for increased user satisfaction (user perspective) that eventually will lead to higher business value of IT (business contribution).

# Chapter 1—Framework for the Governance of Enterprise IT
## Section Two: Content

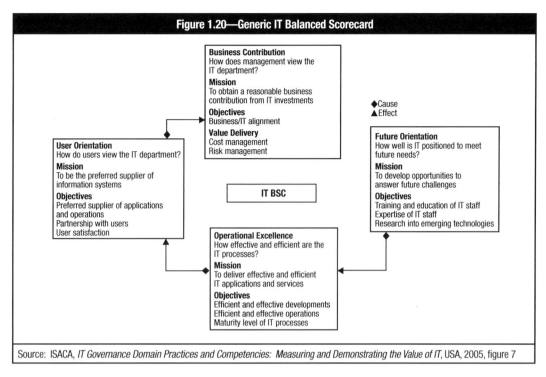

The proposed IT BSC shown in **figure 1.20** links with business through the business contribution perspective (business/IT alignment, value delivery, cost management and risk management). The relationship between IT and business can be more explicitly expressed through a cascade of BSCs. In **figure 1.21**, the relationship between IT BSCs and the business BSC is illustrated. The IT development BSC and the IT operational BSC are both enablers of the IT strategic BSC, which in turn is the enabler of the business BSC. This cascade of scorecards becomes a linked set of measures that will be instrumental in achieving governance of enterprise IT through aligning IT and business strategy and showing how business value is created through IT.

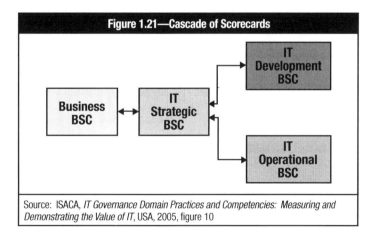

More information regarding outcomes and performance measurement techniques can be found in section 3.7 Outcome and Performance Measurement Techniques.

## 1.11 IT GOVERNANCE REPORTING PROCESSES/MECHANISMS

For effective communication with all stakeholders, it is important to ensure that the enterprise IT performance and conformance measurement and reporting are transparent.

Stakeholders want a say in determining what they expect from information and related technology (what benefits at what acceptable level of risk and at what costs) and what their priorities are in ensuring that expected value is actually being delivered. Some will want short-term returns and others long-term sustainability. Some will be ready to take a high risk that others will not.

These divergent and sometimes conflicting expectations need to be dealt with effectively. In short, stakeholders want:
- To be more involved in setting expectations
- More transparency regarding how expectations will be met
- Assurance that actual results are achieved

This means that stakeholders demand:
- Financial transparency
- Transparency of IT costs, benefits and risk reporting
- Delivery of IT services in line with business requirements

Because IT is a complex and technical topic, it is important to achieve transparency by expressing goals, metrics and performance reports in language meaningful to the stakeholders so that appropriate actions can be taken.

Stakeholder transparency is important to ensure that the communication to stakeholders is effective and timely and the basis for reporting is established to increase performance, identify areas for improvement, and confirm that IT-related objectives and strategies are in line with the enterprise's strategy.

As shown in **figure 1.22**, the COBIT 5 framework provides more detailed guidance and governance practices, in terms of governance of enterprise IT reporting, in line with the Evaluate—Direct—Monitor concept as discussed earlier in this chapter.

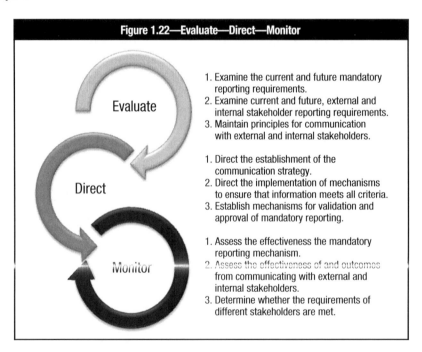

**Chapter 1—Framework for the Governance of Enterprise IT**
**Section Two: Content**

Examples of further specifying the activities for evaluating stakeholder needs, as shown in **figure 1.23**, include:
• IT-related requirements for reporting to regulatory entities are assessed as part of the governance process.
• Different regulatory reporting requirements are cross-referenced to enable harmonization, where possible.
• Geographic reporting needs are considered as part of a holistic stakeholder needs review.

| Figure 1.23—EDM05.01—Evaluate Stakeholder Reporting Requirements ||||| 
|---|---|---|---|---|
| **Governance Practice** | **Inputs** || **Outputs** ||
| EDM05.01 Evaluate stakeholder reporting requirements. Continually examine and make judgement on the current and future requirements for stakeholder communication and reporting, including both mandatory reporting requirements (e.g., regulatory) and communication to other stakeholders. Establish the principles for communication. | **From** | **Description** | **Description** | **To** |
| ^ | EDM02.03 | Actions to improve value delivery | Evaluation of enterprise reporting requirements | MEA01.01 |
| ^ | EDM03.03 | Risk management issues for the board | Reporting and communication principles | MEA01.01 |
| ^ | EDM04.03 | Feedback on allocation and effectiveness of resources and capabilities | ^ | ^ |
| ^ | MEA02.08 | Refined scope | ^ | ^ |
| **Activities** |||||
| 1. Examine and make a judgement on the current and future mandatory reporting requirements relating to the use of IT within the enterprise (regulation, legislation, common law, contractual), including extent and frequency. |||||
| 2. Examine and make a judgement on the current and future reporting requirements for other stakeholders relating to the use of IT within the enterprise, including extent and conditions. |||||
| 3. Maintain principles for communication with external and internal stakeholders, including communication formats and communication channels, and for stakeholder acceptance and sign-off of reporting. |||||
| Source: ISACA, *COBIT 5: Enabling Processes*, USA, 2012, page 48 |||||

**Figure 1.24** gives activities for directing the communication to stakeholders, which should be considered among others:
• Communication guidelines that help establish comprehensive communication plans
• Decision trees to enable the reporting to identify content, medium, timing and distribution
• Escalation paths and reporting mechanisms

| Figure 1.24—EDM05.02—Direct Stakeholder Communication and Reporting ||||| 
|---|---|---|---|---|
| **Governance Practice** | **Inputs** || **Outputs** ||
| EDM05.02 Direct stakeholder communication and reporting. Ensure the establishment of effective stakeholder communication and reporting, including mechanisms for ensuring the quality and completeness of information, oversight of mandatory reporting, and creating a communication strategy for stakeholders. | **From** | **Description** | **Description** | **To** |
| ^ | APO12.04 | Risk analysis and risk profile reports for stakeholders | Rules for validating and approving mandatory reports | MEA01.01 MEA03.04 |
| ^ | ^ | ^ | Escalation guidelines | MEA01.05 |
| **Activities** |||||
| 1. Direct the establishment of the communication strategy for external and internal stakeholders. |||||
| 2. Direct the implementation of mechanisms to ensure that information meets all criteria for mandatory IT reporting requirements for the enterprise. |||||
| 3. Establish mechanisms for validation and approval of mandatory reporting. |||||
| 4. Establish reporting escalation mechanisms. |||||
| Source: ISACA, *COBIT 5: Enabling Processes*, USA, 2012, page 48 |||||

## Chapter 1—Framework for the Governance of Enterprise IT
### Section Two: Content

Examples for activities related to monitoring communication requirements and effectiveness, as shown in **figure 1.25**, include:
- Formal and informal feedback channels are proactively designed into the reporting process.
- Surveys are conducted where applicable.
- Metrics are built into the reporting process to track distribution, receipt, open rates and follow-up activity.

| Figure 1.25—EDM05.03—Monitor Stakeholder Communication | | | | |
|---|---|---|---|---|
| **Governance Practice** | **Inputs** | | **Outputs** | |
| **EDM05.03 Monitor stakeholder communication.** Monitor the effectiveness of stakeholder communication. Assess mechanisms for ensuring accuracy, reliability and effectiveness, and ascertain whether the requirements of different stakeholders are met. | **From** | **Description** | **Description** | **To** |
| | MEA02.08 | • Assurance review report<br>• Assurance review results | Assessment of reporting effectiveness | MEA01.01<br>MEA03.04 |
| **Activities** | | | | |
| 1. Periodically assess the effectiveness of the mechanisms for ensuring the accuracy and reliability of mandatory reporting. | | | | |
| 2. Periodically assess the effectiveness of the mechanisms for, and outcomes from, communication with external and internal stakeholders. | | | | |
| 3. Determine whether the requirements of different stakeholders are met. | | | | |
| Source: ISACA, *COBIT 5: Enabling Processes*, USA, 2012, page 48 | | | | |

To support financial and IT-related transparency, the metrics in **figure 1.26** should be considered.

| Figure 1.26—EDM05 Ensure Stakeholder Transparency Primary IT-related Goals | |
|---|---|
| **IT-related Goal** | **Related Metrics** |
| 03 Commitment of executive management for making IT-related decisions | • Percent of executive management roles with clearly defined accountabilities for IT decisions<br>• Number of times IT is on the board agenda in a proactive manner<br>• Frequency of IT strategy (executive) committee meetings<br>• Rate of execution of executive IT-related decisions |
| 06 Transparency of IT costs, benefits and risk | • Percent of investment business cases with clearly defined and approved expected IT-related costs and benefits<br>• Percent of IT services with clearly defined and approved operational costs and expected benefits<br>• Satisfaction survey of key stakeholders regarding the level of transparency, understanding and accuracy of IT financial information |
| 07 Delivery of IT services in line with business requirements | • Number of business disruptions due to IT service incidents<br>• Percent of business stakeholders satisfied that IT service delivery meets agreed-on service levels<br>• Percent of users satisfied with the quality of IT service delivery |
| Source: ISACA, *COBIT 5: Enabling Processes*, USA, 2012, page 47 | |

To ensure the robustness of the stakeholder reporting process, the metrics in **figure 1.27** should be considered.

| Figure 1.27—EDM05 Ensure Stakeholder Transparency Process Goals and Metrics | |
|---|---|
| **Process Goal** | **Related Metrics** |
| 1. Stakeholder reporting is in line with stakeholder requirements. | • Date of last revision to reporting requirements<br>• Percent of stakeholders covered in reporting requirements |
| 2. Reporting is complete, timely and accurate. | • Percent of reports that are not delivered on time<br>• Percent of reports containing inaccuracies |
| 3. Communication is effective and stakeholders are satisfied. | • Level of stakeholder satisfaction with reporting<br>• Number of breaches of mandatory reporting requirements |
| Source: ISACA, *COBIT 5: Enabling Processes*, USA, 2012, page 47 | |

## 1.12 COMMUNICATION AND PROMOTION TECHNIQUES

For effective GEIT, stakeholder groups need a high level of buy-in to the GEIT process. Knowledge of marketing and communication methods and techniques (for example asserting, persuading, bridging and attracting) plays an important role in helping to define and sell key messages to targeted audiences.

It is quite typical to underestimate the impact of marketing and communication as a critical success factor in the effectiveness of governance of enterprise IT. Today's concept of the customer is anyone who is receptive to information about products or services.

### Importance of Communication and Marketing

In the context of the enterprise's IT organization, customers are internal rather than external (although they can coexist in certain situations). From this viewpoint, it becomes imperative for IT management personnel and those involved in governance of enterprise IT improvement to have a good understanding of, and be able to deploy, suitable marketing and communication approaches to gain the trust and buy-in of stakeholders. Every organization will have its own existing culture and choice of governance of enterprise IT approaches that it wishes to adopt. The road map to follow for cultural change and effective communication will, therefore, be unique to each organization; however, there are common elements, and the methods and techniques employed are generically applicable.

### Communication Strategy and Plan

A robust communication strategy must be seen as fundamental in assisting the successful achievement of IT-related initiatives. This stems from the fact that these initiatives are not about technology alone; rather, they are more about getting people in the organization to adopt and transition to the changes that inevitably result from new or improved processes, which IT helps enable.

In a change management context within the larger space of governance of enterprise IT, a well-crafted and executed communication plan (including the purpose, targeted audience, communication need, message, channels to be used, etc.) will not only inform, but also prepare and persuade, the target audience. It forms an integral part of the overall change management process of any IT-related initiative. Effective communication will ensure that "everyone is on the same page"—that key issues are grasped, objectives are positively accepted by management and staff, and everyone's role is understood.

The communication plan should be based on a well-defined influencing strategy. Behaviors will need to be changed. Therefore, care should be taken to ensure that participants will be motivated and see the benefits of the new approaches as well as understand the consequences of accepting responsibility. If this is not positively communicated, then governance of enterprise IT will not be perceived as part of the corporate mission (with board-level support). Management will resist it as a barrier to getting the job done, a deviation from current priorities or another management fad.

The areas of marketing and communication are no less important than any other area in making governance of enterprise IT effective. An inability to communicate effectively has been one of the major causes of IT failures, with too much technical jargon, lack of business understanding and a poor appreciation of the other party's requirements and issues. Likewise, ineffective marketing will denigrate the success of marketing efforts involved in promoting governance of enterprise IT—lessons learned from structuring and executing successful marketing programs in other domains will thus go a long way toward making these initiatives successful.

### Content of Governance of Enterprise IT Communication Related to Risk

Good practices in governance of enterprise IT point to emphasizing the importance of mitigating IT-related risk when communicating the need for governance of enterprise IT. The following approaches are recommended in the UK National Computing Centre publication on IT governance[26] to ensure that risk has been properly appreciated:

- **Emphasize the business impact of risk associated with misaligned IT strategies, misuse of technology, badly managed operations and ineffective project management. Show how the risk can be mitigated by effective controls:**
  – Use case studies that have impacted the business or other businesses (e.g., virus attacks, critical service outages, projects with "unexpected outcomes") to illustrate how issues might arise.
- **Identify relevant examples of governance providing business benefits beyond the basic requirement of evidencing control:**
  – Use case studies to illustrate how effective governance has identified risk to the business, its objectives and strategy, and brokered an alternative solution.
  – Use case studies to illustrate business benefits as a direct result of effective governance (e.g., reduced costs, improved quality, productivity, reputation and marketing advantages).
- **Use scenario modeling with risk assessment and mitigation:**
  – Consider known and new risk across both business and IT (e.g., external audit requirements).
  – Consider how governance can help mitigate the risk.
  – Calculate a risk factor (likelihood × impact).
  – Consider options: accept, mitigate or assign.
- **Use a common business language for:**
  – Technological risk in financial/economic/business terms
  – Legal/regulatory and contractual implications

More information on risk management and communication can be found in chapter 4 Risk Optimization.

## 1.13 ASSURANCE METHODOLOGIES AND TECHNIQUES

Knowledge and application of assurance methodologies and techniques are the basis for enforcement of the governance initiatives and practices. Without appropriate GEIT monitoring and controls, as prescribed by assurance methodologies and techniques (planning assurance, scoping assurance, executing assurance), the quality and sustainability of initiatives and practices may not meet internal and external requirements.

In general, assurance is defined as a broad term including both the formal audit work (reporting independently towards the audit committee) and practices oriented toward management self-assessments (reporting to management). All these assurance initiatives have similar components and practices, as discussed below.

### Components of Assurance Initiatives

Assurance initiatives in the realm of governance of enterprise IT have a dedicated context and objective—to measure or evaluate a specific subject matter that is the responsibility of another party.[27] For such initiatives, there is generally a stakeholder involved who uses the subject matter, but who has usually delegated operation and custodianship of the subject matter to the responsible party. Therefore, the stakeholder is the end customer of the evaluation and can approve the criteria of the evaluation with the responsible party and the assurance professional. To be called an assurance initiative, five components must be present, as prescribed in the IAASB Assurance Framework[28] and shown in **figure 1.28**:

- A three-party relationship
- Subject matter over which the assurance is to be provided
- Suitable criteria against which the subject matter will be assessed
- A process to execute the assurance engagement
- A conclusion issued by the assurance professional

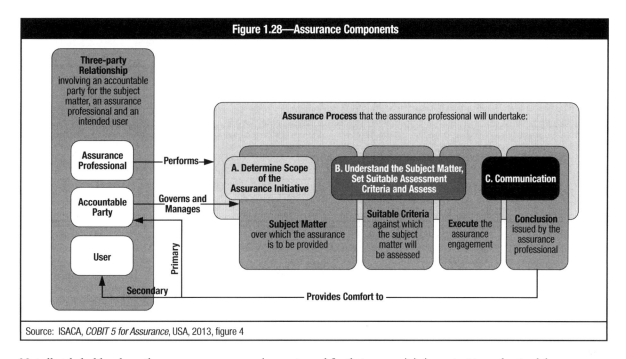

Not all stakeholders have the same assurance requirements, and for that reason, it is important to understand the difference between different types of assurance engagements. These engagements can range from more open self-assessment assignments over internal audit reviews to more standardized external audit assignments. See **figure 1.29**.

| Figure 1.29—Comparison of Assurance Engagement Types | | | |
|---|---|---|---|
| Characteristic | Self-assessment | Internal Audit/Compliance Review | External Audit |
| Independence requirements | • Not required or guaranteed<br>• Objectivity of self-assessors should be encouraged by defining clear responsibilities and correct follow-up. | Should be optimised by the correct composition of the internal audit/compliance department members | Independence of the external auditors should be established, verified and maintained. |
| Interested party (user) | Enabler owners | Executive management, audit committee, operational management and enabler owners involved | Primarily directed toward the board/shareholders, but of importance to the enterprise in general |
| Responsible party (accountable party) | Enabler owners | Management and business enabler owners involved | The board and executive management involved |
| Assurance provider (assurance professional) | Enabler owners | Internal audit/compliance department | External auditor |
| Reporting format and requirements | • Free format<br>• Internal consistency required | Internal consistency required, in line with professional standards | Highly regulated/standardised |
| Governing rules/standards | Standardised approaches based on good practices required | • Standardised approaches based on good practices required<br>• Professional standards and code of ethics to be respected | Adherence to applicable codes of ethics and standards should be established, verified and maintained. |
| Level of trust (reliability) | • Lowest<br>• Depends on the skill and objectivity of the assessor | • Medium<br>• Depends on the skill and expertise of the internal audit/compliance department and on co-operation of the accountable party | Maximal |

Source: ISACA, *COBIT 5 for Assurance*, USA, 2013, figure 8

To fulfill and validate the internal and external requirements of governance and assurance of IT, assurance needs to be conducted on an ongoing basis. The International Standards on Auditing (ISA) Standard 31546 sets out the requirements for the assurance professional to obtain an understanding of internal controls relevant to the audit, which includes the following components:
- The control environment
- The entity's risk assessment process
- The information system (including the related business processes relevant to financial reporting and communications)
- Control activities and monitoring of controls

The ISA states that the assurance professional should consider the need to obtain audit evidence supporting the effective operation of controls directly related to the assertions as well as other indirect controls on which these controls depend, such as underlying general IT controls. Therefore, the minimum requirement for the assurance professional is to understand the information systems underpinning business processes relevant for financial reporting and how the entity has responded to risk arising from IT. Because the use of IT affects the way control activities are implemented in the business and related financial reporting, the assurance professional needs to consider whether the entity has responded adequately to risk arising from IT by establishing effective general IT controls and application controls.

## IT Assurance Road Map

To provide for adequate assurance, it is important to follow a consistent methodology or approach. While the specific approach may be unique to each organization and type of initiative, a fairly common approach that is used is based on three stages: scoping, testing and communicating. The stages and steps of the road map are shown in **figure 1.30**.

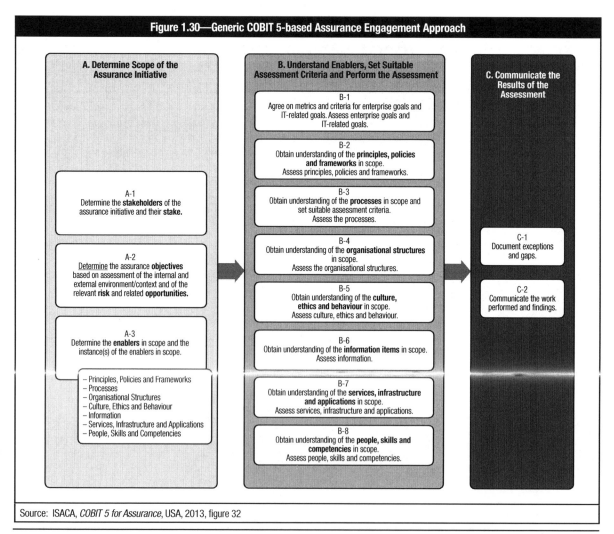

Source: ISACA, *COBIT 5 for Assurance*, USA, 2013, figure 32

Each stage of the road map is briefly explained as follows:
- **Determine the scope of the assurance initiative:** In this step, the stakeholders of the assurance initiative are determined and the objective of the assurance initiative is clarified. An important step in this stage is to scope the assurance universe to the required enablers in the scope.
- **Understand enablers, set assessment criteria and perform the assessment:** In this step, the assurance professional will ensure that he/she fully understands the enablers to be assessed and their related assessment criteria. Next, the core testing steps will be executed to verify control design and operating effectiveness.
- **Communicate the results of the assessment:** In the final step, the assurance professional will document the identified control weaknesses, related business risk and recommendations.

## 1.14 CONTINUOUS IMPROVEMENT TECHNIQUES AND PROCESSES

Applying a continuous improvement life cycle approach provides a method for enterprises to address the complexity and challenges typically encountered during GEIT implementation. There are three interrelated components to the life cycle:
- The core continuous improvement cycle
- The enablement of change
- The management of the program

Applying a continuous improvement life cycle approach provides a method for enterprises to address the complexity and challenges typically encountered during GEIT implementation.

### Components of a Continuous Improvement Cycle

There are three interrelated components to the life cycle, as illustrated in **figure 1.31**: the core GEIT continuous improvement life cycle, the enablement of change (addressing the behavioral and cultural aspects of the implementation or improvement), and the management of the program. In figure 1.31, the initiatives are depicted as continuous life cycles to emphasize the fact that these are not one-off activities, but part of an ongoing process of implementation and improvement that in time become "business as usual," at which time the program can be retired.

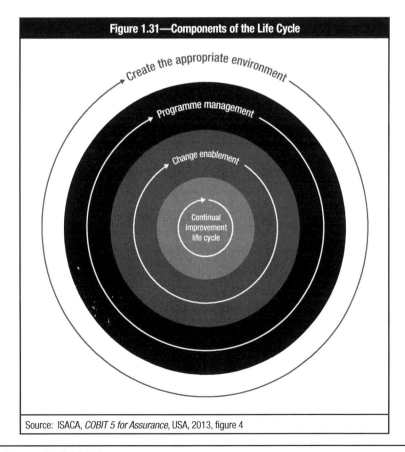

Source: ISACA, *COBIT 5 for Assurance*, USA, 2013, figure 4

## Phases in an Implementation Life Cycle

The seven phases of the implementation life cycle are illustrated in **figure 1.16**. The implementation and improvement program is typically a continuous and iterative one. During the last phase, new objectives and requirements will be identified and a new cycle will be initiated. High-level health checks, assessments and audits often trigger consideration of a GEIT initiative and these results can be used as input to phase 1.

### Phase 1—What Are the Drivers?
Phase 1 identifies current change drivers and creates at executive management levels a desire to change that is then expressed in an outline of a business case. A change driver is an internal or external event, condition or key issue that serves as a stimulus for change. Events, trends (industry, market or technical), performance shortfalls, software implementations and even the goals of the enterprise can act as change drivers. Risk associated with implementation of the program itself will be described in the business case and managed throughout the life cycle. Preparing, maintaining and monitoring a business case are fundamental and important disciplines for justifying, supporting and then ensuring successful outcomes of any initiative, including the improvement of GEIT. They ensure a continuous focus on the benefits of the program and their realization.

### Phase 2—Where Are We Now?
Phase 2 aligns IT-related objectives with enterprise strategies and risk, and prioritizes the most important enterprise goals, IT-related goals and processes. COBIT 5 provides a generic mapping of enterprise goals to IT-related goals to IT processes to help with the selection. Given the selected enterprise and IT-related goals, critical processes are identified that need to be of sufficient capability to ensure successful outcomes. Management needs to know its current capability and where deficiencies may exist. This is achieved by a process capability assessment of the as-is status of the selected processes.

### Phase 3—Where Do We Want to Be?
Phase 3 sets a target for improvement followed by a gap analysis to identify potential solutions. Some solutions will be quick wins and others more challenging, long-term tasks. Priority should be given to projects that are easier to achieve and likely to give the greatest benefit. Longer-term tasks should be broken down into manageable pieces.

### Phase 4—What Needs to Be Done?
Phase 4 plans feasible and practical solutions by defining projects supported by justifiable business cases and developing a change plan for implementation. A well-developed business case will help ensure that the project's benefits are identified and continuously monitored.

### Phase 5—How Do We Get There?
Phase 5 provides for the implementation of the proposed solutions into day-to-day practices and the establishment of measures and monitoring systems to ensure that business alignment is achieved and performance can be measured. Success requires engagement, awareness and communication, understanding and commitment of top management, and ownership by the affected business and IT process owners.

### Phase 6—Did We Get There?
Phase 6 focuses on sustainable transition of the improved governance and management practices into normal business operations and monitoring achievement of the improvements using the performance metrics and expected benefits.

### Phase 7—How Do We Keep the Momentum Going?
Phase 7 reviews the overall success of the initiative, identifies further governance or management requirements and reinforces the need for continuous improvement. It also prioritizes further opportunities to improve GEIT. Program and project management is based on good practices and provides for checkpoints at each of the seven phases to ensure that the program's performance is on track, the business case and risk are updated, and planning for the next phase is adjusted as appropriate. It is assumed that the enterprise's standard approach would be followed. Further guidance on program and project management can also be found in the COBIT 5 process BAI01 *Manage programmes and projects*. Although reporting is not mentioned explicitly in any of the phases, it is a continuous thread through all of the phases and iterations.

The time spent per phase will differ greatly depending on (among other factors) the specific enterprise environment, its maturity, and the scope of the implementation or improvement initiative. However, the overall time spent on each iteration of the full life cycle ideally should not exceed six months, with improvements applied progressively; otherwise, there is a risk of losing momentum, focus and buy-in from stakeholders. The goal is to get into a rhythm of regular improvements.

Larger-scale initiatives should be structured as multiple iterations of the life cycle. Over time, the life cycle will be followed iteratively while building a sustainable approach. This becomes a normal business practice when the phases in the life cycle are everyday activities and continuous improvement occurs naturally.

Well-known methodologies in support of continuous improvement strategies are Six Sigma and TQM. Six Sigma's objective is the implementation of a measurement-oriented strategy focused on process improvement and defect reduction. TQM is a management strategy aimed at embedding awareness of quality in all organizational processes. More information regarding these methods can be found in section 3.6, Continuous Improvement Concepts and Principles.

More information on Methods to Manage Organizational, Process and Cultural Change is discussed in section 1.8.

## ENDNOTES

1. Kotter, John; *Leading Change*, Harvard Business School Press, USA, 1996
2. *Ibid.*
3. International Federation of Accountants, *Enterprise Governance: Getting the Balance Right*, USA, 2003 (*www.ifac.org*). The report provides guidelines to help enterprises be more efficient in their responsibilities for conformance and with value creation and use of resources (performance).
4. Weill, Peter; Jeanne Ross: *IT Governance: How Top Performers Manage IT Decision Rights for Superior Results*, Harvard Business School Press, USA, 2004
5. Van Grembergen, Wim; Steven De Haes; *Enterprise Governance of IT: Achieving Alignment and Value*, 2nd Edition, Springer, USA, 2015
6. De Wit, Bob; Ron Myer; *Strategy: Process, Content, Context*, Cengage Learning EMEA, UK, 2005
7. Van Grembergen, Wim; Steven De Haes; *Implementing Information Technology Governance: Models, Practices and Cases*, IGI Global, USA 2008; *Op cit* Van Grembergen and De Haes
8. ISACA, *Benchmarking and Business Value Assessment of COBIT® 5*, USA 2015
9. *Ibid.*
10. SFIA Foundation, *The Skills Framework for the Information Age (SFIA)*, http://www.sfia-online.org/en
11. European e-Competence Framework, *www.ecompetences.eu*
12. IT Governance Institute, *Board Briefing on IT Governance, 2nd Edition*, USA, 2003, *www.itgi.org*
13. IT Governance Institute; *Governance of the Extended Enterprise: Bridging Business and IT Strategies*, John Wiley & Sons Inc., USA, 2005
14. Fong, Elizabeth; Alan Goldfine; *Information Management Decisions: The Integration Challenge*, US Department of Commerce, National Institute of Standards and Technology (NIST), NIST Special Publication 500-167, USA, 1989, *www.itl.nist.gov*. Quoted in IT Governance Institute; *Governance of the Extended Enterprise: Bridging Business and IT Strategies*, John Wiley & Sons Inc., USA, 2005
15. National Association of State CIOs (NASCIO); *Enterprise Architecture Tool-Kit V2.0*, Adaptive Enterprise Architecture Development Program, USA, 2002, *www.nascio.org*
16. *Op cit* Weill and Ross
17. *Op cit* Kotter
18. *Ibid.*
19. ISACA, *COBIT 5: Enabling Information*, 2013, USA.
20. Weill, Peter; Marianne Broadbent; *Leveraging the New Infrastructure: How Market Leaders Capitalize on Information Technology*, Harvard Business School Press, USA, 1998
21. Kaplan, Robert; David Norton; *The Balanced Scorecard: Translating Strategy Into Action*, Harvard Business School Press, USA, 1996
22. IT Governance Institute, *Board Briefing on IT Governance, 2nd Edition*, USA, 2003, *www.itgi.org*

*Chapter 1—Framework for the Governance of Enterprise IT*
*Section Two: Content*

23. De Haes, Steven; Wim Van Grembergen; "IT Governance Structures, Processes and Relational Mechanisms: Achieving IT/Business Alignment in a Major Belgian Financial Group," in proceedings of the Hawaii International Conference on System Sciences (HICSS), 2005
24. Van Grembergen, Wim; Ronald Saull; Steven De Haes; "Linking the IT Balanced Scorecard to the Business Objectives at a Major Canadian Financial Group," *Journal of Information Technology Cases and Applications*, 2003
25. *Ibid.*
26. UK National Computing Centre, *IT Governance: Developing a Successful Governance Strategy—A Best Practice Guide for Decision Makers in IT*, UK, 2005
27. ISACA, *IT Assurance Guide: Using COBIT®*, USA, 2007, www.isaca.org
28. International Federation of Accountants (IFAC), International Auditing and Assurance Standards Board (IASSB), *International Framework for Assurance Engagements (IAASB Assurance Framework)*, USA, 2004, www.ifac.org/iaasb

# Chapter 2: Strategic Management

## Section One: Overview

| | |
|---|---|
| Domain Definition | 48 |
| Domain Objectives | 48 |
| Learning Objectives | 48 |
| CGEIT Exam Reference | 48 |
| Task and Knowledge Statements | 48 |
| Self-assessment Questions | 51 |
| Answers to Self-assessment Questions | 52 |
| Suggested Resources for Further Study | 53 |

## Section Two: Content

| | | |
|---|---|---|
| 2.1 | An Enterprise's Strategic Plan and How It Relates to Information Technology | 54 |
| 2.2 | Strategic Planning Processes and Techniques | 56 |
| 2.3 | Impact of Changes in Business Strategy on IT Strategy | 65 |
| 2.4 | Barriers to the Achievement of Strategic Alignment | 67 |
| 2.5 | Policies and Procedures Necessary to Support IT and Business Strategic Alignment | 67 |
| 2.6 | Methods to Document and Communicate IT Strategic Planning Processes | 69 |
| 2.7 | Components, Principles and Frameworks of Enterprise Architecture | 72 |
| 2.8 | Current and Future Technologies | 74 |
| 2.9 | Prioritization Processes Related to IT Initiatives | 77 |
| 2.10 | Scope, Objectives and Benefits of IT Investment Programs | 79 |
| 2.11 | IT Roles and Responsibilities and Methods to Cascade Business and IT Objectives to IT Personnel | 81 |
| Endnotes | | 83 |

# Section One: Overview

## DOMAIN DEFINITION
Ensure that IT enables and supports the achievement of enterprise objectives through the integration and alignment of IT strategic plans with enterprise strategic plans.

## DOMAIN OBJECTIVES
The objective of this domain is to ensure that IT enables and supports the achievement of business objectives through the integration of IT strategic plans with business strategic plans and the alignment of IT services with enterprise operations to optimize business processes.

The premise of this domain is that business strategy drives IT strategy and the lack of alignment between them is a major issue that reduces IT's value to the business.

## LEARNING OBJECTIVES
The purpose of this domain is for professionals and executives involved in IT governance to be knowledgeable about why and how business strategy and IT strategy are linked, the concepts behind achieving alignment between them, and how alignment is implemented in practice to enable IT to drive efficient and effective business operations and growth.

## CGEIT EXAM REFERENCE
This domain represents 20 percent of the CGEIT exam (approximately 30 questions).

## TASK AND KNOWLEDGE STATEMENTS

### TASKS
There are six tasks within this domain that a CGEIT candidate must know how to perform. These relate to having to develop, or be part of the development of, an enterprise's IT strategy.

T2.1  Evaluate, direct and monitor IT strategic planning processes to ensure alignment with enterprise goals.
T2.2  Ensure that appropriate policies and procedures are in place to support IT and enterprise strategic alignment.
T2.3  Ensure that the IT strategic planning processes and related outputs are adequately documented and communicated.
T2.4  Ensure that enterprise architecture (EA) is integrated into the IT strategic planning process.
T2.5  Ensure prioritization of IT initiatives to achieve enterprise objectives.
T2.6  Ensure that IT objectives cascade into clear roles, responsibilities and actions of IT personnel.

### KNOWLEDGE STATEMENTS
The CGEIT candidate must have a good understanding of each of the 11 areas delineated by the following knowledge statements. These statements are the basis for the exam. Each statement is defined and its relevance and applicability to this job practice are briefly described as follows:

Knowledge of:

**KS2.1  an enterprise's strategic plan and how it relates to IT**
Business/IT alignment is the fit and integration among business strategy, IT strategy, business structures/operations and IT structures/operations. The idea behind strategic alignment is very comprehensive, but the question is how organizations can achieve this ultimate goal.

**KS2.2  strategic planning processes and techniques**
The foundation of IT strategic alignment is the strategic planning process—defining business strategies from which IT strategies are to be derived. Knowledge of the strategic planning process and techniques will enable development of an IT strategy that is defined and executed in line with business imperatives.

**KS2.3** **impact of changes in business strategy on IT strategy**
The dynamic nature of business requires regular strategy revisions and IT needs to continually readjust to align with the business. Nonaligned IT can have dire consequences for the enterprise as it will inhibit the enterprise to realize its business strategy.

**KS2.4** **barriers to the achievement of strategic alignment**
Strategic alignment is a multifaceted and complex endeavor, often referred to as the alignment challenge. To overcome alignment barriers, it is important to understand the difficulties that organizations experience aligning IT with the business.

**KS2.5** **policies and procedures necessary to support IT and business strategic alignment**
Knowledge of policies and procedures supporting the strategic planning process enable the development of an IT strategy that is defined and executed in line with business imperatives.

**KS2.6** **methods to document and communicate IT strategic planning processes**
To enable an effective and efficient IT strategic planning process, the enterprise requires structured methods to document and communicate the intermediate steps and outcome of the IT strategic planning process. The balanced scorecard is promoted as a very effective means in this regard.

**KS2.7** **components, principles and frameworks of enterprise architecture (EA)**
Enterprise architecture (EA) takes a broad view of business, matching it with associated information and related technologies. It is an important blueprint for IT strategy development and, therefore, components, principles and frameworks related to EA must be understood to effectively align IT to the business.

**KS2.8** **current and future technologies**
The consideration of which technological direction to adopt is key in preparing IT to be an agile enabler for the business. This is especially important for strategic alignment and must be addressed in the context of the role of IT in the future of the business.

**KS2.9** **prioritization processes related to IT initiatives**
The goal of portfolio management is to ensure that an enterprise's overall portfolio of IT-enabled investments is aligned with, and contributes optimal value to, the enterprise's strategic objectives. Enterprise portfolio management activities include establishing and managing resource profiles; defining investment thresholds; evaluating, prioritizing and selecting, deferring or rejecting new investments; managing the overall portfolio; and monitoring and reporting on portfolio performance.

**KS2.10** **scope, objectives and benefits of IT investment programs**
Investment programs are the translation of planned strategy into action (execution). Knowledge of how investment programs are structured helps in understanding the crucial link between strategy blueprints and how they are implemented.

**KS2.11** **IT roles and responsibilities and methods to cascade business and IT objectives to IT personnel**
Engaged stakeholders and personnel will sponsor and facilitate the changes necessary to bring about better strategic alignment. To realize such commitment, it is important to clarify roles and responsibilities and to leverage methods that will help in cascading business and IT objectives to the relevant personnel and stakeholders.

## RELATIONSHIP OF TASK TO KNOWLEDGE STATEMENTS

The task statements are what the CGEIT candidate is expected to know how to do. The knowledge statements delineate what the CGEIT candidate is expected to know to perform the tasks. The task and knowledge statements are mapped in **figure 2.1** insofar as it is possible to do so. Note that although there often is overlap, each task statement will generally map to several knowledge statements.

| Figure 2.1—Task and Knowledge Statements Mapping—IT Strategic Management Domain ||
|---|---|
| **Task Statement** | **Knowledge Statements** |
| T 2.1 Evaluate, direct and monitor IT strategic planning processes to ensure alignment with enterprise goals. | KS2.1 an enterprise's strategic plan and how it relates to IT<br>KS2.2 strategic planning processes and techniques<br>KS2.3 impact of changes in business strategy on IT strategy<br>KS2.4 barriers to the achievement of strategic alignment |
| T2.2 Ensure that appropriate policies and procedures are in place to support IT and enterprise strategic alignment. | K2.5 policies and procedures necessary to support IT and business strategic alignment |
| T2.3 Ensure that the IT strategic planning processes and related outputs are adequately documented and communicated. | K2.6 methods to document and communicate IT strategic planning processes (for example, IT dashboard/balanced scorecard, key indicators) |
| T2.4 Ensure that enterprise architecture (EA) is integrated into the IT strategic planning process. | KS2.7 components, principles and frameworks of enterprise architecture (EA)<br>KS2.8 current and future technologies |
| T2.5 Ensure prioritization of IT initiatives to achieve enterprise objectives. | KS2.9 prioritization processes related to IT initiatives<br>KS2.10 scope, objectives and benefits of IT investment programs |
| T2.6 Ensure that IT objectives cascade into clear roles, responsibilities and actions of IT personnel. | KS2.4 barriers to the achievement of strategic alignment<br>KS2.11 IT roles and responsibilities and methods to cascade business and IT objectives to IT personnel |

## SELF-ASSESSMENT QUESTIONS

CGEIT self-assessment questions support the content in this manual and provide an understanding of the type and structure of questions that have typically appeared on the exam. Questions are written in a multiple-choice format and designed for one best answer. Each question has a stem (question) and four options (answer choices). The stem may be written in the form of a question or an incomplete statement. In some instances, a scenario or a description problem may be included. These questions normally include a description of a situation and require the candidate to answer two or more questions based on the information provided. Many times a question will require the candidate to choose the **MOST** likely or **BEST** answer among the options provided.

In each case, the candidate must read the question carefully, eliminate known incorrect answers and then make the best choice possible. Knowing the format in which questions are asked, and how to study and gain knowledge of what will be tested, will help the candidate correctly answer the questions.

2-1   The **BEST** approach for ensuring the success of IT's contribution to the enterprise is to:

   A. have IT operations report directly to the chief executive officer (CEO).
   B. define the IT strategy based on the enterprise strategy.
   C. make IT responsible and accountable for enterprise performance.
   D. ensure that there is independent oversight of IT performance.

2-2   In an enterprise architecture (EA), which of the following domains should drive the others?

   A. Application
   B. Data
   C. Technology
   D. Business

2-3   Which of the following is the **MOST** significant barrier to achieving strategic alignment between IT and the business?

   A. The role of IT is reactive versus proactive.
   B. The IT organization is decentralized.
   C. There is a lack of structure through which requirements can be understood.
   D. The business has a low reliance on IT (e.g., a cement manufacturer).

## ANSWERS TO SELF-ASSESSMENT QUESTIONS
Correct answers are shown in **bold**.

2-1  A.  Direct reporting of IT operations to the CEO will not ensure alignment.
      **B.  Defining the IT strategy based on the enterprise strategy is the core of strategic alignment—the enablement of enterprise strategy by IT strategy.**
      C.  IT cannot be held responsible or accountable for enterprise performance; particularly because IT is an enabler, not a business objective in itself.
      D.  Independent oversight is one factor that helps align the IT strategy to the enterprise strategy. Independent oversight may consist of audits that investigate and report deficiencies in alignment.

2-2  A.  Applications are part of the higher-level technology infrastructure or delivery systems architecture.
      B.  Data are part of the higher-level technology infrastructure or delivery systems architecture.
      C.  Technology is part of the higher-level technology infrastructure or delivery systems architecture.
      **D.  Business architecture is a key domain component of a business-driven EA.**

2-3  A.  A reactive role is not the most significant barrier to achieving strategic alignment. Depending on the sector, reactive IT can still adequately respond to business strategy and changes.
      B.  A decentralized IT organization is not necessarily a barrier, especially if the business operating model is also decentralized and if proper governance mechanisms are in place.
      **C.  A lack of structure without communication mechanisms is the correct answer because it is difficult for IT to understand business requirements or to provide proactive advice to the business on opportunities created by IT.**
      D.  Low reliance on IT is not as significant a barrier to achieving strategic alignment as a lack of structure.

---

**Note:** For more self-assessment questions, you may also want to obtain a copy of the *CGEIT® Review Questions, Answers & Explanations Manual 4th Edition*, which consists of 250 multiple-choice study questions, answers and explanations.

## SUGGESTED RESOURCES FOR FURTHER STUDY

In addition to the resources cited throughout this manual, the following resources are suggested for further study in this domain (publications in **bold** are stocked in the ISACA Bookstore):

Benson, Robert J.; Thomas L. Bugnitz; William B. Walton; *From Business Strategy to IT Action: Right Decisions for a Better Bottom Line*, John Wiley & Sons Inc., USA, 2004

Chew, Eng K.; Peter Gottschalk (Eds.), *Knowledge Driven Service Innovation and Management: IT Strategies for Business Alignment and Value Creation*, IGI Global, USA, 2013

De Haes, Steven; Wim Van Grembergen; *Enterprise Governance of IT: Achieving Alignment and Value, Second Edition*, Springer, USA, 2015

**ISACA, *COBIT 5*, USA, 2012,** *www.isaca.org/cobit*

**ISACA, *COBIT 5 Implementation*, USA, 2012,** *www.isaca.org/cobit*

**ISACA, *COBIT 5: Enabling Information*, USA, 2013,** *www.isaca.org/cobit*

Kaplan, Robert S.; David P. Norton; *Alignment: Using the Balanced Scorecard to Create Corporate Synergies*, Harvard Business School Press, USA, 2006

Nolan, Richard F.; Warren McFarlan; "Information Technology and the Board of Directors," *Harvard Business Review*, USA, October 2005

Van Grembergen, Wim; De Haes, Steven; *Business Strategy and Applications in Enterprise IT Governance*, IGI Global, USA, 2012

Weill, Peter; Jeanne Ross; *IT Governance: How Top Performers Manage IT Decision Rights for Superior Results*, Harvard Business School Press, USA, 2004

**Weill, Peter; Jeanne Ross; David C. Robertson; *Enterprise Architecture as Strategy: Creating a Foundation for Business Execution*, Harvard Business School Press, USA, 2006**

Wyatt-Haines, Richard; *Align IT: Business Impact Through IT*, John Wiley & Sons Inc., USA, 2007

# Section Two: Content

## 2.1 AN ENTERPRISE'S STRATEGIC PLAN AND HOW IT RELATES TO INFORMATION TECHNOLOGY

Business and IT alignment is the fit and integration among business strategy, IT strategy, business structures and IT structures. The idea behind strategic alignment is very comprehensive, but the question is: How can organizations achieve this ultimate goal?[1] Strategic business-IT alignment remains a top priority for chief executive officers (CEOs) and chief information officers (CIOs) alike.

What does "alignment between business and IT" mean? It comprises two major questions: How is IT aligned with the business? How is the business aligned with IT?

### Strategic Alignment Model

The governance objectives of benefit realization, risk optimization and resource optimization cannot be achieved without effective alignment between business and IT strategy. The alignment between business and IT strategy cannot be achieved by accident. Any such alignment is dynamic and takes different shapes and forms depending on the type of organization.

Henderson and Venkatraman[2] were the first to clearly describe the interrelationship between business strategies and IT strategies in their well-known Strategic Alignment Model (SAM), shown in **figure 2.2**. Many authors used this model for further research. The concept of the SAM is based on two building blocks: strategic fit and functional integration.

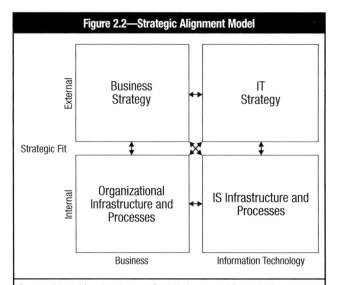

Source: Adapted from Henderson, J.C.; N. Venkatraman; "Strategic Alignment: Leveraging Information Technology for Transforming Organizations," *IBM Systems Journal*, vol. 32, no. 1, 1993. Courtesy of International Business Machines Corporation. ©1993 International Business Machines Corporation.

### Strategic Fit

Strategic fit recognizes that the IT strategy should be articulated in terms of an external domain (how the firm is positioned in the IT marketplace) and an internal domain (how the business and IT processes, including infrastructure, are designed and structured). Strategic fit is equally relevant in the business domain. Two types of functional integration exist: strategic and operational.

## Functional Integration

Strategic functional integration is the link between business strategy and IT strategy reflecting the external components that are important for many companies as IT emerges as a source of strategic advantage. Operational functional integration covers the internal domain and deals with the link between organizational infrastructure and processes and IT infrastructure and processes.

## The Complexity of Strategic Alignment

An important premise of the Strategic Alignment Model is that effective governance of enterprise IT requires a balance among the choices made in all the four domains of **figure 2.2**. Henderson and Venkatraman describe two cross-domain relationships in which business strategy plays the role of driver, and two relationships where IT strategy is the enabler, shown in **figure 2.3**.

- **The strategic execution perspective** is probably the most widely understood because it is the classic, hierarchical view of strategic management. The perspective starts from the premise that business strategy is articulated and that this strategy is the driver for the choices in organizational design and the design in IT infrastructure.
- **The technology transformation perspective** also starts from an existing business strategy, but focuses on the implementation of this strategy through appropriate IT strategy and the articulation of the required IT infrastructure and processes.
- **The competitive potential perspective** allows the adaptation of business strategy through emerging IT capabilities. Starting from the IT strategy, the best set of strategic options for business strategy and a corresponding set of decisions regarding organizational infrastructure and processes are determined.
- **The service level perspective** focuses on how to build a world-class IT service organization. This requires an understanding of the external dimensions of IT strategy with the corresponding internal design of the IT infrastructure and processes.

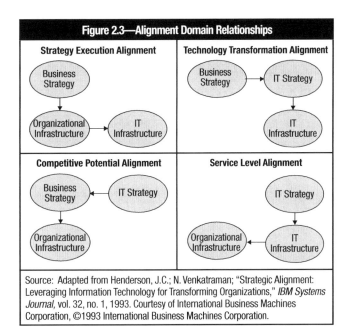

Source: Adapted from Henderson, J.C.; N. Venkatraman; "Strategic Alignment: Leveraging Information Technology for Transforming Organizations," *IBM Systems Journal*, vol. 32, no. 1, 1993. Courtesy of International Business Machines Corporation, ©1993 International Business Machines Corporation.

Henderson and Venkatraman also argue that the external and the internal domains are equally important, but that managers traditionally think of IT strategy in terms of the internal domain because, historically, IT was viewed as a support function that was less essential to the business. In their research results, Henderson and Venkatraman warn of the problems that may surface when a bivariate approach is undertaken with respect to balancing across the four domains:
- IT strategy
- Business strategy
- IT infrastructure
- Organizational infrastructure

For example, when only external issues—IT strategy and business strategy—are considered, a serious underestimation of the importance of internal issues, such as the required redesigning of key business processes, might occur. Therefore, the SAM calls for the recognition of multivariate relationships, which will always take into consideration at least three out of the four defined domains.

## 2.2 STRATEGIC PLANNING PROCESSES AND TECHNIQUES

The foundation of IT strategic alignment is the strategic planning process—defining business strategies from which IT strategies are derived. Knowledge of the strategic planning process and techniques will enable development of IT strategy, defined and executed in line with business imperatives.

Every enterprise operates in a different context; this context is determined by external factors (the market, the industry, geopolitics, etc.) and internal factors (the culture, organization, risk appetite, etc.), and requires a customized governance and management system. Stakeholder needs have to be transformed into an enterprise's actionable strategy.

### The COBIT 5 Goals Cascade and Strategic Planning

The COBIT 5 goals cascade is the mechanism to translate stakeholder needs into specific, actionable and customized enterprise goals, IT-related goals and enabler goals. This translation allows setting specific goals at every level and in every area of the enterprise in support of the overall goals and stakeholder requirements, and thus effectively supports alignment between enterprise needs and IT solutions and services.[3]

The COBIT 5 goals cascade is shown in **figure 2.4**.

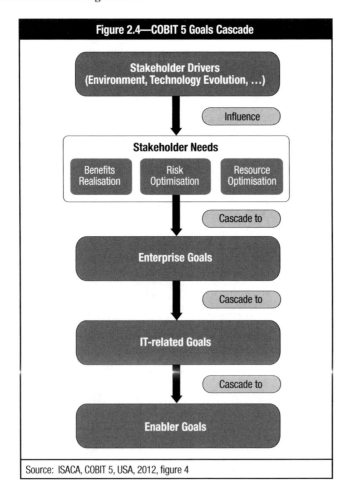

Source: ISACA, COBIT 5, USA, 2012, figure 4

### Step 1. Stakeholder Drivers Influence Stakeholder Needs
Stakeholder needs are influenced by a number of drivers (e.g., strategy changes, a changing business and regulatory environment, and new technologies).

### Step 2. Stakeholder Needs Cascade to Enterprise Goals
Stakeholder needs can be related to a set of generic enterprise goals. These enterprise goals have been developed using the balanced scorecard (BSC)[4] dimensions, and they represent a list of commonly used goals that an enterprise may define for itself. Although this list is not exhaustive, most enterprise-specific goals can be mapped easily onto one or more of the generic enterprise goals.

COBIT 5 defines 17 generic goals, as shown in **figure 2.5**, which includes the following information:
- The BSC dimension under which the enterprise goal fits
- Enterprise goals
- The relationship to the three main governance objectives

"P" stands for primary relationship and "S" for secondary relationship (i.e., a less strong relationship).

| | | Figure 2.5—COBIT 5 Enterprise Goals | | |
|---|---|---|---|---|
| | | **Relation to Governance Objectives** | | |
| **BSC Dimension** | **Enterprise Goal** | **Benefits Realisation** | **Risk Optimisation** | **Resource Optimisation** |
| Financial | 1. Stakeholder value of business investments | P | | S |
| | 2. Portfolio of competitive products and services | P | P | S |
| | 3. Managed business risk (safeguarding of assets) | | P | S |
| | 4. Compliance with external laws and regulations | | P | |
| | 5. Financial transparency | P | S | S |
| Customer | 6. Customer-oriented service culture | P | | S |
| | 7. Business service continuity and availability | | P | |
| | 8. Agile responses to a changing business environment | P | | S |
| | 9. Information-based strategic decision making | P | P | P |
| | 10. Optimisation of service delivery costs | P | | P |
| Internal | 11. Optimisation of business process functionality | P | | P |
| | 12. Optimisation of business process costs | P | | P |
| | 13. Managed business change programmes | P | P | S |
| | 14. Operational and staff productivity | P | | P |
| | 15. Compliance with internal policies | | P | |
| Learning and Growth | 16. Skilled and motivated people | S | P | P |
| | 17. Product and business innovation culture | P | | |
| Source: ISACA, COBIT 5, USA, 2012, figure 5 | | | | |

### Step 3. Enterprise Goals Cascade to IT-related Goals

Achievement of enterprise goals requires a number of IT-related outcomes,[5] which are represented by the IT-related goals. IT-related stands for information and related technology, and the IT-related goals are structured along the dimensions of the IT BSC. COBIT 5 defines 17 IT-related goals, listed in **figure 2.6**.

| | | Figure 2.6—IT-related Goals |
|---|---|---|
| **IT BSC Dimension** | | **Information and Related Technology Goal** |
| Financial | 01 | Alignment of IT and business strategy |
| | 02 | IT compliance and support for business compliance with external laws and regulations |
| | 03 | Commitment of executive management for making IT-related decisions |
| | 04 | Managed IT-related business risk |
| | 05 | Realised benefits from IT-enabled investments and services portfolio |
| | 06 | Transparency of IT costs, benefits and risk |
| Customer | 07 | Delivery of IT services in line with business requirements |
| | 08 | Adequate use of applications, information and technology solutions |
| Internal | 09 | IT agility |
| | 10 | Security of information, processing infrastructure and applications |
| | 11 | Optimisation of IT assets, resources and capabilities |
| | 12 | Enablement and support of business processes by integrating applications and technology into business processes |
| | 13 | Delivery of programmes delivering benefits, on time, on budget, and meeting requirements and quality standards |
| | 14 | Availability of reliable and useful information for decision making |
| | 15 | IT compliance with internal policies |
| Learning and Growth | 16 | Competent and motivated business and IT personnel |
| | 17 | Knowledge, expertise and initiatives for business innovation |
| Source: ISACA, COBIT 5, USA, 2012, figure 6 | | |

The mapping table between IT-related goals and enterprise goals is shown in **figure 2.7**, which shows how each enterprise goal is supported by a number of IT-related goals.

| Figure 2.7—Mapping COBIT 5 Enterprise Goals to IT-related Goals | | | | | | | | | | | | | | | | | | | |
|---|---|---|---|---|---|---|---|---|---|---|---|---|---|---|---|---|---|---|---|
| | | | Enterprise Goal | | | | | | | | | | | | | | | | |
| | | | Stakeholder value of business investments | Portfolio of competitive products and services | Managed business risk (safeguarding of assets) | Compliance with external laws and regulations | Financial transparency | Customer-oriented service culture | Business service continuity and availability | Agile responses to a changing business environment | Information-based strategic decision making | Optimisation of service delivery costs | Optimisation of business process functionality | Optimisation of business process costs | Managed business change programmes | Operational and staff productivity | Compliance with internal policies | Skilled and motivated people | Product and business innovation culture |
| | | | 1. | 2. | 3. | 4. | 5. | 6. | 7. | 8. | 9. | 10. | 11. | 12. | 13. | 14. | 15. | 16. | 17. |
| | | IT-related Goal | Financial | | | | | Customer | | | | | Internal | | | | | Learning and Growth | |
| Financial | 01 | Alignment of IT and business strategy | P | P | S | | | P | S | P | P | S | P | S | P | | | S | S |
| | 02 | IT compliance and support for business compliance with external laws and regulations | | | S | P | | | | | | | | | | | P | | |
| | 03 | Commitment of executive management for making IT-related decisions | P | S | S | | | | | S | S | | S | | P | | | S | S |
| | 04 | Managed IT-related business risk | | | P | S | | P | S | | P | | | | S | | S | S | |
| | 05 | Realised benefits from IT-enabled investments and services portfolio | P | P | | | | S | | S | | | S | S | P | S | | | S |
| | 06 | Transparency of IT costs, benefits and risk | S | | S | | P | | | | | S | P | | P | | | | |
| Customer | 07 | Delivery of IT services in line with business requirements | P | P | S | S | | P | S | P | S | P | P | S | S | | | S | S |
| | 08 | Adequate use of applications, information and technology solutions | S | S | S | | | S | S | | S | S | P | S | | P | | S | S |
| | 09 | IT agility | S | P | S | | | S | | P | | | P | | S | S | | S | P |
| | 10 | Security of information, processing infrastructure and applications | | | P | P | | | P | | | | | | | | P | | |
| Internal | 11 | Optimisation of IT assets, resources and capabilities | P | S | | | | | S | | | P | S | P | S | S | | | S |
| | 12 | Enablement and support of business processes by integrating applications and technology into business processes | S | P | S | | | S | | | S | | S | P | S | S | | | S |

## Figure 2.7—Mapping COBIT 5 Enterprise Goals to IT-related Goals (cont.)

| | | IT-related Goal | 1. Stakeholder value of business investments | 2. Portfolio of competitive products and services | 3. Managed business risk (safeguarding of assets) | 4. Compliance with external laws and regulations | 5. Financial transparency | 6. Customer-oriented service culture | 7. Business service continuity and availability | 8. Agile responses to a changing business environment | 9. Information-based strategic decision making | 10. Optimisation of service delivery costs | 11. Optimisation of business process functionality | 12. Optimisation of business process costs | 13. Managed business change programmes | 14. Operational and staff productivity | 15. Compliance with internal policies | 16. Skilled and motivated people | 17. Product and business innovation culture |
|---|---|---|---|---|---|---|---|---|---|---|---|---|---|---|---|---|---|---|---|
| | | | Financial | | | | | Customer | | | | | Internal | | | | | Learning and Growth | |
| Internal | 13 | Delivery of programmes delivering benefits, on time, on budget, and meeting requirements and quality standards | P | S | S | | | S | | | | S | | S | P | | | | |
| | 14 | Availability of reliable and useful information for decision making | S | S | S | S | | | P | | P | | S | | | | | | |
| | 15 | IT compliance with internal policies | | | S | S | | | | | | | | | | | P | | |
| Learning and Growth | 16 | Competent and motivated business and IT personnel | S | S | P | | | S | | S | | | | | | P | | P | S |
| | 17 | Knowledge, expertise and initiatives for business innovation | S | P | | | | S | | P | S | | S | | S | | | S | P |

Source: ISACA, COBIT 5, USA, 2012, figure 22

### Step 4. IT-related Goals Cascade to Enabler Goals

Achieving IT-related goals requires the successful application and use of a number of enablers. Enablers include processes, organizational structures and information, and for each enabler a set of specific relevant goals can be defined in support of the IT-related goals. Processes are one of the enablers, and **figure 2.8** contains a mapping between IT-related goals and the relevant COBIT 5 processes, which then contain related process goals.

## Figure 2.8—Mapping COBIT 5 IT-related Goals to Processes

| | | | Financial | | | | | | Customer | | Internal | | | | | | | Learning and Growth | |
|---|---|---|---|---|---|---|---|---|---|---|---|---|---|---|---|---|---|---|---|
| | COBIT 5 Process | | 01 Alignment of IT and business strategy | 02 IT compliance and support for business compliance with external laws and regulations | 03 Commitment of executive management for making IT-related decisions | 04 Managed IT-related business risk | 05 Realised benefits from IT-enabled investments and services portfolio | 06 Transparency of IT costs, benefits and risk | 07 Delivery of IT services in line with business requirements | 08 Adequate use of applications, information and technology solutions | 09 IT agility | 10 Security of information, processing infrastructure and applications | 11 Optimisation of IT assets, resources and capabilities | 12 Enablement and support of business processes by integrating applications and technology into business processes | 13 Delivery of programmes delivering benefits, on time, on budget, and meeting requirements and quality standards | 14 Availability of reliable and useful information for decision making | 15 IT compliance with internal policies | 16 Competent and motivated business and IT personnel | 17 Knowledge, expertise and initiatives for business innovation |
| Evaluate, Direct and Monitor | EDM01 | Ensure Governance Framework Setting and Maintenance | P | S | P | S | S | S | P | | S | S | S | S | S | S | S | S | S |
| | EDM02 | Ensure Benefits Delivery | P | | S | | P | P | P | S | | | S | S | S | S | | S | P |
| | EDM03 | Ensure Risk Optimisation | S | S | S | P | | P | S | S | | P | | | | S | S | P | S | S |
| | EDM04 | Ensure Resource Optimisation | S | | S | S | S | S | S | | P | | P | | | S | | | P | S |
| | EDM05 | Ensure Stakeholder Transparency | S | S | P | | | P | P | | | | | | | S | S | S | | S |
| Align, Plan and Organise | APO01 | Manage the IT Management Framework | P | P | S | S | | | S | | P | S | P | S | S | S | | P | P | P |
| | APO02 | Manage Strategy | P | | S | S | S | | P | S | | | S | | S | S | S | | S | P |
| | APO03 | Manage Enterprise Architecture | P | | S | S | S | S | | P | S | P | S | | | S | | | S |
| | APO04 | Manage Innovation | S | | | S | P | | | P | P | | P | S | | S | | | | P |
| | APO05 | Manage Portfolio | P | | S | | P | S | S | S | | | S | | P | | | | S |
| | APO06 | Manage Budget and Costs | S | | S | S | P | P | S | S | | | S | | S | | | | |
| | APO07 | Manage Human Resources | P | S | S | S | | | S | | S | S | P | | P | | | S | P | P |
| | APO08 | Manage Relationships | P | | S | S | S | S | P | S | | | S | P | S | | | | S | P |

Source: ISACA, COBIT 5, USA, 2012, figure 23

### Figure 2.8—Mapping COBIT 5 IT-related Goals to Processes (cont.)

| | | | Financial | | | | | | Customer | | Internal | | | | | | Learning and Growth | |
|---|---|---|---|---|---|---|---|---|---|---|---|---|---|---|---|---|---|---|
| | COBIT 5 Process | | 01 | 02 | 03 | 04 | 05 | 06 | 07 | 08 | 09 | 10 | 11 | 12 | 13 | 14 | 15 | 16 | 17 |
| Align, Plan and Organise | AP009 | Manage Service Agreements | S | | | S | S | S | P | S | S | S | S | | S | P | S | | |
| | AP010 | Manage Suppliers | | S | | P | S | S | P | S | P | S | S | | S | S | S | | S |
| | AP011 | Manage Quality | S | S | | S | P | | P | S | S | | S | | P | S | S | S | S |
| | AP012 | Manage Risk | | P | | P | | P | S | S | S | P | | | P | S | S | S | S |
| | AP013 | Manage Security | | P | | P | | | P | S | S | | P | | | P | | | |
| Build, Acquire and Implement | BAI01 | Manage Programmes and Projects | P | | S | P | P | S | S | S | | | S | | P | | | S | S |
| | BAI02 | Manage Requirements Definition | P | S | S | S | S | | P | S | S | S | S | P | S | S | | | S |
| | BAI03 | Manage Solutions Identification and Build | S | | | S | S | | P | S | | | S | S | S | S | | | S |
| | BAI04 | Manage Availability and Capacity | | | | S | S | | P | S | S | | P | | S | P | | | S |

### Figure 2.8—Mapping COBIT 5 IT-related Goals to Processes (cont.)

| | | | IT-related Goal | | | | | | | | | | | | | | | |
|---|---|---|---|---|---|---|---|---|---|---|---|---|---|---|---|---|---|---|
| | | | Alignment of IT and business strategy | IT compliance and support for business compliance with external laws and regulations | Commitment of executive management for making IT-related decisions | Managed IT-related business risk | Realised benefits from IT-enabled investments and services portfolio | Transparency of IT costs, benefits and risk | Delivery of IT services in line with business requirements | Adequate use of applications, information and technology solutions | IT agility | Security of information, processing infrastructure and applications | Optimisation of IT assets, resources and capabilities | Enablement and support of business processes by integrating applications and technology into business processes | Delivery of programmes delivering benefits, on time, on budget, and meeting requirements and quality standards | Availability of reliable and useful information for decision making | IT compliance with internal policies | Competent and motivated business and IT personnel | Knowledge, expertise and initiatives for business innovation |
| | | | 01 | 02 | 03 | 04 | 05 | 06 | 07 | 08 | 09 | 10 | 11 | 12 | 13 | 14 | 15 | 16 | 17 |
| | COBIT 5 Process | | Financial | | | | | | Customer | | | Internal | | | | | | Learning and Growth | |
| Build, Acquire and Implement | BAI05 | Manage Organisational Change Enablement | S | | S | | S | | S | P | S | | S | | S | P | | | P |
| | BAI06 | Manage Changes | | S | P | S | | P | S | S | P | S | | S | S | S | S | | S |
| | BAI07 | Manage Change Acceptance and Transitioning | | | | S | S | | S | P | S | | | P | S | S | | | S |
| | BAI08 | Manage Knowledge | S | | | S | | | S | S | P | S | S | | | S | | S | P |
| | BAI09 | Manage Assets | | S | | S | | P | S | | S | S | P | | | S | S | | |
| | BAI10 | Manage Configuration | | P | | S | | S | S | S | S | P | | | P | S | | | |
| Deliver, Service and Support | DSS01 | Manage Operations | | S | | P | S | | P | S | S | S | P | | | S | S | S |
| | DSS02 | Manage Service Requests and Incidents | | | | P | | | P | S | | S | | | | S | S | | S |
| | DSS03 | Manage Problems | | S | | P | S | | P | S | S | | P | S | | P | S | | |
| | DSS04 | Manage Continuity | S | S | | P | S | | P | S | S | S | S | | | P | S | S | S |
| | DSS05 | Manage Security Services | S | P | | P | | | S | S | | P | S | | | S | S | | |
| | DSS06 | Manage Business Process Controls | | S | | P | | | P | S | | S | S | | | S | S | S | S |

**Figure 2.8—Mapping COBIT 5 IT-related Goals to Processes** *(cont.)*

| | | | \multicolumn{17}{c}{IT-related Goal} |
|---|---|---|---|---|---|---|---|---|---|---|---|---|---|---|---|---|---|---|---|
| | | | Alignment of IT and business strategy | IT compliance and support for business compliance with external laws and regulations | Commitment of executive management for making IT-related decisions | Managed IT-related business risk | Realised benefits from IT-enabled investments and services portfolio | Transparency of IT costs, benefits and risk | Delivery of IT services in line with business requirements | Adequate use of applications, information and technology solutions | IT agility | Security of information, processing infrastructure and applications | Optimisation of IT assets, resources and capabilities | Enablement and support of business processes by integrating applications and technology into business processes | Delivery of programmes delivering benefits, on time, on budget, and meeting requirements and quality standards | Availability of reliable and useful information for decision making | IT compliance with internal policies | Competent and motivated business and IT personnel | Knowledge, expertise and initiatives for business innovation |
| | | | 01 | 02 | 03 | 04 | 05 | 06 | 07 | 08 | 09 | 10 | 11 | 12 | 13 | 14 | 15 | 16 | 17 |
| Monitor, Evaluate and Assess | MEA01 | Monitor, Evaluate and Assess Performance and Conformance | S | S | S | P | S | S | P | S | S | S | P | | S | S | P | S | S |
| | MEA02 | Monitor, Evaluate and Assess the System of Internal Control | | P | | P | | | S | S | S | | S | | | | S | P | | S |
| | MEA03 | Monitor, Evaluate and Assess Compliance With External Requirements | | P | | P | S | | S | | | | S | | | | | S | | S |

More information on enablers can be found in section 1.4 IT Governance Enablers.

## Value of the COBIT 5 Cascade for Strategic Planning

The goals cascade[6] is important because it allows the definition of priorities for implementation, improvement and assurance of governance of enterprise IT based on (strategic) objectives of the enterprise and the related risk. In practice, the goals cascade:
- Defines relevant and tangible goals and objectives at various levels of responsibility
- Filters the knowledge base of COBIT, based on enterprise goals, to extract relevant guidance for inclusion in the specific implementation, improvement or assurance projects
- Clearly identifies and communicates how enablers are important to achieve enterprise goals

The goals cascade—with its mapping tables between enterprise goals and IT-related goals and between IT-related goals and COBIT 5 enablers (including processes)—does not contain the universal truth, and users should not attempt to use it in a purely mechanistic way, but rather as a guideline. There are various reasons for this, including:
- Every enterprise has different priorities in its goals, and priorities might change over time.
- The mapping tables do not distinguish between size and/or industry of the enterprise. They represent a sort of common denominator of how, in general, the different levels of goals are interrelated.
- The indicators used in the mapping use two levels of importance of relevance, suggesting that there are discrete levels of relevance, whereas, in reality, the mapping will be closer to a continuum of various degrees of correspondence.

From the previous disclaimer, it is obvious that the first step an enterprise should always apply when using the goals cascade is to customize the mapping, taking into account its specific situation. In other words, each enterprise should build its own goals cascade, compare it with COBIT and then refine it.

For example, the enterprise may wish to:
- Translate the strategic priorities into a specific weight of importance for each of the enterprise goals
- Validate the mapping of the goals cascade, taking into account its specific environment, industry, etc.

An example of an application of the goals cascade is provided in **figure 2.9**.

| Figure 2.9—Example—Goals Cascade |
|---|
| An enterprise has defined for itself a number of strategic goals, of which improving customer satisfaction is the most important. From there, it wants to know where it needs to improve in all things related to IT. |
| The enterprise decides that setting customer satisfaction as a key priority is equivalent to raising the priority of the following enterprise goals (from figure 5):<br>• 6. Customer-oriented service culture<br>• 7. Business service continuity and availability<br>• 8. Agile responses to a changing business environment |
| The enterprise now takes the next step in the goals cascade: analysing which IT-related goals correspond to these enterprise goals. A suggested mapping between them is listed in appendix B. |
| From there, the following IT-related goals are suggested as most important (all 'P' relationships):<br>• 01 Alignment of IT and business strategy<br>• 04 Managed IT-related business risk<br>• 07 Delivery of IT services in line with business requirements<br>• 09 IT agility<br>• 10 Security of information, processing infrastructure and applications<br>• 14 Availability of reliable and useful information for decision making<br>• 17 Knowledge, expertise and initiatives for business innovation |
| The enterprise validates this list, and decides to retain the first four goals as a matter of priority. |
| In the next step in the cascade, using the enabler concept (see chapter 5), these IT-related goals drive a number of enabler goals, which include process goals. In appendix C, a mapping is suggested between IT-related goals and COBIT 5 processes. That table allows identification of the most relevant IT-related processes that support the IT-related goals, but, processes alone are not sufficient. The other enablers, such as culture, behaviour and ethics; organisational structures; or skills and expertise are equally important and require a set of clear goals. |
| When this exercise is completed, the enterprise has a set of consistent goals for all enablers that allow it to reach the stated strategic objectives and a set of associated metrics to measure performance. |
| Source: ISACA, COBIT 5, USA, 2012, example 1, page 21 |

## 2.3 IMPACT OF CHANGES IN BUSINESS STRATEGY ON IT STRATEGY

The dynamic nature of business requires frequent strategy revisions, and IT needs to continually readjust to align with the business. Nonaligned IT can have dire consequences for the enterprise as it will inhibit the enterprise to realize its business strategy.

### Agility

Enterprises need to be agile to keep up with their markets, and IT organizations must be agile to stay aligned with their enterprises. As enterprise strategy evolves over time, there must be a constant assessment of the strategic business changes and their impact on the IT organization. Questions to be asked include:
- Does the existing infrastructure support the new business strategies?
- What new capabilities are needed?
- How can existing systems best be leveraged?
- What new systems are needed?

A recent study published by CIONET[7] on the priorities of IT executives pointed at agility to be among the top 10 IT management concerns. However, a recent benchmarking study by ISACA and the University of Antwerp—Antwerp Management School[8] revealed that agility is perceived as one of the weakest achieved goals within IT. This result hints at a "knowing-doing gap"[9]—agility is reported as top-priority on the one hand, but is reported to be "least achieved" on the other hand.

## Enterprise Agility

COBIT 5 refer to the stakeholder's need for enterprise agility in terms of the requirement for "agile responses to a changing business environment." The achievement of such enterprise goal could be measured through:
- Level of board level satisfaction with enterprise responsiveness to new requirements
- Number of critical products and services supported by up-to-date business processes
- Average time required to turn strategic business objectives into an agreed upon and approved initiative

## IT Agility

In order to realize enterprise IT agility, the IT organization of course also needs to be agile. COBIT 5 proposes IT agility to be measured through:
- Level of board level satisfaction with IT's responsiveness to new requirements
- Number of critical business processes supported by up-to-date infrastructure and applications
- Average time required to turn strategic IT objectives into an agreed upon and approved initiative

## Agility Loops

The target to be achieved is a tactically agile IT organization—one that senses and responds to environmental change efficiently and effectively. Agility is an ongoing process, and not a one-time or occasional event. For the entire IT organization to be agile, all members of the IT organization need to understand the need for agility and be committed to this process. There are three work principles involved for the IT organization to be agile. These can be represented as agility loops, as follows:[10]
- Loop 1: Monitoring and deciding
- Loop 2: Improving existing processes
- Loop 3: Creating new processes

Loop 1 encompasses environmental monitoring and responsive decision making. The focus of the environmental monitoring and decision-making process is to identify, analyze and respond to the nonstandard inputs. This is because the most profitable opportunities for better alignment of IT and business often arise from agile responses to new or unexpected events (nonstandard inputs).

Loop 2 focuses on improving existing operations—delivering efficiency. The IT systems that drive the standard operating processes of the enterprise should be as automated and reliable as possible. They are the basic transaction processing systems such as enterprise resource planning (ERP), order management and production scheduling.

Loop 3 focuses on creating new operations—delivering effectiveness. In Loops 2 and 3, data from the environment and customer demands are handled by a set of standardized operating processes. They handle most of the input reliably and efficiently. When the loops are used in combination, the IT organization can sense changes and respond efficiently and effectively.

Examples of events providing nonstandard input are transaction processing volumes increasing or decreasing at unexpected rates, or system processing or operating errors occurring at greater than expected rates. These events signify a need to improve an existing IT operation. Other examples are new competitors entering the market or sales of certain products increasing or decreasing faster than expected. Such events often signify a need to create a new system or work process.

## 2.4 BARRIERS TO THE ACHIEVEMENT OF STRATEGIC ALIGNMENT

Strategic alignment is a multifaceted and complex endeavor, often referred to as the alignment challenge. To overcome alignment barriers, it is important to understand the difficulties that organizations experience aligning business with IT. According to a report by CIONET, IT and business alignment has been a top concern of IT management over the past decade.[11] Weill and Broadbent[12] depict a number of difficulties (barriers) that organizations have experienced while aligning business with IT: expression barriers, specification barriers and implementation barriers.

### Expression Barriers

Expression barriers arise from the organization's strategic context and from senior management behavior, including lack of direction in business strategy. This results in insufficient understanding of and commitment to the organization's strategic focus by operational management.

### Specification Barriers

Specification barriers arise from the circumstances of the organization's IT strategy, such as lack of IT involvement in strategy development and business and IT management conducting two independent monologues. This results in a situation in which business and IT strategies are set in isolation and are not adequately related.

### Implementation Barriers

The nature of the organization's current IT portfolio creates implementation barriers, which arise when there are technical, political or financial constraints on the current infrastructure. A good example of this last barrier is the difficult integration of legacy systems.

## 2.5 POLICIES AND PROCEDURES NECESSARY TO SUPPORT IT AND BUSINESS STRATEGIC ALIGNMENT

Knowledge of policies and procedures supporting the strategic planning process enable the development of an IT strategy that is defined and executed in line with business imperatives.

While some good practices do exist within many enterprises to maximize alignment, ISACA identified in a global survey on alignment that there are a number of concerns including:[13]
- Almost 50 percent of the entities responding to the survey did not have a formalized governance structure designed to ensure IT and business alignment.
- The responsibility for IT strategy is often delegated to management below the board level. In particular, fewer than 25 percent of entities engage board members directly in the IT strategy-setting process.

### Practices Supporting Strategic Alignment

In aligning IT strategy with business strategy, it is important that the IT strategy implementation plan be endorsed by all relevant parties.[14] It is also important that the IT implementation plan be broken down into manageable parts, each with a clear business case incorporating a plan for achieving outcomes and realizing benefits. The board should ensure that the strategy is reviewed regularly in light of technological and operational change. Either the board, or a dedicated IT strategy committee of the board, should drive business alignment by:
- Ensuring that the IT strategy is aligned with the business strategy and that distributed IT strategies are consistent and integrated
- Ensuring that IT delivers against the strategy (delivering on time and within budget, with appropriate functionality and intended benefits—a fundamental building block of alignment and value delivery) through clear expectations and measurement (e.g., balanced business scorecard)
- Balancing investments between systems that support the enterprise, transform the enterprise or that enable the business to grow and compete in new arenas
- Making considered decisions about the focus of IT resources (i.e., their use to break into new markets, drive competitive strategies, increase overall revenue generation, improve customer satisfaction and/or assure customer retention)

Alignment requires planned and purposeful management processes, such as:
- Creating and sustaining awareness of the strategic role of IT at a top management level
- Clarifying the role that IT should play—utility vs. enabler
- Creating IT guiding principles from business maxims (business strategies, ambitions, goals, etc.). For example, "develop partnerships with customers worldwide" can lead to "consolidate customer database and order processing processes."
- Monitoring the business impact of the IT infrastructure and applications portfolio
- Evaluating postimplementation benefits delivered by IT-enabled programs

An effective practice to support strategic alignment is described in the goals cascade of COBIT 5, providing an approach to identify and align enterprise goals and IT-related goals. More information about this goals cascade is discussed in section 2.2 Strategic Planning Processes and Techniques.

## Role of the IT Strategy Committee

The role of IT is shifting from being an enabler to becoming critical for survival. IT strategy committees need to broaden their scope. Not only must they offer advice on strategy when assisting the board in its GEIT responsibilities, but they also need to focus on IT value, risk and performance. In this way, the IT strategy committee can assist the board in effectively assuming accountability for evaluating, directing and monitoring IT.

A more detailed definition of the IT strategy committee is provided in chapter 1, figure 1.13.

## Importance of Policies and Procedures

Policies and procedures reflect management guidance and direction in developing controls over information systems, related resources and IS department processes.

### Policies

Policies are high-level documents that represent the corporate philosophy of an organization. To be effective, policies must be clear and concise. Management must create a positive control environment by assuming responsibility for formulating, developing, documenting, promulgating and controlling policies covering general goals and directives. In addition to corporate policies that set the tone for the organization as a whole, individual divisions and departments should define lower-level policies. To achieve strategic alignment, it is imperative that these divisional and departmental policies are consistent with corporate-level policies.

Policies are one of the governance enablers discussed in COBIT 5. Policies, along with principles, refer to the communication mechanism put in place to convey the governing body's and management's direction and instructions. Policies provide more guidance to put principles into practice. Good policies are effective, efficient and nonintrusive.

### Procedures

Procedures are detailed steps defined and documented for implementing policies. They must be derived from the parent policy and be designed to achieve the spirit (intent) of the policy statement. Procedures must be written in a clear and concise manner so they may be easily and properly understood by those governed by them. Generally, procedures are more dynamic than their respective parent policies. Procedures must reflect the regular changes in business focus and environment. Therefore, frequent reviews and updates of procedures are essential to ensure continuous relevance and alignment with the respective policies.

COBIT 5 stresses the importance of having policies and procedures in one of its management processes, more specifically APO01 *Manage the IT Management Framework*. The details of the management practices are shown in **figure 2.10**.

### Figure 2.10—APO01.08 Maintain Compliance With Policies and Procedures

| Management Practice | Inputs | | Outputs | |
|---|---|---|---|---|
| **APO01.08 Maintain compliance with policies and procedures.** Put in place procedures to maintain compliance with and performance measurement of policies and other enablers of the control framework, and enforce the consequences of non-compliance or inadequate performance. Track trends and performance and consider these in the future design and improvement of the control framework. | **From** | **Description** | **Description** | **To** |
| | DSS01.04 | Environmental policies | Non-compliance remedial actions | MEA01.05 |
| | MEA03.02 | Updated policies, principles, procedures and standards | | |
| **Activities** | | | | |
| 1. Track compliance with policies and procedures. | | | | |
| 2. Analyse non-compliance and take appropriate action (this could include changing requirements). | | | | |
| 3. Integrate performance and compliance into individual staff members' performance objectives. | | | | |
| 4. Regularly assess the performance of the framework's enablers and take appropriate action. | | | | |
| 5. Analyse trends in performance and compliance and take appropriate action. | | | | |
| Source: ISACA, *COBIT 5: Enabling Processes*, USA, 2012, page 56 | | | | |

More information on the IT strategic planning process is discussed in section 2.6 Methods to Document and Communicate IT Strategic Planning Processes.

## 2.6 METHODS TO DOCUMENT AND COMMUNICATE IT STRATEGIC PLANNING PROCESSES

To enable an effective and efficient IT strategic planning process, the enterprise requires structured methods to document and communicate the intermediate steps and outcome of the IT strategic planning process. The balanced scorecard is promoted as a very effective means in this regard.

Strategy has taken on a new urgency as enterprises mobilize intangible and hidden assets to compete in an information-based global economy.[15] The means of value creation have shifted from tangible assets toward intangible assets, and intangible assets generally are not measurable through traditional financial means.

### Business Strategy and the Business Balanced Scorecard

BSCs are often used in translating strategy into action to achieve goals. They have a performance measurement system that goes beyond conventional accounting, which measures those relationships and knowledge-based assets necessary to compete in the information age—customer focus, process efficiency and the ability to learn and grow. Each perspective is needed for a strategy road map and is designed to answer one question about the enterprise's way of doing business:

- **Financial perspective**—To satisfy our stakeholders, what financial objectives must we accomplish?
- **Customer perspective**—To achieve our financial objectives, what customer needs must we serve?
- **Internal process perspective**—To satisfy our customers and stakeholders, in which internal business processes must we excel?
- **Learning perspective**—To achieve our goals, how must our enterprise learn and innovate?

By using the BSC, managers rely on more than short-term financial measures as indicators of the enterprise's performance. They also take into account intangible items, such as level of customer satisfaction, streamlining of internal functions, creation of operational efficiencies and development of staff skills. This unique and more holistic view of business operations contributes to linking long-term strategic objectives with short-term actions.

At the heart of these scorecards is management information supplied by relevant stakeholders and supported by a sustainable reporting system as shown in **figure 2.11**.

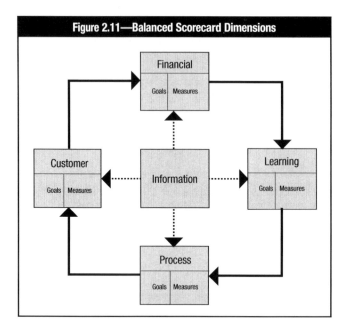

## IT Strategy and the IT Balanced Scorecard

IT not only contributes information to the business scorecards and tools to the different dimensions being measured, but also—because of the criticality of IT itself—needs its own scorecard. Defining clear goals and good measures that unequivocally reflect the business impact of the IT goals is a challenge that needs to be resolved in cooperation with the different governance layers within the enterprise.

Use of an IT BSC is one of the most effective means to aid the board and management in achieving IT and business alignment. The objectives are to establish a vehicle for management reporting to the board, to foster consensus among key stakeholders about IT's strategic aims, to demonstrate the effectiveness and added value of IT, and to communicate about IT's performance, risk and capabilities. This is shown in **figure 2.12**.

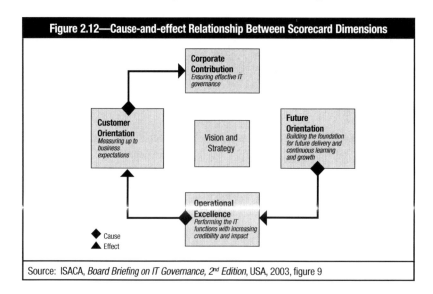

To apply the BSC concepts to the IT function, the four perspectives need to be redefined. An IT BSC template should be developed by considering the following questions:
- **Corporate contribution**—How do business executives view the IT department?
- **Customer orientation**—How do users view the IT department?
- **Operational excellence**—How effective and efficient are the IT processes?
- **Future orientation**—How well is IT positioned to meet future needs?

As stated in ISACA's board briefing publication,[16] use of an IT BSC is one of the most effective means to aid board and management in achieving IT and business alignment and is a best practice for performance measurement and alignment. This is supported by testimonials of executives.[17] To apply this good practice to the IT function as an internal service provider, the four perspectives of the generic BSC should be changed accordingly.

In **figure 2.13**, a generic IT BSC for an IT department is shown. The user orientation perspective represents the user evaluation of IT. The operational excellence perspective represents the IT processes employed to develop and deliver the applications. The future orientation perspective represents the human and technology resources needed by IT to deliver its services over time. The business contribution perspective captures the business value created from the IT investments.

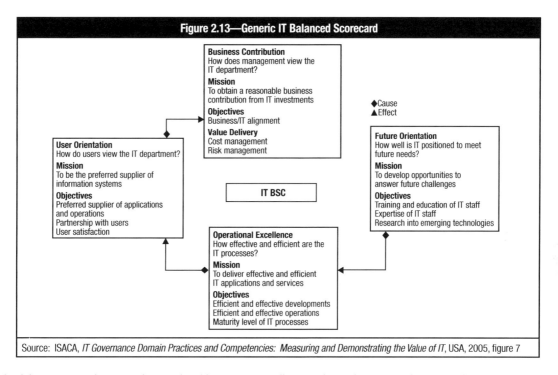

Source: ISACA, *IT Governance Domain Practices and Competencies: Measuring and Demonstrating the Value of IT*, USA, 2005, figure 7

Each of these perspectives must be translated into corresponding metrics and measures that assess the current situation. As noted previously, the cause-and-effect relationships between measures are essential components of the IT BSC, and these relationships are articulated by two types of measures: outcome measures and performance drivers.

Outcome measures, such as programmers' productivity (e.g., number of function points per person per month), need performance drivers, such as IT staff education (e.g., number of education days per person per year), to communicate how the outcomes are to be achieved.

Performance drivers need outcome measures to ensure a way to determine whether the chosen strategy is effective, especially important in cases where a significant investment is made.

These cause-and-effect relationships must be defined throughout the entire scorecard: more and better education of IT staff (future orientation) is an enabler (performance driver) for a better quality of developed systems (operational excellence perspective) that, in turn, is an enabler for increased user satisfaction (user perspective) that, eventually, will lead to higher business value of IT (business contribution).

More information on the IT strategic planning process is discussed in section 2.2 Strategic Planning Processes and Techniques.

More information on the BSC and its applications to IT and GEIT is discussed in section 1.10 IT Governance Monitoring Processes/Mechanisms and section 1.11 IT Governance Reporting Processes/Mechanisms.

## 2.7 COMPONENTS, PRINCIPLES AND FRAMEWORKS OF ENTERPRISE ARCHITECTURE

Enterprise architecture (EA) takes a broad view of business, matching it with associated information and related technologies. It is an important blueprint for IT strategy development and, therefore, components, principles and frameworks related to EA must be understood to effectively align IT to the business.

Architecture can be defined as a representation of a conceptual framework of components and their relationships at a point in time.[18] Architecture discussions have traditionally focused on technology issues. EA takes a broader view of business, matching it with the associated information. It provides the framework for ensuring that enterprisewide goals, objectives and policies are properly and accurately reflected in decision making related to building, implementing or changing information systems. Having appropriate architecture practice in place should also provide reasonable assurance that standards for interprocess communication, data naming, data representation, data structures and information systems will be consistently and appropriately applied across the enterprise.

An organization with effective enterprise architecture has the following virtues:
- Be aware, at management level, about enterprise architecture
- Be aware of business weaknesses and strengths
- Have clear models, understandable by stakeholders
- Improve processes continuously
- Understand the impact of changes; have fewer "unknown unknowns"

### Components of Enterprise Architecture

The widely accepted US National Institute of Standards and Technology (NIST) model for EA[19] is shown in **figure 2.14**.

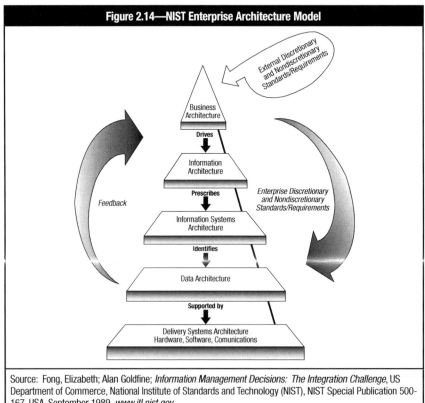

Figure 2.14—NIST Enterprise Architecture Model

Source: Fong, Elizabeth; Alan Goldfine; *Information Management Decisions: The Integration Challenge*, US Department of Commerce, National Institute of Standards and Technology (NIST), NIST Special Publication 500-167, USA, September 1989, *www.itl.nist.gov*

An enterprise is composed of one or more business units that are responsible for a specific business area. The Enterprise Architecture Development Tool-Kit published by the US National Association of State Chief Information Officers (NASCIO)[20] states:

> *Adopting enterprise architecture increases the utility of an enterprise's data by facilitating information sharing between data stores. Committing to an ongoing renewable enterprise architecture process fosters a technology-adaptive enterprise. Enterprise architecture becomes a road map, guiding all future technology investments and identifying and aiding in the resolution of gaps in the entity's business and IT infrastructures.*

EA is the inclusive term used to describe the five layers of architecture:
- Business unit architecture (or business architecture)
- Information architecture
- Information systems architecture (sometimes called solution architecture)
- Data architecture
- Delivery system architecture (sometimes called technology architecture)

More information on these architecture layers can be found in section 1.6 Components, Principles and Concepts Related to Enterprise Architecture.

## COBIT 5 View on Enterprise Architecture

The GEIT COBIT 5 framework also develops guidance in the area of EA. The EA process is positioned as one of the first management processes in the framework, with a process description and purpose statement as shown in **figure 2.15**. The process should support IT-related goals around agility, resource optimization, and the alignment between business and IT strategy.

| Figure 2.15—APO03 Manage Enterprise Architecture ||
|---|---|
| **APO03 Manage Enterprise Architecture** | **Area: Management**<br>**Domain: Align, Plan and Organise** |
| **Process Description**<br>Establish a common architecture consisting of business process, information, data, application and technology architecture layers for effectively and efficiently realising enterprise and IT strategies by creating key models and practices that describe the baseline and target architectures. Define requirements for taxonomy, standards, guidelines, procedures, templates and tools, and provide a linkage for these components. Improve alignment, increase agility, improve quality of information and generate potential cost savings through initiatives such as re-use of building block components. ||
| **Process Purpose Statement**<br>Represent the different building blocks that make up the enterprise and their inter-relationships as well as the principles guiding their design and evolution over time, enabling a standard, responsive and efficient delivery of operational and strategic objectives. ||
| Source: ISACA, *COBIT 5: Enabling Processes*, USA, 2012, page 63 ||

COBIT 5 proposes five key management practices to be adopted around enterprise architecture:
- Develop the enterprise architecture vision.
  - The EA vision provides a first-cut, high-level description of the baseline and target architectures, covering the business, information, data, application and technology domains.
- Define the reference architecture.
  - The reference architecture describes the current and target architectures for the business, information, data, application and technology domains.
- Select opportunities and solutions.
  - Rationalize the gaps between baseline and target architectures, taking both the business and technical perspectives, and logically group them into project work packages.
- Define architecture implementation.
  - Create a viable implementation and migration plan in alignment with the program and project portfolios.
- Provide EA services.
  - The provision of EA services within the enterprise includes guidance to, and monitoring of, implementation projects, formalizing ways of working through architecture contracts, and measuring and communication architecture's value-add and compliance monitoring.

### Information Governance and Management

The EA discussion is closely related to the concepts of information governance and management, as also discussed in the guidance provided in *COBIT 5: Enabling Information*.

According to *COBIT 5: Enabling Information*, information governance ensures:
- That stakeholder needs, conditions and options are evaluated to determine balanced, agreed-on enterprise objectives to be achieved through the acquisition and management of information resources
- Setting the direction of information capabilities through prioritization and decision making
- Monitoring performance and compliance of the information resource against agreed-on direction and objectives

Information governance activities include:
- Communicating information strategies, policies, standards, architecture and metrics
- Tracking and enforcing regulatory compliance and conformance to information policies, standards, architecture and procedures
- Sponsoring, tracking and overseeing the delivery of information management programs

Information management, in turn, is encountered as one or more specific areas of practice in many organizations, with various names, such as data architecture, data administration, database administration, data warehousing, data/information governance, business intelligence and analytics, information architecture, information resource management, EA, and records management. Information management plans, builds, runs and monitors the practices, projects and capabilities that acquire, control, protect, deliver and enhance the value of data and information assets, in alignment with the direction set by the data and information governance body.

## 2.8 CURRENT AND FUTURE TECHNOLOGIES

The consideration of the technological direction to adopt is a key in preparing IT to be an agile enabler for the business. This is especially important for strategic alignment and must be addressed in the context of the role of IT in the future of the business.

COBIT 5 identifies a specific process focused on:
- Maintaining an awareness of IT and related service trends
- Identifying innovation opportunities
- Planning how to benefit from innovation in relation to business needs.

According to the COBIT 5 process APO04 *Manage Innovation*, it is important to analyze what opportunities for business innovation or improvement can be created by emerging technologies, services or IT-enabled business innovation, as well as through existing established technologies and by business and IT process innovation.

These inputs can influence strategic planning and EA decisions, in support of achieving competitive advantage, business innovation, and improved operational effectiveness and efficiency by exploiting IT developments.

The key management practices relating to this innovation process are shown in **figure 2.16**.

More information on enterprise IT architecture can be found in section 2.7 Components, Principles and Frameworks of Enterprise Architecture.

### Figure 2.16—APO04 Manage Innovation Management Practices

**APO04 Manage Innovation**

| Management Practice | Inputs | | Outputs | |
|---|---|---|---|---|
| **APO04.01 Create an environment conducive to innovation.** Create an environment that is conducive to innovation, considering issues such as culture, reward, collaboration, technology forums, and mechanisms to promote and capture employee ideas. | From | Description | Description | To |
| | | | Innovation plan | Internal |
| | | | Recognition and reward programme | APO07.04 |

**Activities**

1. Create an innovation plan that includes risk appetite, the envisioned budget to spend on innovation initiatives, and innovation objectives.
2. Provide infrastructure that can be an enabler for innovation, such as collaboration tools for enhancing work between geographic locations and divisions.
3. Create an environment that is conducive to innovation by maintaining relevant HR initiatives, such as innovation recognition and reward programmes, appropriate job rotation, and discretionary time for experimentation.
4. Maintain a programme enabling staff to submit innovation ideas and create an appropriate decision-making structure to assess and take these ideas forward.
5. Encourage innovation ideas from customers, suppliers and business partners.

| Management Practice | Inputs | | Outputs | |
|---|---|---|---|---|
| **APO04.02 Maintain an understanding of the enterprise environment.** Work with relevant stakeholders to understand their challenges. Maintain an adequate understanding of enterprise strategy and the competitive environment or other constraints so that opportunities enabled by new technologies can be identified. | From | Description | Description | To |
| | Outside COBIT | Enterprise strategy and enterprise SWOT analysis | Innovation opportunities linked to business drivers | APO02.01 |

**Activities**

1. Maintain an understanding of the business drivers, enterprise strategy, industry drivers, enterprise operations and other issues so that the potential value-add of technologies or IT innovation can be identified.
2. Conduct regular meetings with business units, divisions and/or other stakeholder entities to understand current business problems, process bottlenecks, or other constraints where emerging technologies or IT innovation can create opportunities.
3. Understand enterprise investment parameters for innovation and new technologies so appropriate strategies are developed.

| Management Practice | Inputs | | Outputs | |
|---|---|---|---|---|
| **APO04.03 Monitor and scan the technology environment.** Perform systematic monitoring and scanning of the enterprise's external environment to identify emerging technologies that have the potential to create value (e.g., by realising the enterprise strategy, optimising costs, avoiding obsolescence, and better enabling enterprise and IT processes). Monitor the marketplace, competitive landscape, industry sectors, and legal and regulatory trends to be able to analyse emerging technologies or innovation ideas in the enterprise context. | From | Description | Description | To |
| | Outside COBIT | Emerging technologies | Research analyses of innovation possibilities | BAI03.01 |

**Activities**

1. Understand the enterprise's interest and potential for adopting new technology innovations and focus awareness efforts on the most opportunistic technology innovations.
2. Perform research and scanning of the external environment, including appropriate web sites, journals and conferences, to identify emerging technologies.
3. Consult with third-party experts where needed to confirm research findings or as a source of information on emerging technologies.
4. Capture staff members' IT innovation ideas and analyse them for potential implementation.

### Figure 2.16—APO04 Manage Innovation Management Practices (cont.)

**APO04 Manage Innovation (cont.)**

| Management Practice | Inputs | | Outputs | |
|---|---|---|---|---|
| | From | Description | Description | To |
| **APO04.04 Assess the potential of emerging technologies and innovation ideas.** Analyse identified emerging technologies and/or other IT innovation suggestions. Work with stakeholders to validate assumptions on the potential of new technologies and innovation. | | | Evaluations of innovation ideas | BAI03.01 |
| | | | Proof-of-concept scope and outline business case | APO05.03 APO06.02 |
| | | | Test results from proof-of-concept initiatives | Internal |

**Activities**

1. Evaluate identified technologies, considering aspects such as time to reach maturity, inherent risk of new technologies (including potential legal implications), fit with the enterprise architecture, and potential to provide additional value.
2. Identify any issues that may need to be resolved or proven through a proof-of-concept initiative.
3. Scope the proof-of-concept initiative, including desired outcomes, required budget, time frames and responsibilities.
4. Obtain approval for the proof-of-concept initiative.
5. Conduct proof-of-concept initiatives to test emerging technologies or other innovation ideas, identify any issues, and determine whether further implementation or roll-out should be considered based on feasibility and potential ROI.

| Management Practice | Inputs | | Outputs | |
|---|---|---|---|---|
| | From | Description | Description | To |
| **APO04.05 Recommend appropriate further initiatives.** Evaluate and monitor the results of proof-of-concept initiatives and, if favourable, generate recommendations for further initiatives and gain stakeholder support. | | | Results and recommendations from proof-of-concept initiatives | APO02.03 BAI03.09 |
| | | | Analysis of rejected initiatives | APO02.03 BAI03.08 |

**Activities**

1. Document proof-of-concept results, including guidance and recommendations for trends and innovation programmes.
2. Communicate viable innovation opportunities into the IT strategy and enterprise architecture processes.
3. Follow up on proof-of-concept initiatives to measure the degree to which they have been leveraged in actual investment.
4. Analyse and communicate reasons for rejected proof-of-concept initiatives.

| Management Practice | Inputs | | Outputs | |
|---|---|---|---|---|
| | From | Description | Description | To |
| **APO04.06 Monitor the implementation and use of innovation.** Monitor the implementation and use of emerging technologies and innovations during integration, adoption and for the full economic life cycle to ensure that the promised benefits are realised and to identify lessons learned. | | | Assessments of using innovative approaches | APO02.04 BAI03.02 |
| | | | Evaluation of innovation benefits | APO05.04 |
| | | | Adjusted innovation plans | Internal |

**Activities**

1. Assess the implementation of the new technologies or IT innovations adopted as part of IT strategy and enterprise architecture developments and their realisation during programme management of initiatives.
2. Capture lessons learned and opportunities for improvement.
3. Adjust the innovation plan, if required.
4. Identify and evaluate the potential value to be realised from the use of innovation.

Source: ISACA, *COBIT 5: Enabling Processes*, USA, 2012, pages 70-72

## 2.9 PRIORITIZATION PROCESSES RELATED TO IT INITIATIVES

The goal of portfolio management is to ensure that an enterprise's overall portfolio of IT-enabled investments is aligned with, and contributes optimal value to, the enterprise's strategic objectives. Enterprise portfolio management activities include establishing and managing resource profiles; defining investment thresholds; evaluating, prioritizing and selecting, deferring or rejecting new investments; managing the overall portfolio; and monitoring and reporting on portfolio performance.

To maximize the return on IT investments at the enterprise level, various techniques can be helpful, such as preparation of formalized, consistent business cases; use of hurdle rates; attention to portfolio management; and application of metrics such as internal rate of return (IRR), net present value (NPV) and payback period.[21]

For this section, it is important to understand the difference between the concepts of projects, programs and portfolios. These three concepts are discussed in section 3.11 Value Delivery Frameworks.

### Investment Portfolio Categorizations

Depending on the type of IT investment made, the anticipated returns will vary. To analyze this in a systematic manner, a portfolio approach (i.e., IT portfolio management) is used. There are two popular portfolio categorization paradigms used:

- **The market research company META Group has defined three categories of IT-related spending or investment:**
  - Run the business—This is the spending necessary to maintain existing operations at the existing level, including initiatives to observe regulatory compliance.
  - Grow the business—Examples of this category is the spending necessary to provide additional automation to improve efficiency or the consolidation of data centers to reduce costs and increase competitiveness.
  - Transform the business—This is the spending associated with the introduction of new areas of business, the expansion into new markets or any other radical transformation project designed to lead to significantly enhanced revenues and profits.
- **Peter Weill of the Massachusetts Institute of Technology Center for Information Systems Research (CISR) has suggested the following four-part categorization:**[22]
  - Transactional investments—These investments provide the IT to process the basic, repetitive transactions of the business (e.g., mortgage processing, claims processing or account management). Their main purposes are to increase efficiency and reduce costs.
  - Informational investments—These investments provide the information systems for managing and controlling the enterprise. Investments in this category typically include systems for management and financial control, decision making, planning, communication and accounting.
  - Strategic investments—These investments are usually designed to add real value to the business by increasing competitive advantage, enabling entry into new markets, or otherwise increasing or enhancing revenue streams. Examples include a new system to support an Internet-enabled banking initiative or a cable-TV-enabled insurance marketing channel.
  - Infrastructure investments—Investments in infrastructure can be costly and of long duration, but they may not, in themselves, generate any directly quantifiable financial benefits, although the business applications that depend upon the infrastructure will benefit. Examples include implementation of a new or upgraded systems management product or the implementation of a new operating system.

These are just two examples of the categorization classifications used or suggested by different industry experts. There are clear similarities across the classifications as well as different nuances within each. Each enterprise must select and adopt the categorization scheme that is most relevant. It is important to understand the purposes of having some form of categorization scheme for IT investments. These purposes include:
- **A greater ability to construct and monitor a balanced portfolio of IT investments**—It is probable that a healthy, growing enterprise will have investments in all categories. A proper mix is essential to ensure that risk is understood and managed, growth is encouraged and focus continues to be placed on essential "keeping the lights on" activities as well as the longer-term strategic investments.
- **The ability to better define risk and return targets for investments**—For example, it is probable that a strategic investment will carry a higher risk and, therefore, an expectation of higher return than an informational investment, which will almost certainly be low-risk and will, therefore, be undertaken with an expectation of a lower return.

## IT-enabled Investment Programs

No matter which categorization approach is adopted, within each category of IT investment portfolio in the enterprise, there would be a number of programs. Programs are cast to focus on the achievement of business benefit outcomes and typically are made up of a number of linked projects. Effective program management planning involves sequencing activities or projects according to their interlinkages and contribution to business outcomes, and taking into consideration assumptions or risk. ISACA's publication on getting started with value management[23] outlines an approach to program management planning called the Results Chain™ technique. The technique depicts the interlinkages between the various elements in a change program and is conceptually illustrated in **figure 2.17**.

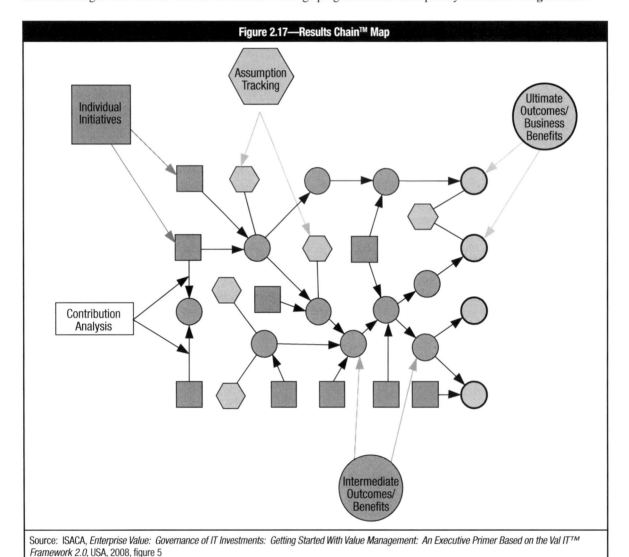

Source: ISACA, *Enterprise Value: Governance of IT Investments: Getting Started With Value Management: An Executive Primer Based on the Val IT™ Framework 2.0*, USA, 2008, figure 5

The road map for change programs can be effectively developed using the previous technique or other similar planning techniques. Once this road map is developed, attributes to each of the elements can be better defined (e.g., roles and responsibilities, measures for tracking progress, and benefits management for the change program). Key to the development of the road map is achieving consensus among the relevant stakeholders to commit to the change program as defined by the road map. There is another similar analysis technique called the Benefits Dependency Network that is published by the Cranfield School of Management.[24]

Another useful technique used in program management is stage-gating. It was first popularized by Robert Cooper in the 1980s as an approach to make product development more effective. While it serves as a blueprint for managing the new product processes, it is equally effective in managing a multiproject program. Mapping the entire program process as a series of predetermined steps, or stages, helps to break the overall program initiative into a more manageable and logical set of tasks. Each stage consists of a set of certain cross-functional and parallel activities which must be successfully completed prior to obtaining management approval to proceed to the next stage. The entrance to each stage is called a gate, and these gates, which are normally decision-making meetings, control the process and serve as:
- Quality control
- Go checkpoints, kill checkpoints, readiness checks, must-meet criteria, should-meet criteria
- A marker for the action plan for the next stage

An application of stage-gating is the UK Government Gateway Review process.

More information on portfolio management and benefits realization can be found in section 3.2 Basic Principles of Portfolio Management and section 3.8 Procedures to Manage and Report the Status of IT Investments.

## 2.10 SCOPE, OBJECTIVES AND BENEFITS OF IT INVESTMENT PROGRAMS

Investment programs are the translation of planned strategy into action (execution). Knowledge of how investment programs are structured helps in understanding the crucial link between strategy blueprints and how they are implemented.

Optimizing investments requires the ability to evaluate and compare investments, objectively select those with the highest potential to create value, and manage all of the investments to maximize value. Realizing value from IT-enabled investments requires more than delivering IT solutions and services—it also requires changes to some or all of the following: the nature of the business itself; business processes, skills and competencies; and organization, all of which must be included in the business case for the investment.

### Current Practice in Business Case Development

The seeds of success or failure are sown in the business case. Recent research from the University of Antwerp Management School (UAMS)[25] clearly positioned the entire portfolio management process, including business case development, as one of the key practices for achieving better alignment between business and IT and, by extension, business value. However, enterprises generally are not good at developing or using complete and comprehensive business cases. A 2006 Cranfield University School of Management study[26] found that while 96 percent of respondents develop business cases for most investments involving IT, however:
- 69 percent were not satisfied with the business case development process.
- 68 percent were not satisfied with the identification and structuring of benefits.
- 81 percent were not satisfied with the evaluation and review of results.
- 38 percent admitted that benefits claims were exaggerated to get the business cases approved.

### Business Case Components

The business case contains a set of assumptions on how value will be created, assumptions that should be well tested to ensure that the expected outcomes are achieved. The business case should also be based on qualitative and quantitative indicators that substantiate these assumptions and provide decision makers with insight supporting future investment decisions.

The business case consists of the major input resources as well as the three workstreams driving the outcome (see **figure 2.18**). These workstreams include delivering technical capabilities (e.g., a customer relationship management [CRM] application), operational capabilities (e.g., users have access to complete customer information) and business capabilities (e.g., information is used to support cross-selling). Each of these workstreams needs to be documented with data to support the investment decision and portfolio management processes: initiatives, costs, risk, assumptions, outcomes and metrics.

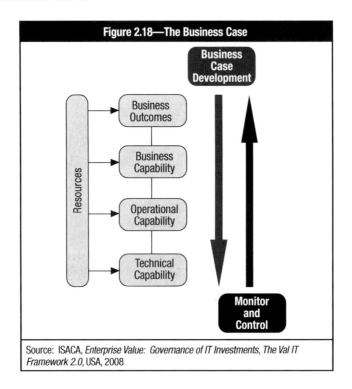

Source: ISACA, *Enterprise Value: Governance of IT Investments, The Val IT Framework 2.0*, USA, 2008

At a minimum, the business case should include the following:
- The business benefits targeted, their alignment with business strategy and who in the business functions will be responsible for securing them
- The business changes needed to create additional value
- The investments needed to make the business changes
- The investments required to change or add new IT services and infrastructure
- The ongoing IT and business costs of operating in the changed way
- The risk inherent in the above, including any constraints or dependencies
- Who will be accountable for the successful creation of optimal value
- How the investment and value creation will be monitored throughout the economic life cycle, and the metrics to be used

## Business Cases as Operational Tools

The business case should be developed from a strategic perspective—from the top down—starting with a clear understanding of the desired business outcomes and progressing to a detailed description of critical tasks and milestones as well as key roles and responsibilities. The business case is not a static document supporting a one-time use; rather, it is a dynamic operational tool that must be continually updated to reflect the current view of the future so that the viability of the program can be maintained.

In most enterprises today, the business case is generally seen as a necessary evil or a bureaucratic hurdle to get over to obtain required financial and other resources. The focus is on the technology project, and the costs of the technology, with only a cursory discussion of benefits or changes that the business might need to make to create or sustain value from use of the technology. Business cases are also all too often treated as "one-off" documents that are rarely looked at again once the required resources have been obtained—other than, possibly, at a postimplementation review.

This approach to business cases can cause challenges down the road. A well-developed and intelligently used business case for a business change program is actually one of the most valuable tools available to management—the quality of the business case and the processes involved in its creation and use throughout the economic life cycle of an investment has an enormous impact on creating and sustaining value. It describes a proposed journey from initial ideas to realizing expected outcomes for beneficiaries (i.e., those whose money is being invested and for whom the return should be secured) and other affected stakeholders.

The business case is an operational tool that must be continually updated to reflect the current reality so as to provide the basis for informed decision making—not just for initial commitment of resources, but for managing the investment through its full economic life cycle. In so doing, the business case must include answers to the following questions[27]—answers based on relevant, current and accurate business-focused information:
- **Are we doing the right things?**—What is proposed? For what business outcome? How do the projects within the program contribute?
- **Are we doing them the right way?**—How will it be done? What is being done to ensure that it will fit with other current or future capabilities?
- **Are we getting them done well?**—What is the plan for doing the work? What resources and funds are needed?
- **Are we getting the benefits?**—How will the benefits be delivered? What is the value of the program?

In summary, a business case must be:
- At the business program level
- Complete and comprehensive, including the full scope of change required in achieving the desired outcomes
- A "living" operational document that is kept up to date
- Used to manage the program through its full economic life cycle

More information on the business case can be found in section 3.3 Benefit Calculation Techniques and section 3.8 Procedures to Manage and Report the Status of IT Investments.

## 2.11 IT ROLES AND RESPONSIBILITIES AND METHODS TO CASCADE BUSINESS AND IT OBJECTIVES TO IT PERSONNEL

Engaged stakeholders and personnel will sponsor and facilitate the changes necessary to bring about better strategic alignment. To realize such commitment, it is important to clarify roles and responsibilities and to leverage methods that will help in cascading business and IT objectives to the relevant personnel and stakeholders.

There are six prescribed activities to undertake when selling the value proposition of the IT strategy to key stakeholders:[28]
- Illustrating and quantifying the IT strategy
- Continuous communication
- Focus on explanation and training
- Using a participatory style of decision-making process
- Mastering the "operational" art
- Risk considerations at the CIO level

### Illustrating and Quantifying the IT Strategy
Illustrating and quantifying the IT strategy is explained in sections 2.1 through 2.3.

### Continuous Communication
Failing to properly and consistently communicate the strategy is tantamount to having no strategy at all. Securing feedback and participation—both integral elements of strategy execution—depends on the CIO's proactive communication and advocacy. The execution of strategy is all about getting things done. And to get things done, there is a need for the active cooperation and participation of different people, all of whom probably have slightly different agendas. The CIO must constantly explain what is in the IT strategy for them and why they should actively, vocally support the IT strategy. The CIO also listens to people's responses and makes adjustments to IT strategy when necessary. It must be remembered that the business goal (the destination) usually remains steady over a two- to

four-year period, but the way to reach that goal (the strategy) can change as the situation unfolds. Listening to how people respond to the message is an important part of sound communication. The message needs to evolve so that people buy in to it. They buy in and become supporters of the strategy when they know that what they have to say is being heard.

## Focus on Explanation and Training

While communication must take place enterprisewide, there is a need for the CIO to focus a special level of explanation and training on the small subset of people building or using the systems called for in the IT strategy. With such people, the CIO moves beyond communicating the strategy and listening to feedback. What makes the difference, here, is spending considerable time one-on-one and in these small group settings explaining the details of the tactics being used and why those tactics can deliver success, and making sure that people get the training they need so they can perform the tasks that the tactics demand of them. When they understand the tactics and believe those tactics can bring success, they will buy in to the project wholeheartedly. Then the CIO steps back and lets them do their jobs. When people do not buy in to a project, they require constant supervision and cajoling to get anything done. An unwillingness to participate shows that the tactics employed may well be flawed. It is worth remembering that tactics are composed of sequences of techniques. Every job and profession has a core set of techniques; in the IT profession, some of the core techniques are process mapping, data modeling, and object-oriented design and programming. It is advisable to make sure to define what techniques are required for the tactics to be used, and to make sure that people receive adequate training in those techniques.

## Using a Participatory Style of Decision-making Process

If the CIO handles the communication, explanation and training phases effectively, he or she will generate more opportunities to secure commitment and participation, leading to a more open decision-making process—without the CIO abdicating responsibility for setting and carrying out the IT strategy and/or assuming authority for certain decisions.

People expect the CIO to make the tough calls when there is no consensus among lieutenants, when there is not enough information and when time is short. But, otherwise, people usually do not like a dictatorial style of leadership. The best, most competent people want an active voice in the decision-making process. As the leader of the decision-making process, the CIO's role is to see that timely and accurate information is available and that people get a chance to examine it, ask questions and voice their opinions. The CIO's role is to ask questions that focus people's attention on the important issues. It is important to keep people from wandering off the subject, bemoaning past mistakes and discussing personalities instead of issues. When the CIO acts as a participatory leader, consensus decisions usually emerge that combine the collective wisdom of the entire management team. It is important to encourage people to take ownership of their decisions and act on them without constant oversight. Five elements have been identified that create an effective decision-making process, all of which must be present to produce the most effective, participatory decision-making environment:
- A functioning project management office (PMO)
- Relevant data displayed in easily understandable dashboard summaries
- Regular weekly meetings
- Obligation to dissent
- Trust

## Mastering the "Operational" Art

Strategy defines a way to reach a desired destination; tactics focus on executing the projects necessary to reach the destination. Before tactics can be addressed, the projects must be chosen—an exercise that relies on the CIO's ability to see opportunities and risk from an operational perspective and his or her appreciation of their potential. CIOs can choose from many potential projects to implement their strategies, but how do they select the project with the best strategic fit? Time teaches successful CIOs how to master this operational art. Mastering the operational art has been described as knowing when a tactical move can deliver strategic results. Effective CIOs develop a keen sense of the operational art. They see opportunities to employ available means to achieve breakthrough results. Where others see only obstacles, masters of the operational art see openings.

### Risk Considerations at the CIO Level

Understanding opportunity and risk, and capitalizing on them, calls for a CIO to be bold—to know the difference between a smart, calculated risk and a foolish gamble. A smart, calculated risk is an action that is not a certain success, but one that has potential to deliver extraordinary rewards compared to the risk taken. A foolish gamble is one that delivers a small reward at the cost of risk bearing dire consequences. CIOs need to clearly understand the potential upside of a project in comparison to the magnitude of its downside. Unless there is literally nothing to lose, it is best never to take a risk where the magnitude of the downside could overwhelm one's ability to recover and exit from the situation intact enough to try again, another day.

A BSC can also be helpful in translating IT strategy into action. More information on the BSC can be found in section 1.10 IT Governance Monitoring Processes/Mechanisms section 1.11 IT Governance Reporting Processes/Mechanisms, and section 2.6 Methods to Document and Communicate IT Strategic Planning Processes.

## ENDNOTES

[1] Van Grembergen, Wim; Steven De Haes; *Enterprise Governance of Information Technology: Achieving Strategic Alignment and Value, Springer*, USA, 2009

[2] Henderson J.C.; N. Venkatraman; "Strategic Alignment: Leveraging Information Technology for Transforming Organizations," *IBM Systems Journal*, vol. 32, no. 1, 1993

[3] ISACA, COBIT 5, USA, 2012, www.isaca.org/cobit

[4] Kaplan, Robert S.; David P. Norton; *The Balanced Scorecard: Translating Strategy Into Action*, Harvard University Press, USA, 1996

[5] IT-related outcomes are not the only intermediate benefit required to achieve enterprise goals. All other functional areas in an organization, such as finance and marketing, also contribute to the achievement of enterprise goals; however, within the context of COBIT 5, only IT-related activities and goals are considered.

[6] The goals cascade is based on research performed by the University of Antwerp Management School, IT Alignment and Governance Research Institute in Belgium.

[7] Derkson, Barry; Jerry Luftman; *Key European IT Management Trends for 2015*, CIONET, Belgium 2015

[8] ISACA, *Benchmarking and Business Value Assessment of COBIT® 5*, USA, 2015

[9] Pfeffer, Jeffrey; Robert I. Sutton; *The Knowing-doing Gap: How Smart Companies Turn Knowledge Into Action*, Harvard Business School Press, USA, 2000

[10] Stenzel, Joe; *CIO Best Practices: Enabling Strategic Value with Information Technology, 2nd Edition*, John Wiley & Sons, USA, 2012

[11] Derksen, Barry; Jerry Luftman, *Key European IT Management Trends for 2015*, CIONET, Belgium, 2015

[12] Weill, Peter; Marianne Broadbent; *Leveraging the New Infrastructure: How Market Leaders Capitalize on Information Technology*, Harvard Business School Press, 1998

[13] IT Governance Institute, *IT Governance Domains, Practices and Competencies: IT Alignment: Who Is in Charge?*, USA, 2005

[14] IT Governance Institute, *Board Briefing on IT Governance, 2nd Edition*, USA, 2003

[15] Ibid.

[16] Ibid.

[17] IT Governance Institute; *IT Governance Domain Practices and Competencies: Measuring and Demonstrating the Value of IT*, USA, 2005

[18] IT Governance Institute; *Governance of the Extended Enterprise: Bridging Business and IT Strategies*, John Wiley & Sons Inc., USA, 2005

[19] Fong, Elizabeth; Alan Goldfine; *Information Management Decisions: The Integration Challenge*, US Department of Commerce, National Institute of Standards and Technology (NIST), NIST Special Publication 500-167, USA, September 1989 (www.itl.nist.gov). Quoted in IT Governance Institute; *Governance of the Extended Enterprise: Bridging Business and IT Strategies*, John Wiley & Sons Inc., USA, 2005

[20] National Association of State Chief Information Officers (NASCIO); *Enterprise Architecture Tool-Kit V2.0*, Adaptive Enterprise Architecture Development Program, USA, 2002. www.nascio.org

[21] IT Governance Institute, *IT Governance Domains, Practices and Competencies: Optimising Value Creation From IT Investments*, USA, 2005

[22] *Op cit* Weill
[23] ISACA, *Enterprise Value: Governance of IT Investments, Getting Started With Value Management*, USA, 2008. Results Chain™ by Fujitsu.
[24] Peppard, Joe; John Ward; *Unlocking Sustained Business Value From IT Investments*, Cranfield School of Management, UK, 2003, www.som.cranfield.ac.uk/som/
[25] De Haes, Steven; Wim Van Grembergen, "An Exploratory Study Into the Design of an IT Governance Minimum Baseline Through Delphi Research," *Communications of the Association of Information Systems*, vol. 22, 2008
[26] Ward, John, "Delivering Value From Information Systems and Technology Investments: Learning From Success," Forum (monthly newsletter of Cranfield School of Management, UK), August, 2006
[27] Based on the "Four Ares" as described by John Thorp in his book *The Information Paradox—Realizing the Business Benefits of Information Technology*, written jointly with Fujitsu, first published in 1998 and revised in 2003, McGraw-Hill, Canada.
[28] *Op cit* Stenzel

# Chapter 3: Benefits Realization

## Section One: Overview

| | |
|---|---|
| Domain Definition | 86 |
| Domain Objectives | 86 |
| Learning Objectives | 86 |
| CGEIT Exam Reference | 86 |
| Task and Knowledge Statements | 86 |
| Self-assessment Questions | 90 |
| Answers to Self-assessment Questions | 91 |
| Suggested Resources for Further Study | 92 |

## Section Two: Content

| | | |
|---|---|---|
| 3.1 | IT Investment Management Processes, Including the Economic Life Cycle of Investments | 93 |
| 3.2 | Basic Principles of Portfolio Management | 95 |
| 3.3 | Benefit Calculation Techniques | 100 |
| 3.4 | Process and Service Measurement Techniques | 102 |
| 3.5 | Processes and Practices for Planning, Development, Transition, Delivery, and Support of Solutions and Services | 105 |
| 3.6 | Continuous Improvement Concepts and Principles | 109 |
| 3.7 | Outcome and Performance Measurement Techniques | 110 |
| 3.8 | Procedures to Manage and Report the Status of IT Investments | 112 |
| 3.9 | Cost Optimization Strategies | 116 |
| 3.10 | Models and Methods to Establish Accountability Over IT Investments | 117 |
| 3.11 | Value Delivery Frameworks | 126 |
| 3.12 | Business Case Development and Evaluation Techniques | 132 |
| | Endnotes | 136 |

# Section One: Overview

## DOMAIN DEFINITION
Ensure that information technology (IT)-enabled investments are managed to deliver optimized business benefits and that benefit realization outcome and performance measures are established, evaluated and progress is reported to key stakeholders.

## DOMAIN OBJECTIVES
The objective of benefits realization is to ensure that IT and the business fulfill their value management responsibilities, particularly that:
- IT-enabled business investments achieve the promised benefits and deliver measurable business value
- Required capabilities (solutions and services) are delivered on time and within budget
- IT services and other IT assets continue to contribute to business value.

The premise of benefits realization is that there is strong concern at board and senior management levels that the high expenditures on IT-related initiatives are not realizing the business benefits they promise. Studies and surveys also indicate high levels of loss from ill-planned and ill-executed initiatives. The focus on value has become more prevalent while the capability maturity of value management practices in most enterprises has remained low.

## LEARNING OBJECTIVES
The purpose of this chapter is to ensure that professionals and executives involved in governance understand, define and execute appropriate value management practices for the enterprise. This includes formulating and implementing a road map to introduce and improve value management practices, benefits realization planning and monitoring progress through metrics and with appropriate remediation measures where necessary.

## CGEIT EXAM REFERENCE
This domain represents 16 percent of the CGEIT exam (approximately 24 questions).

## TASK AND KNOWLEDGE STATEMENTS

### TASKS
There are eight tasks within this domain that a CGEIT candidate must know how to perform. These relate to managing or contributing to the development of, a systematic, analytical and continuous value governance process:

T3.1   Ensure that IT-enabled investments are managed as a portfolio of investments.
T3.2   Ensure that IT-enabled investments are managed through their economic life cycle to achieve business benefit.
T3.3   Ensure that business ownership and accountability for IT-enabled investments are established.
T3.4   Ensure that IT investment management practices align with enterprise investment management practices.
T3.5   Ensure that IT-enabled investment portfolios, IT processes and IT services are evaluated and benchmarked to achieve business benefit.
T3.6   Ensure that outcome and performance measures are established and evaluated to assess progress toward the achievement of enterprise and IT objectives.
T3.7   Ensure that outcome and performance measures are monitored and reported to key stakeholders in a timely manner.
T3.8   Ensure that improvement initiatives are identified, prioritized, initiated and managed based on outcome and performance measures.

## KNOWLEDGE STATEMENTS

The CGEIT candidate must have a good understanding of each of the twelve areas delineated by the following knowledge statements. These statements are the basis for the exam. Each statement is defined and its relevance and applicability to this job practice are briefly described as follows:

Knowledge of:

**KS3.1  IT investment management processes, including the economic life cycle of investments**
The goal of IT investment management (IM) is to ensure that the enterprise's individual IT-enabled investments contribute to optimal value. When organizational leaders commit to investment management they improve their ability to:
- Identify business requirements
- Develop a clear understanding of candidate investment programs
- Analyze alternative approaches to implementing the programs
- Define each program and document, and maintain a detailed business case for it, including the benefits' details, throughout the full economic life cycle of the investment
- Assign clear accountability and ownership, including those for benefits realization
- Manage each program through its full economic life cycle, including retirement
- Monitor and report on each program's performance

**KS3.2  basic principles of portfolio management**
The goal of portfolio management (PM) is to ensure that an enterprise secures optimal value across its portfolio of IT-enabled investments. An executive commitment to portfolio management helps enterprises:
- Establish and manage resource profiles
- Define investment thresholds
- Evaluate, prioritize, and select, defer, or reject new investments
- Manage and optimize the overall investment portfolio
- Monitor and report on portfolio performance

**KS3.3  benefit calculation techniques**
The imperatives of ensuring that IT-enabled investments provide the value being promised requires proper analysis using established and appropriate techniques. The COBIT 5 and Information Technology Infrastructure Library (ITIL) frameworks provide the value management practices and IT service life cycle perspectives, respectively, to update and contextualize traditional investment analysis techniques and approaches.

**KS3.4  process and service measurement techniques**
High performance is directly related to effective performance management practices and tools (such as maturity models, benchmarking, metrics). Greater visibility of the detailed performance of operations and investments enables enterprises to influence performance and improve agility.

**KS3.5  processes and practices for planning, development, transition, delivery and support of solutions and services**
With the heightened complexity of IT solutions and services to execute the enterprise's strategy and vision, and the rapid rise in its pivotal contribution to business automation and value delivery, the perspective of IT solution and service delivery has assumed a preeminence, evidenced by the fast-growing service management movement—the IT Service Management Forum (itSMF). There is now an evolving body of good practices (ITIL) and an international standard (International Organization for Standardization [ISO] 20000) to detail specific IT service management processes and methodologies. In terms of solutions development, the system development life cycle (SDLC) is probably one of the best-known IT practices and is the mainstay of IT development since the very beginning of IT enablement of business processes.

**KS3.6 continuous improvement concepts and principles**
Continuous process improvement and associated techniques are highly useful for mapping GEIT to organizational imperatives, especially the improvement of process capability maturity. Knowledge enables and aligns GEIT improvement within the enterprise.

**KS3.7 outcome and performance measurement techniques**
The purpose of performance measurement is to uncover, communicate and evolve organizational performance drivers. The choice of metrics communicates to stakeholders what is important, and this affects what gets done. Choosing metrics that answer critical management questions improves management's visibility into key processes.

**KS3.8 procedures to manage and report the status of IT investments**
Understanding the relevance of business cases to manage and report on IT investments is of primary importance to all management levels across both the business and IT parts of the enterprise—from the chief executive officer (CEO) and the C-suite to those directly involved in and responsible for the selection, procurement, development, implementation, deployment and benefits realization processes.

**KS3.9 cost optimization strategies**
Cost optimization is typically a constant focus in today's business environment when expenditures on IT-related activities and resources have escalated dramatically. Additional factors, such as the drive to enhance efficiency, responding to competition or facing an economic downturn, may cause additional attention to be placed on cost optimization strategies. In the context of value delivery, cost optimization is necessary to ensure that IT services enable the business to create the required business value using resources (people, applications, infrastructure and information) to deliver the appropriate capabilities at the optimal cost.

**KS3.10 models and methods to establish accountability over IT investments**
There is a need for the business to take ownership of, and be accountable for, governing the use of IT in creating value from IT-enabled business investments. This implies a crucial shift in the minds of the business and IT, moving away from managing IT as a "cost" toward managing IT as an "asset" to create business value. As Weill and Ross describe in their 2009 book titled *IT Savvy: What Top Executives Must Know to Go from Pain to Gain*:[1] "If senior managers do not accept accountability for IT, the company will inevitable throw its IT money to multiple tactical initiatives with no clear impact on the organizational capabilities. IT becomes a liability instead of a strategic asset."

**KS3.11 value delivery frameworks**
The lack of value governance is the main cause of the waste in IT investments. Surveys have consistently revealed that 20 to 70 percent of large-scale investments in IT-enabled business change are wasted, challenged or fail to bring a return to the enterprise. ISACA's global surveys published in 2006[2] and 2008[3] also generally support these findings. As one survey measuring costs and value shows, it is obvious that ineffective or immature value governance practices are responsible for the poor showing that, in many enterprises, less than eight percent of the IT budget is actually spent on initiatives that create value for the enterprise.[4] Value delivery frameworks, such as Val IT, can help organizations to find the appropriate guidance in implementing their value governance practices and processes.

**KS3.12 business case development and evaluation techniques**
Most enterprises see a business case as a necessary evil or a bureaucratic hurdle to get over to obtain required financial and other resources. The focus is on the technology project, and the costs of the technology, with only a cursory discussion of benefits or changes that the business might need to make to create or sustain value from use of the technology. Business cases are also all too often treated as "one-off" documents that are rarely looked at again once the required resources have been obtained—other than, possibly, at a postimplementation review. This approach to business cases can cause challenges down the road. A well-developed and intelligently used business case for a business change program is actually one of the most valuable tools available to management—the quality of the business case and the processes involved in its creation and use throughout the economic life cycle of an investment has an enormous impact on creating and sustaining value.

## RELATIONSHIP OF TASK TO KNOWLEDGE STATEMENTS

The task statements are what the CGEIT candidate is expected to know how to do. The knowledge statements delineate what the CGEIT candidate is expected to know to perform the tasks. The task and knowledge statements are approximately mapped in **figure 3.1** insofar as it is possible to do so. Note that although there often is overlap, each task statement will generally map to several knowledge statements.

| Figure 3.1—Task and Knowledge Statements Mapping—Benefits Realization Domain ||
|---|---|
| **Task Statement** | **Knowledge Statements** |
| T3.1 Ensure that IT-enabled investments are managed as a portfolio of investments. | KS3.2 basic principles of portfolio management |
| T3.2 Ensure that IT-enabled investments are managed through their economic life cycle to achieve business benefit. | KS3.1 IT investment management processes, including the economic life cycle of investments<br>KS3.3 benefit calculation techniques (for example, earned value [EV], total cost of ownership [TCO], return on investment [ROI]<br>KS3.11 value delivery frameworks (for example, Val IT) |
| T3.3 Ensure that business ownership and accountability for IT-enabled investments are established.<br>T3.4 Ensure that IT investment management practices align with enterprise investment management practices. | KS3.10 models and methods to establish accountability over IT investments<br>KS3.1 IT investment management processes, including the economic life cycle of investments<br>KS3.11 value delivery frameworks (for example, Val IT) |
| T3.5 Ensure that IT-enabled investment portfolios, IT processes and IT services are evaluated and benchmarked to achieve business benefit. | KS3.4 process and service measurement techniques (for example, maturity models, benchmarking, key performance indicators [KPIs])<br>KS3.5 processes and practices for planning, development, transition, delivery and support of IT solutions and services<br>KS3.6 continuous improvement concepts and principles<br>KS3.12 business case development and evaluation techniques |
| T3.6 Ensure that outcome and performance measures are established and evaluated to assess progress toward the achievement of enterprise and IT objectives. | KS3.7 outcome and performance measurement techniques (for example, service metrics, key performance indicators [KPIs])<br>KS3.8 procedures to manage and report the status of IT investments<br>KS3.12 business case development and evaluation techniques |
| T3.7 Ensure that outcome and performance measures are monitored and reported to key stakeholders in a timely manner. | KS3.2 basic principles of portfolio management<br>KS3.7 outcome and performance measurement techniques (for example, service metrics, key performance indicators [KPIs])<br>KS3.8 procedures to manage and report the status of IT investments |
| T3.8 Ensure that improvement initiatives are identified, prioritized, initiated and managed based on outcome and performance measures. | KS3.5 processes and practices for planning, development, transition, delivery and support of IT solutions and services<br>KS3.6 continuous improvement concepts and principles<br>KS3.7 outcome and performance measurement techniques (for example, service metrics, key performance indicators [KPIs])<br>KS3.8 procedures to manage and report the status of IT investments<br>KS3.9 cost optimization strategies (for example, outsourcing, adoption of new technologies) |

## SELF-ASSESSMENT QUESTIONS

CGEIT self-assessment questions support the content in this manual and provide an understanding of the type and structure of questions that have typically appeared on the exam. Questions are written in a multiple-choice format and designed for one best answer. Each question has a stem (question) and four options (answer choices). The stem may be written in the form of a question or an incomplete statement. In some instances, a scenario or a description problem may be included. These questions normally include a description of a situation and require the candidate to answer two or more questions based on the information provided. Many times a question will require the candidate to choose the **MOST** likely or **BEST** answer among the options provided.

In each case, the candidate must read the question carefully, eliminate known incorrect answers and then make the best choice possible. Knowing the format in which questions are asked, and how to study and gain knowledge of what will be tested, will help the candidate correctly answer the questions.

3-1 The **PRIMARY** benefit of managing IT-enabled investments using investment management practices is to:

    A. enable decision making about discretionary and nondiscretionary investments.
    B. optimize the value of these investments.
    C. avoid getting into risky investments.
    D. realize investment benefits.

3-2 The **BEST** use of a business case for IT-enabled investments is as a:

    A. static document supporting the initial justification of the investment.
    B. measure of the financial performance of the investment.
    C. strategic document used over the life of the investment.
    D. checklist to monitor the business outcomes of the investment.

3-3 After conducting a project performance evaluation, early project cancellation is a **BEST** practice because it:

    A. mitigates against project failure (preventing failing projects from fulfilling their eventual outcome).
    B. recovers the budgeted investment funds.
    C. encourages only the most profitable projects to survive.
    D. implies strict levels of business case development and decision making.

# ANSWERS TO SELF-ASSESSMENT QUESTIONS

Correct answers are shown in **bold**.

3-1   A.   Leading practices in investment management do not advocate this manner of decision making. They have established that the basis for decision making of IT-enabled investments is not discretionary/nondiscretionary, but rather on the investments' relative value (benefits less costs adjusted for risk).
  **B.   Optimizing the value of these investments is the key benefit of using investment management practices for managing IT-enabled investments.**
  C.   This is incorrect from two viewpoints: Investment management practices do not advocate avoiding risky investments, but advocate that all investments should be carefully evaluated and managed; and this is not a valid standpoint as a primary benefit.
  D.   Investment management practices do not ensure benefit realization for investments. Benefit realization of investments is enabled by ensuring good solution and service delivery, coupled with good program management.

3-2   A.   A business case is not a static document; it should be updated throughout the life cycle of the investment. Using the business case as a static document will prevent the business case from being used to guide the management of the investment, specifically the investment activity (with benefit, cost and risk implications) throughout its economic life.
  B.   Use of a business case to measure financial performance only is a limited use of the business case and excludes other criteria, such as business process improvements or customer satisfaction.
  **C.   Business cases are best used as strategic documents detailing the desired business outcomes and describing the critical tasks and milestones for the investment activity throughout its economic life, together with key roles and responsibilities.**
  D.   Although business cases may include checklists to monitor the business outcomes of the investment this is only a limited application of a business case.

3-3   **A.   Canceling a project as soon as it becomes apparent that it cannot be delivered satisfactorily is a best practice because it is a sign of strong management and good governance. For nondiscretionary projects (which may not be cancelled), they should be re-scoped to yield a satisfactory performance evaluation.**
  B.   Early project cancellation will not recover expended funds.
  C.   Early project cancellation would be a limiting practice because some projects may need to be undertaken, regardless (such as potentially loss-making compliance-related, and nondiscretionary projects).
  D.   This is incorrect because it implies early project cancellation is a sign of project failure and weakness.

---

**NOTE:** For more self-assessment questions, you may also want to obtain a copy of the *CGEIT® Review Questions, Answers & Explanations Manual 4th Edition*, which consists of 250 multiple-choice study questions, answers and explanations.

## SUGGESTED RESOURCES FOR FURTHER STUDY

In addition to the resources cited throughout this manual, the following resources are suggested for further study in this domain (publications in **bold** are stocked in the ISACA Bookstore):

De Haes, Steven; Dirk Gemke; John Thorp; Wim Van Grembergen; "KLM's Enterprise Governance of IT Journey: From Managing IT Costs to Managing Business Value," *MIS Quarterly Executive*, vol. 10, 2011, pp. 109-120

De Haes, Steven; Van Grembergen, Wim; *Enterprise Governance of IT: Achieving Alignment and Value, Second Edition*, Springer, USA, 2015

Harvard Business School Press; *Harvard Business Review on the Business Value of IT*, USA, 1999

**ISACA, *The Business Case Guide: Using Val IT™ Framework 2.0*, USA, 2008, www.isaca.org/valit**

**ISACA, COBIT 5, USA, 2012, www.isaca.org/cobit**

**ISACA, *Enterprise Value: Governance of IT Investments, The Val IT™ Framework 2.0*, USA, 2010, www.isaca.org/valit**

Keen, Jack M.; Bonnie Digrius; *Making Technology Investments Profitable: ROI Road Map from Business Case to Value Realization, Second Edition*, John Wiley and Sons Inc., USA, 2011

Letavec, Craig; *Strategic Benefits Realization: Optimizing Value Through Programs, Portfolios and Organizational Change Management*, J. Ross Publishing, USA, 2014

Maes, Kim; Steven De Haes; Wim Van Grembergen; "The Business Case as an Operational Management Instrument—A Process View," *ISACA Journal*, vol. 4, 2014.

McCormack, Ade; *The IT Value Stack: A Boardroom Guide to IT Leadership*, John Wiley & Sons Inc., USA, 2007

Thorp, John; *The Information Paradox: Realizing the Business Benefits of Information Technology*, McGraw-Hill Reyerson, USA, 1998

# Section Two: Content

## 3.1 IT INVESTMENT MANAGEMENT PROCESSES, INCLUDING THE ECONOMIC LIFE CYCLE OF INVESTMENTS

Investment management is the process of managing the organizational assets through the end-to-end asset life cycle to meet specific benefit goals set by senior management. Because IT is an important asset, it is imperative for management to develop, define and mandate IT investment management processes to ensure that IT investments meet benefit goals.

The goal of IT investment management is to ensure that the enterprise's individual IT-enabled investments contribute to optimal value. When organizational leaders commit to investment management they improve their ability to:[5]
- Identify business requirements
- Develop a clear understanding of candidate investment programs
- Analyze alternative approaches to implementing the programs
- Define each program and document, and maintain a detailed business case for it, including the benefit details, throughout the full economic life cycle of the investment
- Assign clear accountability and ownership, including those for benefits realization
- Manage each program through its full economic life cycle, including retirement
- Monitor and report on each program's performance

There are three key components of investment management. The first is the business case, which is essential to selecting the right investment programs and managing them during their execution. The second is program management, which governs all processes that support execution of the programs. The third is benefits realization—the set of tasks required to actively manage the realization of program benefits. Each of these components is described in greater detail in the following sections.

### The Business Case

A comprehensive business case is critical to the outcome of the program, yet few enterprises are adept at developing and documenting them. A 2006 Cranfield University School of Management study[6] found that while 96 percent of respondents developed business cases for most investments involving IT, 69 percent were not satisfied with the effectiveness of the practice.

The business case contains a set of assumptions on how value will be created, assumptions that should be well tested to ensure that the expected outcomes are achieved. The business case should also be based on qualitative and quantitative indicators that substantiate these assumptions and provide decision makers with insight supporting future investment decisions.

More information on the business case and related concepts can be found in section 2.10 Scope, Objectives and Benefits of IT Investment Programs, section 3.3 Benefit Calculation Techniques, section 3.8 Procedures to Manage and Report the Status of IT Investments and section 3.12 Business Case Development and Evaluation Techniques.

## Program Management

Realizing business value is not about acquiring technology, but about using IT in conjunction with associated changes in the nature of the business, business processes, individuals' work and competencies, and organizational structures. All changes, and the capabilities required to enable the changes, must be understood, defined, monitored and managed as a comprehensive program of business change in which IT plays a necessary, but not solely sufficient, part. Effective program management requires maintaining a constant focus on the desired business outcomes, the full scope of initiatives required to achieve the outcomes, the relationship between the initiatives and how they individually and collectively contribute to the outcomes, and any assumptions that are being made related to those contributions or to the outcomes themselves. This requires that the IT function and other parts of the business work closely together, each with clearly understood roles and responsibilities, and shared accountabilities.

More information on program management and related concepts can be found in section 2.9 Prioritization Processes Related to IT Initiatives.

## Benefits Realization

Benefits realization is enormously important for several reasons. One is that not all benefits are equal. Val IT distinguishes between two types of benefits. The first are "business benefits," which contribute directly to value, as defined earlier. The second are "intermediate benefits," which do not directly create value, even though they might be beneficial for one or more groups of stakeholders. For example, improvements in specific types of customer service that do not contribute to increased profits would be considered intermediate benefits. Another reason is that benefits do not just happen and rarely happen according to plan. A focus on benefits realization helps address these challenges by actively managing investments across their full economic life cycle—from proposal to profit or improved service performance. Benefits realization ensures that intermediate benefits—such as improvements in customer service—contribute to business benefits—such as additional profits. Benefits realization further ensures that the realization of business benefits is unfolding at levels of return sufficient enough to merit the resources being expended to achieve the benefits. In the absence of effective benefits realization, optimal value will not be created, or, worse, value may be eroded or destroyed.

Benefits are typically delivered through extensive changes to business practices and decision making. There is a growing consensus that organizational factors are far more critical to successful information systems implementation than technical considerations.

## Full Economic Life Cycle Management

The ongoing management through the full economic life cycle is where most enterprises cut corners. In most cases, the program will be considered closed after completing the activities in the program plan and delivering the required business and IT capabilities. In general, the benefits and the expected value, as set in the business case, will not be realized until some later time, long after the delivery of IT and business capabilities. It is only then that the program and, subsequently, the business case will have proven that they delivered the expected benefits. While the timing of program and business case closure—which likely will not be the same—may vary in different organizations and for different types of investments, it is important to understand that the full economic life cycle of an investment decision includes the following, as illustrated in **figure 3.2**:
- **Investment phase**—Developing the necessary capabilities
- **Adoption phase**—Implementing the capabilities
- **Value creation phase**—Achieving the expected level of performance and moving the delivered capabilities into the active service portfolio
- **Value sustainment phase**—Assuring that the assets resulting from the investment continue to create value, which may well include additional investments required to sustain value
- **Retirement phase**—Decommissioning the resulting assets

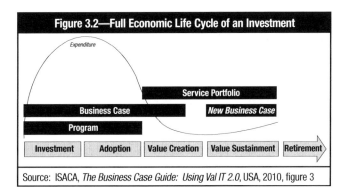

Figure 3.2—Full Economic Life Cycle of an Investment

Source: ISACA, *The Business Case Guide: Using Val IT 2.0*, USA, 2010, figure 3

## 3.2 BASIC PRINCIPLES OF PORTFOLIO MANAGEMENT

The goal of portfolio management is to ensure that an enterprise secures optimal value across its portfolio of IT-enabled investments. An executive commitment to portfolio management helps enterprises:[7]
- Establish and manage resource profiles
- Define investment thresholds
- Evaluate, prioritize, and select, defer, or reject new investments
- Manage and optimize the overall investment portfolio
- Monitor and report on portfolio performance

IT-enabled business investment programs need to be managed as part of the overall portfolio of investments so that all of the enterprise's investments can be selected and managed on a common basis. The programs in the portfolio must be clearly defined, evaluated, prioritized, selected and managed actively throughout their full economic life cycles to optimize value for individual programs and the overall portfolio. This includes optimizing the allocation of the finite investment resources available to the enterprise, the management of risk, the early identification and correction of problems (including program cancellation, if appropriate), and board-level investment portfolio oversight.

Good portfolio management should lead to a balanced portfolio of IT-enabled investments. It recognizes that there are different categories of investment with differing levels of complexity and degrees of freedom in allocating funds. Examples of such categories could include, but are not limited to:
- Investments focused on innovating business models or products
- Investments focused on growing the business
- Investments focused on improving operational processes and activities
- Mandatory investment due to legal and compliance requirements

Evaluation criteria with appropriate weightings should be established for each category within the investment portfolio. The decision to include a program in the portfolio is not a one-time commitment. The portfolio composed of potential and approved investments should be managed actively on a continuing basis and not considered just when approval is sought. Depending on the relative performance of active programs and the opportunity offered by potential programs within the portfolio—as well as changes to the internal and external business environment—the composition of the portfolio may be adjusted by management.

The COBIT 5 framework prescribes the following key management practices for effective portfolio management:[8]
- APO05.01 Establish the target investment mix.
- APO05.02 Determine the availability and sources of funds.
- APO05.03 Evaluate and select programmes to fund.
- APO05.04 Monitor, optimise and report on investment portfolio performance.
- APO05.05 Maintain portfolios.
- APO05.06 Manage benefits achievement.

More information regarding these management practices is provided in the following sections.

- **APO05.01 Establish the target investment mix.** Review and ensure clarity of the enterprise and IT strategies and current services. Define an appropriate investment mix based on cost, alignment with strategy, and financial measures such as cost and expected ROI over the full economic life cycle, degree of risk, and type of benefit for the programs in the portfolio. Adjust the enterprise and IT strategies where necessary, **figure 3.3.**

| Figure 3.3—APO05.01—Establish the Target Investment Mix ||||||
|---|---|---|---|---|
| Management Practice | Inputs ||  Outputs ||
| APO05.01 Establish the target investment mix. Review and ensure clarity of the enterprise and IT strategies and current services. Define an appropriate investment mix based on cost, alignment with strategy, and financial measures such as cost and expected ROI over the full economic life cycle, degree of risk, and type of benefit for the programmes in the portfolio. Adjust the enterprise and IT strategies where necessary. | From | Description | Description | To |
| | EDM02.02 | Investment types and criteria | Defined investment mix | Internal |
| | APO02.05 | • Strategic road map<br>• Risk assessment initiatives<br>• Definition of strategic initiatives | Identified resources and capabilities required to support strategy | Internal |
| | APO06.02 | Prioritisation and ranking of IT initiatives | Feedback on strategy and goals | APO02.05 |
| | APO09.01 | Definitions of standard services | | |
| | BAI03.11 | Service definitions | | |
| Activities |||||
| 1. Validate that IT-enabled investments and current IT services are aligned with enterprise vision, enterprise principles, strategic goals and objectives, enterprise architecture vision, and priorities. |||||
| 2. Obtain a common understanding between IT and the other business functions on the potential opportunities for IT to drive and support the enterprise strategy. |||||
| 3. Create an investment mix that achieves the right balance amongst a number of dimensions, including an appropriate balance of short- and long-term returns, financial and non-financial benefits, and high- and low-risk investments. |||||
| 4. Identify the broad categories of information systems, applications, data, IT services, infrastructure, IT assets, resources, skills, practices, controls and relationships needed to support the enterprise strategy. |||||
| 5. Agree on an IT strategy and goals, taking into account the inter-relationships between the enterprise strategy and the IT services, assets and other resources. Identify and leverage synergies that can be achieved. |||||
| Source: ISACA, *COBIT 5: Enabling Processes*, USA, 2012, page 74 |||||

- **APO05.02 Determine the availability and sources of funds.** Determine potential sources of funds, different funding options and the implications of the funding source on the investment return expectations, **figure 3.4.**

| Figure 3.4—APO05.02—Determine the Availability and Sources of Funds ||||||
|---|---|---|---|---|
| Management Practice | Inputs || Outputs ||
| APO05.02 Determine the availability and sources of funds. Determine potential sources of funds, different funding options and the implications of the funding source on the investment return expectations. | From | Description | Description | To |
| | | | Funding options | APO02.05 |
| | | | Investment return expectations | EDM02.01<br>APO02.04<br>APO06.02<br>BAI01.06 |
| Activities |||||
| 1. Understand the current availability and commitment of funds, the current approved spending, and the actual amount spent to date. |||||
| 2. Identify options for obtaining additional funds for IT-enabled investments, internally and from external sources. |||||
| 3. Determine the implications of the funding source on the investment return expectations. |||||
| Source: ISACA, *COBIT 5: Enabling Processes*, USA, 2012, page 75 |||||

- **APO05.03 Evaluate and select programmes to fund.** Based on the overall investment portfolio mix requirements, evaluate and prioritize program business cases, and decide on investment proposals. Allocate funds and initiate programs, **figure 3.5**.

| Figure 3.5—APO05.03—Evaluate and Select Programmes to Fund ||||||
|---|---|---|---|---|---|
| **Management Practice** | **Inputs** ||| **Outputs** ||
| **APO05.03 Evaluate and select programmes to fund.** Based on the overall investment portfolio mix requirements, evaluate and prioritise programme business cases, and decide on investment proposals. Allocate funds and initiate programmes. | From | Description | Description | To ||
| | EDM02.01 | • Evaluation of investment and services portfolios<br>• Evaluation of strategic alignment | Programme business case | APO06.02<br>BAI01.02 ||
| | EDM02.02 | Investment types and criteria | Business case assessments | APO06.02<br>BAI01.06 ||
| | APO03.01 | Architecture concept business case and value proposition | Selected programmes with ROI milestones | EDM02.01<br>BAI01.04 ||
| | APO04.04 | Proof-of-concept scope and outline business case | | ||
| | APO06.02 | Budget allocations | | ||
| | APO06.03 | • Budget communications<br>• IT budget and plan | | ||
| | APO09.01 | Identified gaps in IT services to the business | | ||
| | APO09.03 | Service level agreements (SLAs) | | ||
| | BAI01.02 | • Programme benefit realisation plan<br>• Programme mandate and brief<br>• Programme concept business case | | ||
| **Activities** ||||||
| 1. Recognise investment opportunities and classify them in line with the investment portfolio categories. Specify expected enterprise outcome(s), all initiatives required to achieve the expected outcomes, costs, dependencies and risk, and how all would be measured. ||||||
| 2. Perform detailed assessments of all programme business cases, evaluating strategic alignment, enterprise benefits, risk and availability of resources. ||||||
| 3. Assess the impact on the overall investment portfolio of adding candidate programmes, including any changes that might be required to other programmes. ||||||
| 4. Decide which candidate programmes should be moved to the active investment portfolio. Decide whether rejected programmes should be held for future consideration or provided with some seed funding to determine whether the business case can be improved or discarded. ||||||
| 5. Determine the required milestones for each selected programme's full economic life cycle. Allocate and reserve total programme funding per milestone. Move the programme into the active investment portfolio. ||||||
| 6. Establish procedures to communicate the cost, benefit and risk-related aspects of these portfolios to the budget prioritisation, cost management and benefit management processes. ||||||
| Source: ISACA, *COBIT 5: Enabling Processes*, USA, 2012, page 75 ||||||

- **APO05.04 Monitor, optimise and report on investment portfolio performance.** On a regular basis, monitor and optimize the performance of the investment portfolio and individual programs throughout the entire investment life cycle, **figure 3.6.**

### Figure 3.6—APO05.04—Monitor, Optimise and Report on Investment Portfolio Performance

| Management Practice | Inputs | | Outputs | |
|---|---|---|---|---|
| **APO05.04 Monitor, optimise and report on investment portfolio performance.** On a regular basis, monitor and optimise the performance of the investment portfolio and individual programmes throughout the entire investment life cycle. | **From** | **Description** | **Description** | **To** |
| | EDM02.01 | Evaluation of investment and services portfolios | Investment portfolio performance reports | EDM02.03 APO09.04 BAI01.06 MEA01.03 |
| | EDM02.03 | • Actions to improve value delivery<br>• Feedback on portfolio and programme performance | | |
| | APO04.06 | Evaluation of innovation benefits | | |
| | BAI01.06 | Stage-gate review results | | |
| **Activities** | | | | |
| 1. Review the portfolio on a regular basis to identify and exploit synergies, eliminate duplication between programmes, and identify and mitigate risk. | | | | |
| 2. When changes occur, re-evaluate and reprioritise the portfolio to ensure that the portfolio is aligned with the business strategy and the target mix of investments is maintained so the portfolio is optimising overall value. This may require programmes to be changed, deferred or retired, and new programmes to be initiated. | | | | |
| 3. Adjust the enterprise targets, forecasts, budgets and, if required, the degree of monitoring to reflect the expenditures to be incurred and enterprise benefits to be realised by programmes in the active investment portfolio. Incorporate programme expenditures into chargeback mechanisms. | | | | |
| 4. Provide an accurate view of the performance of the investment portfolio to all stakeholders. | | | | |
| 5. Provide management reports for senior management's review of the enterprise's progress towards identified goals, stating what still needs to be spent and accomplished over what time frames. | | | | |
| 6. Include in the regular performance monitoring information on the extent to which planned objectives have been achieved, risk mitigated, capabilities created, deliverables obtained and performance targets met. | | | | |
| 7. Identify deviations for:<br>• Budget control between actual and budget<br>• Benefit management of:<br>  – Actual vs. targets for investments for solutions, possibly expressed in terms of ROI, NPV or internal rate of return (IRR)<br>  – The actual trend of service portfolio cost for service delivery productivity improvements | | | | |
| 8. Develop metrics for measuring IT's contribution to the enterprise, and establish appropriate performance targets reflecting the required IT and enterprise capability targets. Use guidance from external experts and benchmark data to develop metrics. | | | | |
| Source: ISACA, *COBIT 5: Enabling Processes*, USA, 2012, page 76 | | | | |

- **APO05.05 Maintain portfolios.** Maintain portfolios of investment programmes and projects, IT services and IT assets, **figure 3.7**.

| Figure 3.7—APO05.05—Maintain Portfolios ||||| 
|---|---|---|---|---|
| **Management Practice** | **Inputs** || **Outputs** ||
| APO05.05 Maintain portfolios. Maintain portfolios of investment programmes and projects, IT services and IT assets. | From | Description | Description | To |
| | BAI01.14 | Communication of programme retirement and ongoing accountabilities | Updated portfolios of programmes, services and assets | APO09.02 BAI01.01 |
| | BAI03.11 | Updated service portfolio | | |
| **Activities** |||||
| 1. Create and maintain portfolios of IT-enabled investment programmes, IT services and IT assets, which form the basis for the current IT budget and support the IT tactical and strategic plans. |||||
| 2. Work with service delivery managers to maintain the service portfolios and with operations managers and architects to maintain the asset portfolios. Prioritise portfolios to support investment decisions. |||||
| 3. Remove the programme from the active investment portfolio when the desired enterprise benefits have been achieved or when it is clear that benefits will not be achieved within the value criteria set for the programme. |||||
| Source: ISACA, *COBIT 5: Enabling Processes*, USA, 2012, page 76 |||||

- **APO05.06 Manage benefits achievement.** Monitor the benefits of providing and maintaining appropriate IT services and capabilities, based on the agreed-on and current business case, **figure 3.8**.

| Figure 3.8—APO05.06—Manage Benefits Achievement ||||| 
|---|---|---|---|---|
| **Management Practice** | **Inputs** || **Outputs** ||
| APO05.06 Manage benefits achievement. Monitor the benefits of providing and maintaining appropriate IT services and capabilities, based on the agreed-on and current business case. | From | Description | Description | To |
| | BAI01.04 | Programme budget and benefits register | Benefit results and related communications | EDM02.01 APO09.04 BAI01.06 |
| | BAI01.05 | Results of benefit realisation monitoring | Corrective actions to improve benefit realisation | APO09.04 BAI01.06 |
| **Activities** |||||
| 1. Use the agreed-on metrics and track how benefits are achieved, how they evolve throughout the life cycle of programmes and projects, how they are being delivered from IT services, and how they compare to internal and industry benchmarks. Communicate results to stakeholders. |||||
| 2. Implement corrective action when achieved benefits significantly deviate from expected benefits. Update the business case for new initiatives and implement business process and service improvements as required. |||||
| 3. Consider obtaining guidance from external experts, industry leaders and comparative benchmarking data to test and improve the metrics and targets. |||||
| Source: ISACA, *COBIT 5: Enabling Processes*, USA, 2012, page 77 |||||

When analyzing the inputs/outputs in the previous figures on portfolio management, it is clear that COBIT 5 proposes that appropriate guidance from the board and executive committee is an important enabler for successful portfolio management. More specifically, the governance process EDM02 *Ensure benefits delivery* is shown to be an important input to portfolio management. This stresses the importance of the role of the board of directors and executive committee in securing optimal value from IT-enabled initiatives. The board and the executive committee need to take accountability for the following areas:

- **EDM02.01 Evaluate value optimisation.**
  - Continually evaluate the portfolio of IT-enabled investments, services and assets to determine the likelihood of achieving enterprise objectives and delivering value at a reasonable cost. Identify and make judgement on any changes in direction that need to be given to management to optimise value creation.

- **EDM02.02 Direct value optimisation.**
  – Direct value management principles and practices to enable optimal value realisation from IT-enabled investments throughout their full economic life cycle.
- **EDM02.03 Monitor value optimisation.**
  – Monitor the key goals and metrics to determine the extent to which the business is generating the expected value and benefits to the enterprise from IT-enabled investments and services. Identify significant issues and consider corrective actions.

## 3.3 BENEFIT CALCULATION TECHNIQUES

The imperatives of ensuring that IT-enabled investments provide the value being promised requires proper analysis using established and appropriate techniques. The COBIT 5 and ITIL frameworks provide the value management practices and IT service life cycle perspectives, respectively, to update and contextualize traditional investment analysis techniques and approaches.

It is challenging to measure the benefits accruing from IT-enabled investment. Most of the benefits delivered from IT investment are indirect. Therefore, financial and nonfinancial benefit measurement techniques should be used for measuring the benefits of IT.

### The Investment Cycle and the Service Management Cycle

The premise for life cycle management has two important contextual references:
- **Management of IT-enabled investments in enterprises**—Should follow a life cycle paradigm. One of the seven principles of the Val IT framework[9] states that IT-enabled investments will be managed through their full economic life cycle. As part of this principle, business cases must be kept current from the initiation of an investment, until any resulting service is retired. This principle recognizes that there will always be some degree of uncertainty and that variability over time—in costs, risk, benefits, strategy and organizational and external changes—must be taken into account in determining whether funding should be continued, increased, decreased or stopped.
- **Development of the IT service life cycle as part of IT service management (ITSM)**—This is exemplified in the ITIL V3 framework.[10] As business dependency on IT has grown dramatically over recent years, the practice of service management has grown correspondingly in importance. The ISO 20000 standard for ITSM is now in place and has gained recognition globally. The ITIL framework has seen increasing adoption as a means to meet compliance requirements for the standard.

Within these contextual references, the value of IT-enabled investments to the business has to be maximized. In the Val IT framework, value is defined as benefits less cost, adjusted for risk. (Discussion of the Val IT framework can be found in chapters 1 and 3 of this manual.) Therefore, the application of CBA techniques should be scoped considering this definition, and within the two context references given previously.

The second context reference mentioned above can be explained by leveraging the IT service life cycle as defined by the ITIL framework.[11] This life cycle contains five elements, each of which relies on service principles, processes, roles and performance measures:
- Service strategy (SS)
- Service design (SD)
- Service transition (ST)
- Service operation (SO)
- Continuous service improvement (CSI)

Each part of the life cycle exerts influence on the other and relies on the other for inputs and feedback. In this way, a constant set of checks and balances throughout the service life cycle ensures that, as business demand changes with business need, the services can adapt and respond effectively to the changes.

At the heart of the service life cycle is the key principle—all services must provide measurable value to business objectives and outcomes. ITIL service management focuses on business value as its prime objective. Each practice revolves around ensuring that everything a service provider does to manage IT services for the business customer can be measured and quantified in terms of business value. This has become very important today because IT organizations must operate themselves as businesses to demonstrate a clear return on investment and equate service performance with business value to the customer.

## Financially Oriented Cost-benefit Techniques

CBA involves comparing the costs with the benefits of the IT-enabled investment that can be directly and indirectly attributed to the investment. Some financially-oriented and commonly used cost-benefit techniques for IT-enabled investment decision making include the following:

- **Payback period**—This is a simple technique in which the time period necessary to recoup the initial investment is calculated and used to evaluate an investment and/or a set of mutually exclusive investments.
- **NPV/IRR**—These are based on the well-known corporate financial management principle of the time value of money. The principle states that the longer a return is deferred into the future, the lower its current value. So returns that will be realized further into the future are worth less than those realized sooner. As a result, cash inflows from an IT-enabled investment must be discounted and the present value of the investment is used to evaluate whether to invest.
- **ROI**—This is a relatively simple calculation that provides decision-making information in terms of a ratio. The ratio of expected profit to the initial investment cost is compared to the opportunity cost of capital; if the return is greater than the opportunity cost of capital, the investment should be undertaken.
- **Breakeven analysis**—This is often used in several ways to evaluate IT-enabled investments; however, it is most often used by comparing the present value of the costs with the present value of the benefits of the IT-enabled investment.

Financially-oriented cost benefit analyses are most widely used in the business case to justify IT-enabled investments. However, they fall short in capturing the total value add of any IT-enabled investment because many benefits of IT-enabled investments are difficult to measure (intangible). Hence, nonfinancially oriented cost-benefit techniques should supplement the financially-oriented cost-benefit techniques.

## Nonfinancially Oriented Cost-benefit Techniques

Nonfinancial CBA involves a comparative examination of the costs and benefits of a project. Such analysis tries to overcome the problem of financial ROI by finding some surrogate measure for intangible costs or benefits that can be expressed in monetary terms. For example, if one of the objectives of introducing an IT investment is to increase customer satisfaction, the benefit may be expressed in terms of reducing the cost of returned products and reducing the number of customer complaints.

The approach attempts to deal with two problems: the difficulty of quantifying the value of benefits that do not directly accrue to the investor in the project and the difficulty of identifying the benefits or costs which do not have an obvious market value or price (i.e., intangible factors). Therefore, a nonfinancial CBA method is useful where the costs and benefits are intangible, but the method requires the existence of a broad agreement on the measures used to attach a value to the intangibles.

Emphasis on intangible factors in the evaluation of IT-enabled investments should be a part of any complete CBA. There is a need to look beyond cost savings in evaluating IT benefits. Such benefits as improved customer service, improved product quality and better flexibility are often the result of IT-enabled investments, but are hard to quantify in a convincing manner.

Some nonfinancial cost-benefit techniques for IT-enabled investment decision making include the following:[12]
- **Organizational flexibility**—Organizational flexibility is one way of understanding the business value of IT. The concept of flexibility is operationalized using the ideas of stimulus, response and ease of response (time, cost, scope). When combined with Michael Porter's value chain model, in which each part of the value chain is thought of as consisting of multiple processes, it is argued that areas may be identified where flexibility may add value to enterprises. Examples of IT-driven organizational flexibility are presented in distinct areas of the value chain (logistics, operations, marketing and sales) as well as across different areas of the value chain. Among the advantages of this approach is that organizational flexibility may be viewed as a source of competitive advantage and that IT-enabled investments can add significant business value by enhancing organizational flexibility.

- **Information economics**—Information economics is a composite CBA technique, tailored to cope with the particular intangibles and uncertainties found in information systems projects. It is driven by the notion that information has economic value and is a system of weights and measures that quantifies intangible benefits and ranks proposed projects by their expected contribution to business objectives.

  Information economics breaks IT-enabled investment analysis into three decision factors: tangible CBA, intangible benefit analysis and intangible risk analysis. In practice, tangible and intangible corporate objectives are laid out and assigned a relative weight. Proposed systems then receive scores in each business objective and risk category, based on their potential impact on that objective. Information economics retains ROI calculations for those benefits and costs which can be directly ascertained through a conventional cost-benefit process.

  The decision-making process used in the information economics methodology is based on a ranking and scoring technique of intangibles and risk factors associated with the IT-enabled investment. It identifies IT performance measures and uses them to rank the economic impact of all the changes on the enterprise's performance caused by the introduction of the IT. Surrogate measures are often used for most intangible and risk factors that are hard to estimate.

  One technique used in information economics is value linking, which involves assessment of benefits achieved in other departments in calculating the contribution of IT. Another technique is value restructuring, which assumes that if a function exists in an enterprise, then it has some value. This suggests consideration of IT investments as analogous to that of the research and development department or the legal department of an enterprise.

  The strength of the information economics method is that it links the quantification and comparison approaches with qualification approaches. The limitation of information economics, however, is the focus on simple, idealized settings that can be modeled with applicable mathematical models, often requiring many simplifying assumptions. Clearly, real-world information systems involve complex relationships, variables and parameters—even when rigorous models can be formulated, they cannot be solved analytically.

  Parker and Benson, the developers of the information economics model in 1989, advocated that for enterprises to gain a competitive edge, the way IT is financially justified should also change. They argued that financial CBA is not adequate for evaluation of IT applications, except when dealing only with cost-avoidance issues. Information economics, which measures and justifies IT on the basis of business performance, is positioned as a better method.

  More information on the related concepts of portfolio management and business cases can be found in section 2.9 Prioritization Processes Related to IT Initiatives, section 2.10 Scope, Objectives and Benefits of IT Investment Programs, section 3.1 IT Investment Management Processes, Including the Economic Life Cycle of Investments and section 3.2 Basic Principles of Portfolio Management.

## 3.4 PROCESS AND SERVICE MEASUREMENT TECHNIQUES

High performance is directly related to effective performance management practices and tools (such as maturity models, benchmarking, metrics). Greater visibility of the detailed performance of operations and investments enables enterprises to influence performance and improve agility.

Leading enterprises take great care in designing IT performance reports that are concise, easy to understand and tailored to various management needs and audiences. Executive managers, in particular, often require data presentations and displays that focus on bottom-line performance results. When presented in this manner, they can quickly digest information, focus on problem areas, seek pertinent follow-up data and be more efficient in making or recommending project or program decisions.

## Balanced Scorecard and Metrics

"Use of an IT balanced scorecard is one of the most effective means to aid to board and management to achieve IT and business alignment."[13] In these words, the IT Governance Institute promotes the IT BSC as a good practice for performance measurement and alignment. This is supported by testimonials of several executives, such as:

> "The major advantage of the IT BSC is that it provides a systematic translation of the strategy into critical success factors and metrics, which materializes the strategy" (chief information officer [CIO] of a financial organization)[14]

> "The Balanced Scorecard gives a balanced view of the total value delivery of IT to the business. It provides a snapshot of where your IT organization is at a certain point in time. Most executives, like me, do not have the time to drill down into the large amount of information." (vice president of an insurance organization)[15]

To apply this good practice to the IT function as an internal service provider, the four perspectives of the generic BSC should be changed accordingly. In **figure 3.9**, a generic IT BSC for an IT department is shown. The user orientation perspective represents the user evaluation of IT. The operational excellence perspective represents the IT processes employed to develop and deliver the applications. The future orientation perspective represents the human and technology resources needed by IT to deliver its services over time. The business contribution perspective captures the business value created from the IT investments.

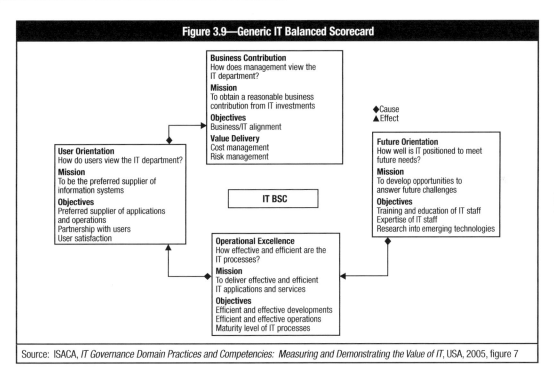

More information on metrics, BSC and benchmarking can be found in section 1.10 IT Governance Monitoring Processes/Mechanisms and section 1.11 IT Governance Reporting Processes/Mechanisms.

## Process Capability Model

The COBIT 5 product set includes a process capability model, based on the internationally recognized ISO/IEC 15504 Software Engineering—Process Assessment standard. This model provides a means to measure the performance of any of the governance (evaluate, direct and monitor [EDM]-based) processes or management (plan, build, run and monitor [PBRM]-based) processes and will allow areas for improvement to be identified.

The ISO/IEC 15504 standard specifies that process capability assessments can be performed for various purposes and with varying degrees of rigor. Purposes can be internal, with a focus on comparisons between enterprise areas and/or process improvement for internal benefit, or they can be external, with a focus on formal assessment, reporting and certification.

The COBIT 5 process capability approach can be summarized as shown in **figure 3.10**.

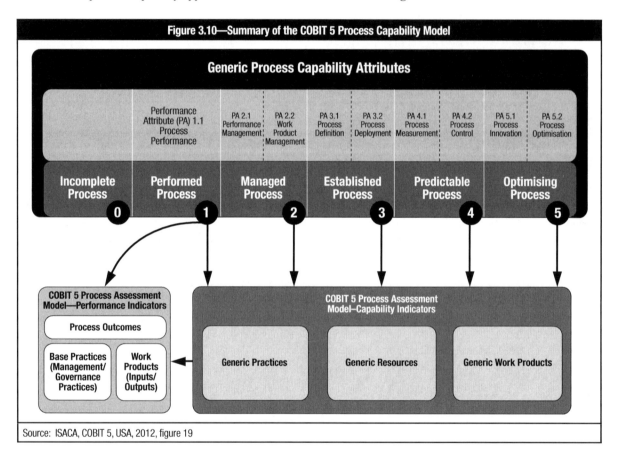

Source: ISACA, COBIT 5, USA, 2012, figure 19

Process capability is expressed in terms of process attributes grouped into capability levels. The capability level of a process is determined on the basis of the achievement of specific process attributes (e.g., process definition) according to ISO/IEC 15504-2:2003.

There are six levels of capability that a process can achieve, including an "incomplete process" designation if the practices in it do not achieve the intended purpose of the process:
- **0 Incomplete process**—The process is not implemented or fails to achieve its process purpose. At this level, there is little or no evidence of any systematic achievement of the process purpose.
- **1 Performed process (one attribute)**—The implemented process achieves its process purpose.
- **2 Managed process (two attributes)**—The previously described performed process is now implemented in a managed fashion (planned, monitored and adjusted) and its work products are appropriately established, controlled and maintained.
- **3 Established process (two attributes)**—The previously described managed process is now implemented using a defined process that is capable of achieving its process outcomes.

- **4 Predictable process (two attributes)**—The previously described established process now operates within defined limits to achieve its process outcomes.
- **5 Optimizing process (two attributes)**—The previously described predictable process is continuously improved to meet relevant current and projected business goals.

Each capability level can be achieved only when the level below has been fully achieved. For example, a process capability level 3 (established process) requires the process definition and process deployment attributes to be largely achieved, on top of full achievement of the attributes for a process capability level 2 (managed process).

There is a significant distinction between process capability level 1 and the higher capability levels. Process capability level 1 achievement requires the process performance attribute to be largely achieved, which actually means that the process is being successfully performed and the required outcomes obtained by the enterprise. The higher capability levels then add different attributes to it. In this assessment scheme, achieving a capability level 1, even on a scale to 5, is already an important achievement for an enterprise. Note that each individual enterprise shall choose (based on cost-benefit and feasibility reasons) its target or desired level, which very seldom will happen to be one of the highest.

The ISO/IEC 15504 standard specifies that process capability assessments can be performed for various purposes and with varying degrees of rigor. Purposes can be internal, with a focus on comparisons between enterprise areas and/or process improvement for internal benefit, or they can be external, with a focus on formal assessment, reporting and certification.

## 3.5 PROCESSES AND PRACTICES FOR PLANNING, DEVELOPMENT, TRANSITION, DELIVERY, AND SUPPORT OF SOLUTIONS AND SERVICES

With the heightened complexity of IT solutions and services to execute the enterprise's strategy and vision, and the rapid rise in its pivotal contribution to business automation and value delivery, the perspective of IT solution and service delivery has assumed a preeminence, evidenced by the fast-growing service management movement—the IT Service Management Forum (itSMF). There is now an evolving body of good practices (ITIL) and an international standard (ISO 20000) to detail specific processes and methodologies. In terms of solutions development, the SDLC is probably one of the best-known IT practices and is the mainstay of IT development since the beginning of IT enablement of business processes.

Currently, enterprises are dependent on IT to satisfy their corporate aims, meet their business needs and deliver value to customers.[16]

For this to happen in a manageable, accountable and repeatable way, the business must ensure that high-quality IT services are provided that are:
- Matched to business needs and user requirements
- Compliant with legislation
- Effectively and efficiently sourced and delivered
- Continually reviewed and improved

### ITIL and ISO 20000 as Reference Frameworks

The origins of IT as a services discipline can be traced back to the late 1980s, when the UK Central Computer and Telecommunications Agency (CCTA, later to become part of the UK Office of Government Commerce [OGC]) developed the initial concepts. This was the beginning of the IT Infrastructure Library (ITIL) framework, now in its ITIL 2011 version. In 2000, the British Standards Institute developed a standard (BS 15000) for ITSM, and this was later evolved to become the international ISO 20000 standard, now in its second version. Achievement of the standard certifies organizations as having passed auditable practices and processes in ITSM. The reference body of knowledge for service delivery best practices is ITIL. It is a comprehensive framework detailed over five volumes in the current version.

Like most standards, ISO 20000 is primarily used as a demonstration of compliance to accepted good practice. In addition to the central elements of good ITSM best practice, it also requires service providers to implement the Plan-Do-Check-Act (PDCA) methodology (Deming's quality circle) and apply it to their service management processes. This ensures continual service improvement by the service provider, so that the organization's processes develop, mature and adapt to their customers' requirements, errors and omissions are avoided, and those problems that have been dealt with do not recur.

### Service Management Processes

IT service delivery practices and processes have been well defined through ITIL and the ISO 20000 standard and continue to evolve. These practices and processes are applicable to IT service provider organizations, whether as an internal department or division or as an external service provider. The service management processes are shown in **figure 3.11**.

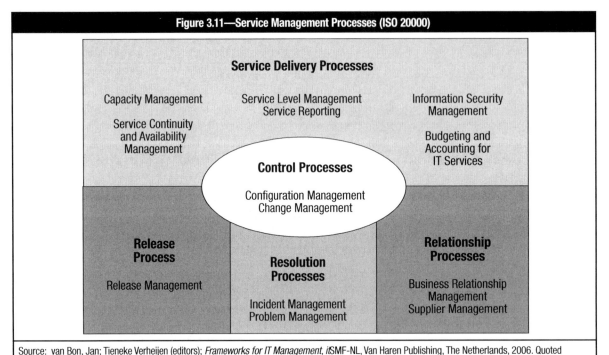

Source: van Bon, Jan; Tieneke Verheijen (editors); *Frameworks for IT Management, it*SMF-NL, Van Haren Publishing, The Netherlands, 2006. Quoted from: International Organization for Standardization (ISO); *Information Technology—Service Management, ISO/IEC 20000-1:2005 and 20000-2:2005*, Switzerland, 2005

There are four group types of key processes that are all related to the fifth group type—control processes. For all processes, the standard defines objectives and specifications. The various processes specified for service delivery are stated in the standard as follows:

- **Capacity Management**—The objective is to ensure that the service provider has, at all times, sufficient capacity to meet the current and future agreed-on demands of the customer's business needs.
- **Service Continuity and Service Availability Management**—The objective is to ensure that agreed-on service continuity and availability commitments to customers can be met in all circumstances.
- **Service Level Management**—The objective is to define, agree on, record and manage levels of service.
- **Service Reporting**—The objective is to produce agreed-on, timely, reliable, accurate reports for informed decision making and effective communication.
- **Information Security Management**—The objective is to manage information security effectively within all service activities.
- **Budgeting and Accounting for IT Services**—The objective is to budget and account for the cost of service provision.

Next to the service delivery processes, ISO 20000 also identifies:
- **Release processes**—Key to enabling the smooth implementation of new services, or major changes to existing services
- **Resolution Processes**—Incident management and problem management
- **Relationship Processes**—Supporting business relationship management and supplier management in the end-to-end supply chain
- **Control Processes**—Configuration management, change management, and release and deployment management

## ITIL 2011

In ITIL 2011, the service delivery module of ITIL covered the more forward-looking delivery aspects of service provision and consisted of the following processes: SLM, financial management for IT services, capacity management, IT service continuity management and availability management.[17] These processes were principally concerned with developing plans for improving the quality of IT services delivered.

In ITIL 2011, the most significant development has been the move from a process-based framework to a more comprehensive structure reflecting the life cycle of IT services. An illustration frequently used to view the operational phases—design, transition and operation—shows the spokes of a wheel, with strategy at the hub and continual service improvement all around the rim (**figure 3.12**).

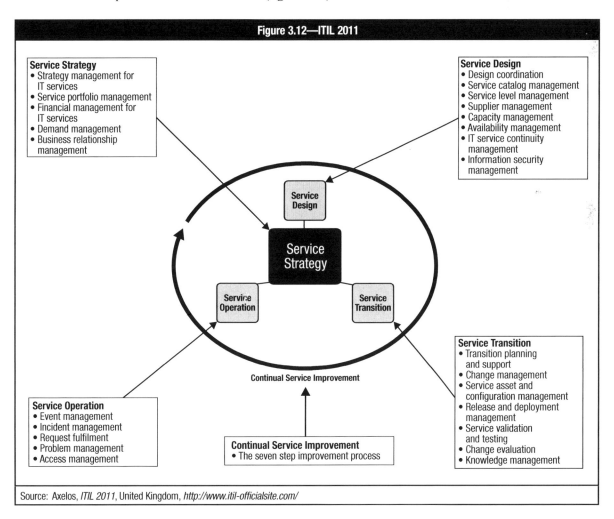

Source: Axelos, *ITIL 2011*, United Kingdom, http://www.itil-officialsite.com/

## System Development Life Cycle and Agile Development

The SDLC is probably one of the best-known IT practices and is the mainstay of IT development since the beginning of IT enablement of business processes. The SDLC can be divided into 10 phases during which defined IT work products are created or modified. The tenth phase occurs when the system is disposed of and the task performed is either eliminated or transferred to other systems. Not every project will require that the phases be sequentially executed; however, the phases are interdependent. Depending on the size and complexity of the project, phases may be combined or may overlap. **Figure 3.13** shows the various phases of the SDLC, along with a brief description of the purpose of, and activity in, each phase. In the context of management and control, the SDLC phases serve as a programmatic guide to project activity and provide a flexible, but consistent, way to conduct projects to a depth matching the scope of the project. Each SDLC phase objective is typically described with key deliverables, a description of recommended tasks and a summary of related control objectives for effective management. It is critical for the project manager to establish and monitor control objectives during each SDLC phase while executing projects.

| Figure 3.13—Phases in System Development Life Cycle (SDLC) | |
|---|---|
| Initiation | Begins when a sponsor identifies a need or an opportunity. Concept proposal is created. |
| System Concept Development | Defines the scope or boundary of the concepts. Includes systems boundary document, cost-benefit analysis, risk management plan and feasibility study. |
| Planning | Develops a project management plan and other planning documents. Provides the basis for acquiring the resources needed to achieve a solution. |
| Requirements Analysis | Analyzes user needs and develops user requirements. Create a detailed functional requirements document. |
| Design | Transforms detailed requirements into complete, detailed systems design document. Focuses on how to deliver the required functionality. |
| Development | Converts a design into a complete information system. Includes acquiring and installing systems environment; creating and testing databases, preparing test case procedures; preparing test files, coding, compiling, refining programs; performing test readiness review and procurement activities. |
| Integration and Test | Demonstrates that developed system conforms to requirements as specified in the functional requirements document. Conducted by quality assurance staff and users. Produces test analysis reports. |
| Implementation | Includes implementation preparation, implementation of the system into a production environment, and resolution of problems identified in the integration and test phases. |
| Operations and Maintenance | Describes tasks to operate and maintain systems in a production environment. Includes postimplementation and in-process reviews. |
| Disposition | Describes end-of-system activities. Emphasis is given to proper preparation of data. |
| Source: US Department of Justice; *Information Resources Management*, chapter 1, figure 1-1, USA, 2003 | |

Agile development approaches propose alternative methods for software development. Assuming that all requirements cannot be articulated upfront, agile approaches, such as the Scrum methodology, propose a more iterative and incremental approach instead of the sequential approach of the SDLC. The "Agile Manifesto" introduced agile development in 2001. Since then, the Agile Movement—with all its values, principles, methods, practices, tools, champions and practitioners, philosophies, and cultures—has significantly changed the landscape of the modern software engineering and commercial software development in the Internet era. To support such agile development approaches, specific governance and control requirements need to be established in the organization.

## The Link Between Solution Delivery and Service Management

It is important to understand the difference between the processes required to build new business solutions (solution delivery) and the service management processes required when solutions are moved into production. As seen in a recent case study described by De Haes, et al., this split between solution delivery and service management, as explained at KLM (and called internally "the bicycle"), is shown in **figure 3.14**.[18] It represents a program portfolio cycle (solution delivery) and a service portfolio cycle (service management).

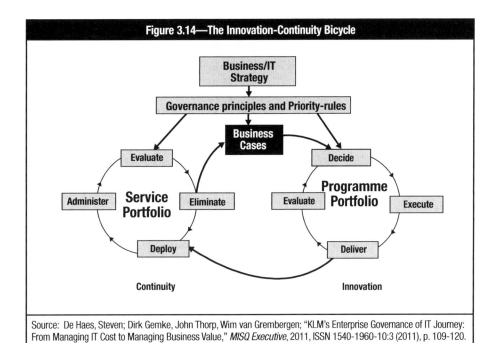

Figure 3.14—The Innovation-Continuity Bicycle

Source: De Haes, Steven; Dirk Gemke, John Thorp, Wim van Grembergen; "KLM's Enterprise Governance of IT Journey: From Managing IT Cost to Managing Business Value," *MISQ Executive*, 2011, ISSN 1540-1960-10:3 (2011), p. 109-120. Used with permission.

As visualized, the business/IT strategy drives the definition and application of the governance principles and priority rules and the definition of business cases. The approved business cases are managed in the program (innovation cycle), which, after delivery, become operational services being deployed and administered in the service (continuity) portfolio. As a result of ongoing evaluation, services may continue with no change, re-enter the innovation cycle through a new business case or be eliminated (retired).

## 3.6 CONTINUOUS IMPROVEMENT CONCEPTS AND PRINCIPLES

Continuous process improvement and associated techniques are highly useful for mapping GEIT to organizational imperatives, especially the improvement of process maturity. Knowledge of these paradigms enables and align GEIT improvement within the enterprise.

There are several frameworks which embody, partially or wholly, the generic concept of continuous improvement, and they fall broadly within the realm of quality management.

### Six Sigma

Six Sigma is an IT-appropriate process improvement methodology, although the fundamental objective is to reduce errors to fewer than 3.4 defects per million executions regardless of the process. Given the wide variation in IT deliverables, roles and tasks within IT operational environments, IT leadership should determine whether it is reasonable to expect delivery at a Six Sigma level. Six Sigma is a data-driven approach that supports continuous improvement. It is business-output-driven in relation to customer specification and focuses on dramatically reducing process variation using Statistical Process Control (SPC) measures. Because Six Sigma requires data, it is important to start capturing data as soon as possible after activity execution. Even questionable data have value because the approach provides the opportunity to analyze why the data do not make sense.

Six Sigma's objective is the implementation of a measurement-oriented strategy focused on process improvement and defect reduction. A Six Sigma defect is defined as anything outside customer specifications. There are two primary submethodologies within Six Sigma: DMAIC and DMADV. The DMAIC process is an improvement method for existing processes for which performance does not meet expectations or for which incremental improvements are desired. The DMADV process focuses on the creation of new processes.

## Total Quality Management

Total quality management (TQM) is a management strategy aimed at embedding awareness of quality in all organizational processes. It is a set of systematic activities carried out by the entire enterprise to effectively and efficiently achieve company objectives and provide products and services with a level of quality that satisfies customers at the appropriate time and price. At the core of TQM is a management approach to long-term success through customer satisfaction. In a TQM effort, all members of an enterprise participate in improving processes, products, services and the culture in which they work. Quality management for IT services is a systematic way of ensuring that all the activities necessary to design, develop and implement IT services that satisfy the requirements of the organization and of users take place as planned and that the activities are carried out in a cost-effective manner.

## Plan-Do-Check-Act

Plan-Do-Check-Act (PDCA) is an iterative four-step management method used in business for the control and continuous improvement of processes and products. It is also known as the Deming circle/cycle/wheel, Shewhart cycle, control circle/cycle, or Plan–Do–Study–Act (PDSA). The steps in each successive PDCA cycle are:
- **Plan**—Establish the objectives and processes necessary to deliver results in accordance with the expected output (the target or goals). By establishing output expectations, the completeness and accuracy of the specification is also a part of the targeted improvement. When possible, start on a small scale to test possible effects.
- **Do**—Implement the plan, execute the process and make the product. Collect data for charting and analysis in the following Check and Act steps.
- **Check**—Study the actual results (measured and collected in the Do step) and compare against the expected results (targets or goals from the Plan step) to ascertain any differences. Look for deviation in implementation from the plan, and also look for the appropriateness/completeness of the plan to enable the execution, i.e., the Do step. Charting data can make it much easier to see trends over several PDCA cycles and to convert the collected data into information. Information is what you need for the next step, Act.
- **Act**—Request corrective actions on significant differences between actual and planned results. Analyze the differences to determine their root causes. Determine where to apply changes that will include improvement of the process or product. When a pass through these four steps does not result in the need to improve, the scope to which PDCA is applied may be refined to plan and improve with more detail in the next iteration of the cycle, or attention needs to be placed in a different stage of the process.

## 3.7 OUTCOME AND PERFORMANCE MEASUREMENT TECHNIQUES

The purpose of performance measurement is to uncover, communicate and evolve organizational performance drivers. The choice of measures communicates to stakeholders what is important, and this affects what gets done. Choosing measures that answer critical management questions improves management's visibility into key processes.

In effective performance management approaches, metrics are not just used for assigning accountabilities or to comply with reporting requirements. They are used to create and facilitate action to improve performance and, therefore, governance of IT.

### The Need for Metrics

Metrics and performance information need to be linked to strategic management processes. An effective performance management system produces information that delivers the following benefits:
- It is an early warning indicator of problems and the effectiveness of corrective action.
- It provides input to resource allocation and planning. It can help enterprises prepare for future conditions that are likely to impact program and support function operations and the demands for products and services, such as decreasing personnel or financial resources or changes in work load. Use of metrics can give organizations lead times for needed resource adjustments, if these conditions are known in advance.
- It provides periodic feedback to employees, customers and stakeholders about the quality, quantity, cost and timeliness of products and services.

Perhaps most important is that metrics build a common results language among all decision makers. Selected measures define what is important to an enterprise, what it holds itself accountable for, how it defines success and how it structures its improvement efforts.

In terms of defining what to measure, effective performance metrics typically concentrate on a few vital, meaningful indicators that are economical, quantitative and usable for the desired results.

If there are too many metrics, enterprises may become too intent on measurement and lose focus on improving results. A guiding principle is to measure that which matters most. Historically, IT has never lacked in the measurement area. In fact, many IT organizations usually measure far too many things that have little or no value.

## Metrics Defined in COBIT

In the context of GEIT, goals and metrics are defined by the COBIT framework at three levels (see **figure 2.5**):[19]
- Enterprise goals as metrics that define the organizational context and objectives and how to measure them
- IT-related goals and metrics that define what the business expects from IT and how to measure it
- Process goals and metrics that define what the IT-related process must deliver to support IT's objectives and how to measure it

In these three levels, it is important to make a distinction between outcome measures and performance drivers. Outcome measures indicate whether goals have been met. These can be measured only after the fact and, therefore, are sometimes called lag indicators. Using the COBIT framework, outcome measures inform management whether an IT function, process or activity has achieved its goals. The outcome measures of the IT functions are often expressed in terms of information criteria, as follows:
- Availability of information needed to support the business needs
- Absence of integrity and confidentiality risk
- Cost efficiency of processes and operations
- Confirmation of reliability, effectiveness and compliance

As a measure, indicators for performance achievement (called performance indicators or lead indicators) indicate whether goals are likely to be met. In contrast to outcome measures, performance indicators can be used for measurement before the outcomes are clear and, therefore, are sometimes called lead indicators. Using the COBIT framework, performance indicators define measures that determine how well the business, IT function or IT process is performing in enabling the goals to be reached. They are lead indicators of whether goals will likely be reached, thereby driving the higher-level goals. They often measure the availability of appropriate capabilities, practices and skills, and the outcome of underlying activities. For example, a service delivered by IT is a goal for IT, but a performance indicator and a capability for the business. This is why performance indicators are sometimes referred to as performance drivers, particularly in BSCs.

More information related to metrics and the balanced scorecard can be found in section 1.10 IT Governance Monitoring Processes/Mechanisms and section 2.6 Methods to Document and Communicate IT Strategic Planning Processes.

## Metrics Defined in ITIL

In the context of continuous service improvement (CSI) for IT, the ITIL framework[20] specifies three types of metrics that an enterprise will need to collect to support CSI activities as well as other process activities, as follows:
- **Technology metrics**—These are often associated with component and application-based metrics such as performance, availability, etc.
- **Process metrics**—These are captured in the form of critical success factors (CSFs), KPIs and activity metrics for the service management processes. Process metrics can help determine the overall health of a process. Four key questions that KPIs can help answer concern quality, performance, value and compliance in following the process. The implementation of CSI uses these metrics as input in identifying improvement opportunities for each process.
- **Service metrics**—These are the results of the end-to-end service. Component metrics are used to compute the service metrics.

ITIL defines a metric as a scale of measurement defined in terms of a standard, i.e., in terms of a well-defined unit. The quantification of an event through the process of measurement relies on the existence of an explicit or implicit metric, which is the standard to which measurements are referenced. ITIL specifies metrics as a system of parameters or ways to quantitatively assess a process that is to be measured, along with the processes to carry out the measurement. Metrics, therefore, define what is to be measured and are usually specialized by the subject area, in which case they are valid only within a certain domain and cannot be directly benchmarked or interpreted outside that domain. Generic metrics can, however, be aggregated across subject areas or business units of an enterprise.

## SMART Metrics

The acronym for specific, measurable, attainable, realistic and timely (SMART) is commonly used as a basis for goal setting; however, this applies as well to the preferred design and selection characteristics of measures and metrics. In the context of measures and metrics for IT performance measurement, the meaning of SMART is usually interpreted for a business expectation of IT, IT process or IT activity as follows:

- **Does the measure or metric provide any useful insight into a specific dimension?**—If it does not, then information provided by that measure or metric will not be valued.
- **Is the measure or metric measurable?**—It is fundamental that the measure or metric be measurable; otherwise, it will not facilitate comparison or trend analysis.
- **Is the measure or metric achievable or attainable?**—This relates to the measure or metric being used for target setting. Set it too low, and it does not provide a challenge and is not valued; set it too high, and it can fail to motivate because it is perpetually beyond reach. To be valued, the target measure or metric must be set realistically.
- **Is the measure or metric realistic?**—These are the criteria that determine whether the measure or metric provides useful information in the context within which it is used. Too often, IT metrics are provided that have little or no use to the business.
- **Is the measure or metric timely?**—This relates to the frequency and timeliness of the measure or metric. It should be reflective of the actual situation so that analysis and action can be undertaken.

In addition, measures or metrics should follow the keep-it-simple principle. Having complex and/or many measures or metrics tends to confuse and detract from incisive analysis and effective remedial action.

As an example, the metrics defined in the COBIT framework have been developed with the following characteristics in mind:
- **High insight-to-effort ratio**—The insight into performance and the achievement of goals as compared to the effort to capture them
- **Comparable internally**—Percent against a base or numbers over time
- **Comparable externally**—Irrespective of enterprise size or industry
- **Better to have a few good metrics than a longer list of lower-quality metrics**—Possibly even just one very good metric that could be influenced by different means
- **Easy to measure**—Not to be confused with targets

## 3.8 PROCEDURES TO MANAGE AND REPORT THE STATUS OF IT INVESTMENTS

The seeds of success or failure for IT-related investments are sown in the business case. Recent research from the University of Antwerp Management School (UAMS)[21] clearly positioned the entire portfolio management process, including business case development, as one of the key practices for achieving better alignment between business and IT and, by extension, business value. However, enterprises generally are not good at developing or using complete and comprehensive business cases. A 2006 Cranfield University School of Management study[22] found that 96 percent of respondents develop business cases for most investments involving IT. However:
- 69 percent were not satisfied with the business case development process.
- 68 percent were not satisfied with the identification and structuring of benefits.
- 81 percent were not satisfied with the evaluation and review of results.
- 38 percent admitted that benefits claims were exaggerated to get the business cases approved.

Understanding the relevance of business cases to manage and report on IT investments is of primary importance for all management levels across both the business and IT parts of the enterprise—from the CEO and the C-suite to those directly involved in and responsible for the selection, procurement, development, implementation, deployment and benefits realization processes.

IT investments represent a profound conundrum within many enterprises. There are no other investments within a company that occupy such a large and growing expenditure, yet lack disciplined management, processes and performance measurements. However, a majority of enterprises are aggressively scrutinizing the amount of investment allocated to IT in an effort to cut costs, achieve economies of scale and drive shareholder value to get more and do more with less. The primary focus on IT investments is on short-term projects and priorities with near-term benefits, delaying and, in many cases, eliminating long-term strategic investments that are crucial for the growth and transformation of the enterprise.

## Business Case Life Cycle

The full life cycle of the business case encompasses three fundamental aspects:
1. **Obtain complete, comprehensive and accurate information at an appropriate level of detail.** The business case should provide a complete and shared understanding of the expected business outcomes (intermediate lead and end lag outcomes) of an investment. It should describe the assumptions taken, how the business outcomes and the validity of the assumptions will be measured, and the full scope of initiatives required in achieving the expected outcomes. These initiatives should include any required changes to the nature of the enterprise's business model, business processes, people skills and competencies, enabling technology, and organizational structure required to achieve outcomes. This means not only investments in technology and business capital expenditures (capex), but also operational and technology expenditures (opex) for realizing and completing the business transformation—for example, cleaning up legacy systems, changing business processes via training on the job, adding new hires and implementing new processes.

   Key risk to the successful completion of individual initiatives (i.e., delivery risk) and the achievement of the desired outcomes (i.e., benefit risk) also need to be identified and documented, together with mitigation actions. The business case should contain all of the information needed for analyzing the strategic alignment and financial and nonfinancial benefits and risk of the investment and for determining its relative value. It should be derived within the context of best- and worst-case scenarios, when appropriate.

   At minimum, the business case should include the following:
   - **The reason for the investment**—The opportunity or problem that the investment is intended to address
   - **The recommended solution/approach**—Including alternatives considered and proposed timetable
   - **The business benefits targeted**—Their alignment with business strategy, how they will be measured and who in the business functions will be responsible for securing them
   - **The initial investment and ongoing costs**—Both the IT and business costs of operating in the changed way
   - **The business changes**—Needed to create and realize sustained additional value and the investments needed to make the changes
   - **The risk inherent in the approach**—Including delivery risk (the risk of not being able to deliver required capabilities) and benefit risk (the risk of the organization not being able to make and sustain the changes required to use the capabilities to create and sustain value)
   - **The governance approach for the investment**—How the investment and value creation will be monitored throughout the economic life cycle, the metrics to be used and who will be ultimately accountable for the successful creation of optimal value

   The business case should, as appropriate, include high-level summaries of and links to:
   - The detailed program plan (including individual project plans)
   - The resourcing plan
   - The financial plan (including cost-benefit analysis)
   - The benefits realization plan (including the benefits register)
   - The (organizational) change management plan
   - The risk management plan (including the risk register)

The information in the business case should be validated by appropriate plausibility checks. These should include the appraisal of the logic behind the claimed contribution of initiatives (and intermediate outcomes), the outcomes and associated benefits. The appraisal is best supported by empirical evidence (derived from evaluation of previous investments), especially concerning the logic of contribution and assumptions.

2. **Continually update the business case as (internal/external) events occur that (could) influence the business case.** Forecasting future effects of IT-enabled investments involves making assumptions about internal and external conditions. Even with the best of processes, a business case is still no more than a snapshot or best guess at a point in time. The business case process involves much more than providing that initial snapshot to determine whether to proceed with an investment. It involves updating the business case as circumstances change or more information is available so that the business case can be used as an operational tool to manage the investment.

Any investment should be managed by taking into account that there always is risk, and that risk should be monitored and controlled throughout the life cycle of the investment by an iterative process of risk identification, risk assessment and risk treatment. Personnel should be encouraged to anticipate, identify and report risk. Risk management and response plans should be maintained and ready for use. Reports on project risk monitoring should be part of progress evaluations and, therefore, updates of the business case.

The business case should be continually updated as the projected costs or benefits of the investments change, when risk changes, or in preparation for regular reviews.

3. **Use the business case as a management tool to support informed decision making throughout the full economic life cycle of the investment (decision).** Once an investment is approved, the business case is the primary tool to monitor and manage the delivery of the required capabilities and the desired outcomes through the full economic life cycle of the investment.

As an integrated part of the enterprise portfolio, the investment should be actively managed. If the investment is not performing as expected, or business requirements have changed, the approach or desired outcomes may need to be adjusted or the investment may have to be canceled. Postmortem analysis of all major investments must be done to learn from success and failure and to continuously improve the portfolio quality.

This ongoing management through the full economic life cycle is where many enterprises cut corners. In most cases, the program will be considered closed after completing the activities in the program plan and delivering the required business and IT capabilities. In general, the benefits and the expected value, as set in the business case, will not be realized until some later time, long after the delivery of IT and business capabilities. It is only then that the program and, subsequently, the business case will have proven that they delivered the expected benefits. While the timing of program and business case closure—which likely will not be the same—may vary in different organizations and for different types of investments, it is important to understand that the full economic life cycle of an investment decision includes the following, as illustrated in **figure 3.2**:
- **Investment phase**—Developing the necessary capabilities
- **Adoption phase**—Implementing the capabilities
- **Value creation phase**—Achieving the expected level of performance and moving the delivered capabilities into the active service portfolio
- **Value sustainment phase**—Assuring that the assets resulting from the investment continue to create value, which may well include additional investments required to sustain value
- **Retirement phase**—Decommissioning the resulting assets

In a recent study at the University of Antwerp—Antwerp Management School, the researchers concluded that many business cases are weakly developed or are not implemented in a timely manner after the investment has formally been approved. Looking at this issue from a process perspective, they developed a conceptual model of a business-case process that can facilitate continuous business-case use throughout the investment life cycle, as seen in **figure 3.15**.

### Figure 3.15—The Business Case Process Aligned With an Investment Life Cycle

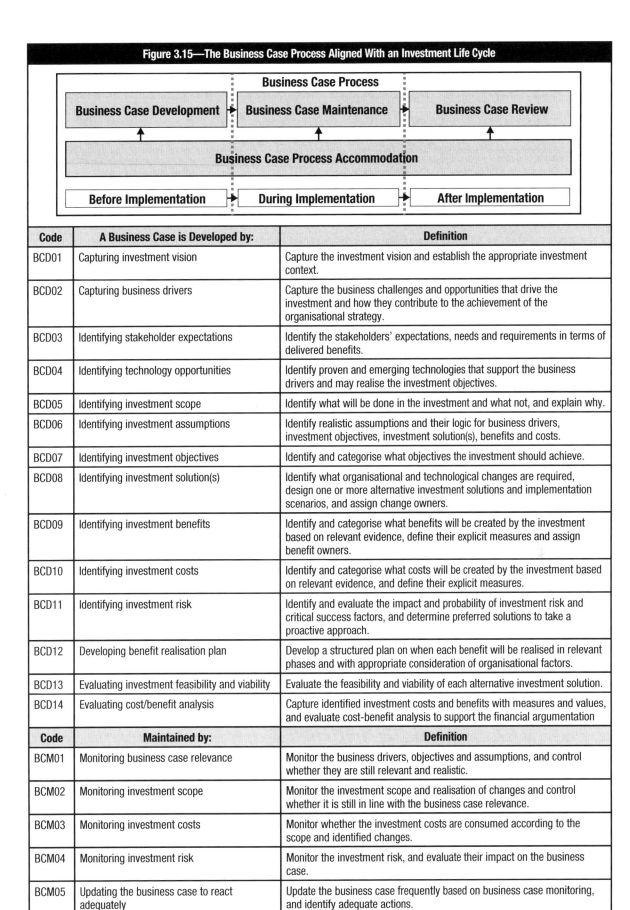

| Code | A Business Case is Developed by: | Definition |
|---|---|---|
| BCD01 | Capturing investment vision | Capture the investment vision and establish the appropriate investment context. |
| BCD02 | Capturing business drivers | Capture the business challenges and opportunities that drive the investment and how they contribute to the achievement of the organisational strategy. |
| BCD03 | Identifying stakeholder expectations | Identify the stakeholders' expectations, needs and requirements in terms of delivered benefits. |
| BCD04 | Identifying technology opportunities | Identify proven and emerging technologies that support the business drivers and may realise the investment objectives. |
| BCD05 | Identifying investment scope | Identify what will be done in the investment and what not, and explain why. |
| BCD06 | Identifying investment assumptions | Identify realistic assumptions and their logic for business drivers, investment objectives, investment solution(s), benefits and costs. |
| BCD07 | Identifying investment objectives | Identify and categorise what objectives the investment should achieve. |
| BCD08 | Identifying investment solution(s) | Identify what organisational and technological changes are required, design one or more alternative investment solutions and implementation scenarios, and assign change owners. |
| BCD09 | Identifying investment benefits | Identify and categorise what benefits will be created by the investment based on relevant evidence, define their explicit measures and assign benefit owners. |
| BCD10 | Identifying investment costs | Identify and categorise what costs will be created by the investment based on relevant evidence, and define their explicit measures. |
| BCD11 | Identifying investment risk | Identify and evaluate the impact and probability of investment risk and critical success factors, and determine preferred solutions to take a proactive approach. |
| BCD12 | Developing benefit realisation plan | Develop a structured plan on when each benefit will be realised in relevant phases and with appropriate consideration of organisational factors. |
| BCD13 | Evaluating investment feasibility and viability | Evaluate the feasibility and viability of each alternative investment solution. |
| BCD14 | Evaluating cost/benefit analysis | Capture identified investment costs and benefits with measures and values, and evaluate cost-benefit analysis to support the financial argumentation |
| **Code** | **Maintained by:** | **Definition** |
| BCM01 | Monitoring business case relevance | Monitor the business drivers, objectives and assumptions, and control whether they are still relevant and realistic. |
| BCM02 | Monitoring investment scope | Monitor the investment scope and realisation of changes and control whether it is still in line with the business case relevance. |
| BCM03 | Monitoring investment costs | Monitor whether the investment costs are consumed according to the scope and identified changes. |
| BCM04 | Monitoring investment risk | Monitor the investment risk, and evaluate their impact on the business case. |
| BCM05 | Updating the business case to react adequately | Update the business case frequently based on business case monitoring, and identify adequate actions. |

| \multicolumn{3}{c}{Figure 3.15—The Business Case Process Aligned With an Investment Life Cycle *(cont.)*} |
|---|---|---|
| Code | Maintained by: | Definition |
| BCR01 | Identifying objective evaluation criteria | Identify and communicate objective criteria with predefined weighting that helps to evaluate the investment effectiveness and efficiency. |
| BCR02 | Evaluating investment effectiveness | Monitor benefits realisation, and evaluate the contribution of investment objectives and changes. |
| BCR03 | Evaluating investment efficiency | Evaluate the effort and costs that were consumed to realise the investment. |
| Code | A Business Case Process Accommodated by: | Definition |
| BCPA01 | Establishing an adaptable business case approach | Establish an adaptable business case approach according to investment, and accept a growing maturation and granularity through its development and usage. |
| BCPA02 | Establishing business case templates, training and guidance | Establish standard business case templates and tools and accommodate training and guidance on what constitute business case practices and how to employ them adequately. |
| BCPA03 | Establishing maximum objectivity in business case usage | Maximise objectivity to support well-founded and comparable decision making without influence from politics, lobbying or institutional powers. |
| BCPA04 | Establishing simple and dynamic business case usage | Describe and employ business case practices and its content in a simple, straightforward and dynamic manner to encourage their usage. |
| BCPA05 | Establishing business case practices as a standard approach | Establish and evangelise business case practices as a standard way of working. |
| BCPA06 | Ensuring business case practice improvements | Ensure business case practice improvements further through experience and continuous learning. |
| BCPA07 | Ensuring communication and involvement with stakeholders | Ensure clear communication and active involvement with all stakeholders in order to gain insight, commitment and ownership. |
| BCPA08 | Ensuring stakeholder confirmation | Ensure formal confirmation from relevant stakeholders on the (updated) business case to increase their commitment. |
| BCPA09 | Evaluating business cases regularly | Evaluate all business case documents in order to make well-founded decisions to approve, continue or stop the investment. |

Source: Maes, Kim; Steven De Haes; Wim Van Grembergen; "The Business Case as an Operational Management Instrument—A Process View," *ISACA Journal*, vol. 4, 2014

## 3.9 COST OPTIMIZATION STRATEGIES

Cost optimization is typically a constant focus in today's business environment when expenditures on IT-related activities and resources have escalated dramatically. Additional factors, such as the drive to enhance efficiency, responding to competition or facing an economic downturn may cause additional attention to cost optimization strategies. In the context of value delivery, cost optimization is necessary to ensure that IT services enable the business to create the required business value using resources (people, applications, infrastructure and information) to deliver the appropriate capabilities at the optimal cost.

According to Gartner, there are four discrete levels at which cost optimization can be exercised,[23] categorized in two lower levels (IT procurement and cost savings within IT) and two upper levels (joint business/IT cost savings and enable innovation/business restructuring). As stated by Gartner:[24] "The two lower levels are focused on the reduction of cost within IT: the two upper levels involve IT and the business teaming up to reduce operating costs." The broad definitions of the four discrete levels are as follows, as defined by Gartner,[25] listed from lowest to highest:
- **IT procurement**—"True partnerships with IT vendors mean that each party benefits in the good times and makes joint sacrifices in times of economic uncertainty. Each year, IT organizations spend billions of dollars for hardware, software, IT services and telecommunications services. The manner in which IT organizations approach procurement issues will affect how much they can reduce spending to meet business goals. In the IT Procurement level of the framework, we provide guidance on getting the best pricing for your IT purchases, best practices for contract negotiations and renegotiation, terms and conditions, selection of service providers, alternative delivery and acquisition models, and other issues related to procurement of IT input."

- **Cost savings within IT**—"A priority for many IT organizations will be to identify opportunities to reduce baseline IT costs, not just move them to another budget center. Where IT organizations focus is where they will be successful with costs savings. In the Cost Savings Within IT level of the framework, we provide specific cost-saving opportunities by technology and service area, ranges or specific percentages of savings that are typical, things you have to do to achieve these savings, alternatives to the current status quo, and pros and cons related to possible cost-saving actions."
- **Joint business and IT cost savings**—"Consider that the average IT budget is roughly 3% of revenue, while total operating expenses are 80% to 90% of revenue. If your enterprise is looking to reduce costs in 2009, join IT with the business to reduce costs in business operating expenses. In the Joint Business and IT Cost Savings level of the framework, we provide specific guidance on implementing cost-saving technologies in conjunction with the business that may include investments to reduce cost long term, or alternatives to current technologies. We also explore energy-saving strategies with technology, modernization, total cost of ownership, and alternative delivery and acquisition models."
- **Enable innovation and business restructuring**—"As economic uncertainty passes, cost optimization will refocus on efforts to implement long-term process improvement and enable business structuring and innovation. Research at this level of the cost-optimization framework will focus on IT and business process improvement to reposition the business to competitive advantage using IT or, in some cases, help the business "find a business model" that enables it to survive the economic downturn."

Gartner recommends that "IT leaders should use Gartner's Four Levels of Cost Optimization framework as an organizing structure in which to track and balance cost optimization programs and communicate the impact of cost optimization to the business."

In terms of approaches to cost optimization, a business plan for cost optimization should formulated and executed jointly with the business. On the IT side, specific approaches might include:
- **Use of best-practice methodologies, frameworks and standards**—Especially the ITIL/ISO 20000 processes of capacity management and configuration management
- **Use of benchmarking to compare against other IT organizations**—Can lead to further and significant cost savings compared to levels without benchmarking
- **Re-evaluation/renegotiation of IT outsourcing**—To develop transition scenarios toward alternative delivery models such as managed services and Software as a Service (SaaS). The rationale for considering outsourcing can go beyond cost-optimization. The benefits include service quality improvements, scalability, better risk management and freeing internal resources to focus on core, value-adding activities. It is no longer an enterprise's ownership of capabilities that matters, but rather its ability to control and make the most of critical capabilities, whether or not they reside on the enterprise's balance sheet.

More information on related matter of outsourcing can be found in section 5.4 Outsourcing and Offshoring Approaches That May Be Employed to Meet the Investment Program and Operational Level Agreements and Service Level Agreements.

## 3.10 MODELS AND METHODS TO ESTABLISH ACCOUNTABILITY OVER IT INVESTMENTS

In academic and professional literature, the term "IT governance" is more commonly used than "governance of enterprise IT." However, due to the focus on "IT" in the naming of the concept, most discussion around, and implementation of, GEIT occurs within the IT area. Yet, it is clear that business value from IT investments cannot be realized by the IT function, but will always be created by the business through its use of IT. For example, there will be no business value created when IT delivers a new customer relationship management (CRM) application on time, on budget and to specification, if the business has not made the necessary changes to the business model, business processes, organizational structure, people competencies, and the reward system required to effectively integrate the new IT system into its business operations. IT-enabled investments should therefore always be treated as business programs, composed of a collection of business and IT projects delivering all the capabilities required to create and sustain business value.[26]

This discussion clarifies the need for the business to take ownership of, and be accountable for, governing the use of IT in creating value from IT-enabled business investments. It also implies a crucial shift in the minds of the business and IT, moving away from managing IT as a "cost" toward managing IT as an "asset" to create business value. As Weill and Ross describe in their 2009 book titled *IT Savvy: What Top Executives Must Know to Go from Pain to Gain*:[27] "If senior managers do not accept accountability for IT, the company will inevitable throw its IT money to multiple tactical initiatives with no clear impact on the organizational capabilities. IT becomes a liability instead of a strategic asset."

A common and critical dilemma confronting enterprises today is how to ensure that they realize value from their large-scale investments in IT and IT-enabled change. IT-enabled investments can bring huge rewards, but only with the right value management approaches and appropriate established accountabilities. ISACA's Val IT framework offers a broad set of good practices supporting the adoption of such value management processes and approaches. This section describes how the Dutch airline company KLM introduced value management for their IT-enabled investments, analyzed through the lens of Val IT. As such, this section will provide insight to practitioners regarding how to introduce better value management approaches and accountabilities based on Val IT.

### Val IT as a Framework for GEIT and Value Management

A recent important framework addressing GEIT, with a specific focus on value management and creation, is ISACA's Val IT framework. This framework starts from the premise that value creation out of IT investments is a business responsibility in the first place. To support business people in organizing and developing these responsibilities, Val IT presents a set of 22 IT-related business processes and associated key management practices, management guidelines and maturity models, as shown in **figure 3.16**.

Val IT presents 22 processes categorized in three domains: value governance (VG), portfolio management (PM) and investment management (IM). Value governance addresses the structures and processes required to ensure that value management practices are embedded in the organization. Value governance deals with the engagement of the leadership (VG1), the definition and implementation of value management practices (VG2) and the integration of the latter into the organization's financial management processes (VG4). It also addresses the fact that portfolio types and criteria need to be defined by the business (VG3), that effective governance monitoring should be established over the value management practices (VG5) and that there should be a continuous improvement cycle through implementing lessons learned (VG6). It is clear that these processes are defined at a higher level in Val IT and encompass necessary conditions to enable a value-based approach in portfolio and investment management.

Portfolio management addresses the processes required to manage the entire portfolio of IT-enabled investments. This domain starts with stating that the strategic direction of the organization should be clarified and that the target portfolio mix should be defined (PM1). Also, the available resources in terms of funding (PM2) and human resources (PM3) need to be inventoried. Based on detailed business cases arising from the investment management processes (processes IM1–IM5), investment programs are selected and moved into the active portfolio (PM4). The performance of this active portfolio needs to be continuously monitored, reported on (PM5) and optimized (PM6), based on performance reports coming out of the investment management processes.

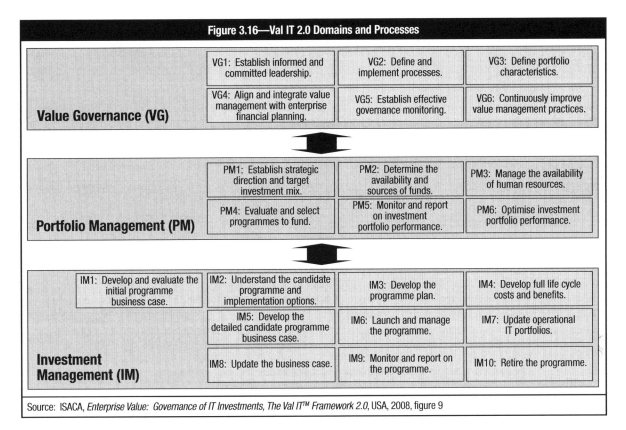

Source: ISACA, *Enterprise Value: Governance of IT Investments, The Val IT™ Framework 2.0*, USA, 2008, figure 9

The processes in the IM domain are situated at the level of one single IT-enabled investment. The first five processes in investment management are about the emergence of new investment opportunities in the organization (IM1) and the development of detailed business cases (IM5) for the approved opportunities, including analyses of alternative courses (IM2) of action, definition of a detailed program plan (IM3) and full cost-benefit analysis (IM4). After approval of detailed business cases (PM4), investment programs need to be launched (IM6), monitored (IM8) and, if required, business cases need to be updated (IM9). All investment programs need to be retired (IM10), bringing programs to an orderly closure when there is agreement that the desired business value has been achieved or when it is clear it will not be achieved. Also, changes to operational IT portfolios as a result of the investment program need to be incorporated in the portfolios of IT services, assets or resources (IM7).

More information on Val IT can be found in section 3.12 Business Case Development and Evaluation Techniques.

### Case Study: Analyzing IT Value Management at the Dutch Airline Company KLM Through the Lens of Val IT

Although KLM did not specifically use Val IT to introduce value management, the company was involved in the development of Val IT version 2.0 and, as a result, there was some knowledge sharing in both directions between KLM and the Val IT development team. In this section, we analyze KLM's value management approaches through the lens of Val IT. First, the case company and context is briefly discussed, followed by a more detailed discussion on their value governance, portfolio management and investment management processes. Each time, a mapping is made to the Val IT processes. The focus of the case description is on understanding how a real-life organization implemented the intent of these good practices.[28]

### The Case Company: KLM

The airline company KLM was founded in 1919 and has its home base and hub in Amsterdam Schiphol Airport (Netherlands). KLM currently employs over 33,000 people worldwide and manages a fleet of about 200 aircraft. In 2004, KLM merged with Air France, after which both companies continued to operate as separate airlines, each with their own identity and brand and each benefiting from each other strengths. In terms of financial turnover after the merger, Air France-KLM is the world's largest airline group, transports the most passengers and is the world's second-largest cargo transporter. In 2009, Air France-KLM operated flights to 255 destinations in 115 countries on four continents.

This case focuses on the KLM activities within the Air France-KLM group. The KLM Executive Committee is composed of the CEO, chief financial officer (CFO), managing director and the executive vice presidents (EVPs) of the major business units and services (commercial, in-flight services, operations, ground services, cargo, engineering and maintenance, IT and human resources [HR]). In 2009-2010, KLM IT employed close to 1,000 (internal and external) full-time employees (FTEs), with an IT budget of around 300 million. As shown in **figure 3.17**, KLM IT is organized around IT development activities, IT operations activities and the CIO office, addressing aspects of the enterprise/IT architecture, IT strategy, value and portfolio management, sourcing strategy, and risk and security. The mission of the IT department is to "create business value by delivering reliable IT services to the business processes, and innovative IT solutions to enable and support business changes." The following strategic goals for IT support this mission:
- IT is a world class information services provider and will be able to deliver the best value to the company.
- The IT cost levels will be at a competitive industry level.
- The IT architecture and infrastructure will enable the growth ambitions of Air France-KLM.

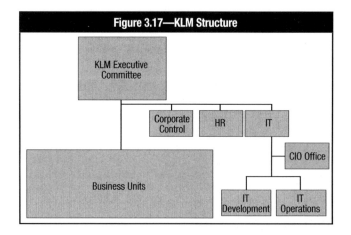

IT is a business-critical enabler for KLM; however, at the same time, it can be a source of both success and discontent. In 2001, the balance had tilted toward discontent due to a lack of trust in what was perceived as a very costly and unresponsive IT department. This occurred in a business climate that was increasingly challenging, and which became dramatically more so after the 9/11 terrorist attacks. After that event, KLM's CEO seized the opportunity to make a structural break with the past, and reexamine and transform KLM's business and GEIT.

The EVP of the operations control center was appointed as the new CIO. It was felt that having the CIO coming out of the "real business" would help in getting the GEIT discussion out of the IT area and have it put on the business executive's agenda. The newly appointed CIO received three clear priorities:
- Provide the reasons why, or why not, to outsource IT.
- Create a business/IT board to organize joint success.
- Design simple governance principles to restore control enabling steering by the EVPs and the CIO.

To respond to these requirements, the CIO office was established as a support function to the CIO, consolidating a number of already existing, loosely coupled and different functions, such as an IT strategy office, program management, and business/IT liaison roles. In the words of the vice president (VP) of the CIO office:

> *In the scenario that we would outsource IT, both IT operations and development would mainly be sourced outside KLM, but the activities of the of CIO office would be kept internally, as it governs IT strategy, architecture, security, business/IT alignment, etc. The goal of the CIO office is to enable effective IT, in support of business needs.*

### Value Governance at KLM

It was decided that, ahead of the first priority stated above, the primary focus should be to introduce better governance principles and practices (priority 3). A project titled "IT: A Collaborative Effort" was launched, focused on enabling all stakeholders to better understand the cost and value of IT, which in turn would enable them to make more informed decisions on what and how to potentially outsource (priority 1). In support of priority 2, a business/IT board was established, composed of the CEO, CIO and all business unit EVPs, meeting every quarter to discuss and decide on strategic issues involving IT.

With regard to priority 3, the CIO office, in collaboration with the business, designed a set of principles that would significantly simplify IT-related governance. The starting premise was that these principles should put the business in full control of all IT demand and IT spending. In support of these principles, a number of governance practices were introduced in the business and IT organizations, including the establishment of the business/IT board and demand management functions for each business domain. These governance principles and practices were introduced as "the only way of working" between business and IT for all business units and activities. These practices also supported the creation of portfolio management processes driven by the business units. The portfolio management processes evolved from being IT resource- and supply-driven toward business demand-driven with an innovative and rigorous approach to evaluation and selection.

The definition of the first draft set of governance principles and practices was mainly driven by the CIO office. These principles were later refined with the involved business parties and are now shared in the organization through the intranet. According to the director of value management and alliances (member of the CIO office):

> *These principles and practices are still challenged from time to time. Our position is that we are always open for discussion for each of these principles and practices, but up till now, we have each time in the end reconfirmed them.*

The stated principles and practices apply for all business units and are presented in internal KLM presentations as shown in **figure 3.18**. The involved parties acknowledge that this list does not really distinguish between principles and practices and presents them in a mixed way, but it was felt to be a pragmatic and practical list that was workable for KLM. The CIO office developed more detailed background information and internal documentation to explain the impact and consequences of each of these principles and practices.

**Figure 3.18—Governance Principles and Practices**

1. For the business, there should be no difference between working with an internal or external IT provider.
2. Differentiate between *what* and *how* (and *why*).
3. Improve the demand function by creating a business demand office per business domain.
4. Improve the supply function by creating an innovation organizer and a service manager per business domain.
5. Create monthly decision meetings of *what* and *how* (management and IT).
6. Focus on the costs that can be influenced in full and those that can be influenced in part: split between innovation and continuity.
7. Each innovation (investment) has one business owner to whom all costs are charged.
8. Each service (continuity) has one business owner to whom all costs are charged.
9. Create a top-down budget framework and simplified budget process.
10. Activity-based costing is applied to process primary cost to product cost.

Referring back to Val IT, the goal of the Val IT VG domain is to ensure that value management practices are embedded in the enterprise, enabling it to secure optimal value from its IT-enabled investments. Val IT proposes six processes in this domain, as shown in **figure 3.19**. Mapping these processes to the above description makes clear that the adoption of some of these processes is very nicely illustrated at KLM. KLM's definition of the governance practices and principles (**figure 3.18**) ensures informed and committed leadership (VG1), appropriate governance monitoring (VG5) and the implementation of value management processes (VG2). Also, some of these principles address specific issues, such as VG4 being covered in principles 9 and 10.

| Figure 3.19—Illustration of VAL IT Processes at KLM | |
|---|---|
| Val IT Management Processes | Illustrated at KLM |
| **Value Governance** | |
| VG1 Establish informed and committed leadership. | X |
| VG2 Define and implement processes. | X |
| VG3 Define portfolio characteristics. | |
| VG4 Align and integrate value management with enterprise financial planning. | X |
| VG5 Establish effective governance monitoring. | X |
| VG6 Continuously improve value management practices. | |

### Portfolio and Investment Management at KLM

The previous governance principles and practices were needed as key building blocks in support of having effective portfolio and investment management processes driven by the business units. The design of these portfolio and investment management processes was done by the portfolio management office (part of the CIO office) and is shown in **figure 3.20**. Three approval stages are defined, going from "idea selection" to "program go" and "investment approval." For each of these phases, clear decision thresholds were defined. For investments between €150,000 and €500,000, the EVP, director of finance and control and business development office (BDO) of a business unit can approve the go/no-go decision in each phase; investments above €500,000 are approved by the business unit investment committee (BIC), comprising the business unit COO, EVP, director of finance and control and BDO; and investments above €5,000,000 are approved by the executive committee (EC).

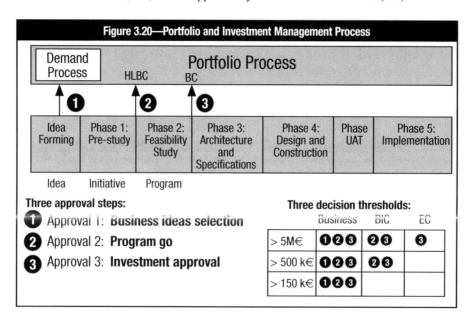

The initial phase addresses the initiation of the investment proposals or idea generation. In this phase, all business ideas are gathered and captured by the BDOs (demand process) and turned into potential initiatives for which a high-level business case (HLBC) will be developed. These HLBCs include descriptive information, classifications and high-level cost and benefits estimates and risk. The VP of BDO for passenger operations clarifies:

> *It is often hard to quantify some benefits at this stage. For example, the cost avoided of an aircraft not needing to land on another location because of better support systems. But still, we try to make as good as possible educated estimations.*

If an initiative is approved, it is turned into a program for which a full business case is developed based on a detailed feasibility study. To enable common and comparable business cases, a business case template was developed as a mandatory instrument for all investments above €150,000.

In order to be able to prioritize all of the business cases, it is crucial to know what the organization's business drivers are. The director of value management and alliances makes clear:

> *Our experience was that it was often difficult to obtain a clear list of business priorities from a business unit. However, we needed these priorities to enable the selection of 'the right things' and for that reason we used a methodology to help us and the business in making these business priorities transparent.*

To enable this process, the business drivers of a business unit are captured and ranked by the CIO office through interviews with the business unit executives. Next, for each incoming investment proposal, the contribution to each of these ranked business drivers is determined, ranging from "low" toward "extreme." The result of this exercise is an initial portfolio containing a ranked, but still unconstrained, list of all investment proposals at business unit level. The VP of BDO for passenger operations explains the importance of this process:

> *These priorities are the basis to build a 'business plan' for the BDO of a specific business unit, describing all the things that the BDO office of a business unit can be held accountable for. I have even turned this business plan into a video clip on YouTube, to demonstrate to all our business and IT stakeholders our commitment for the next year.*

After this prioritization, total demand of all business units typically exceeds the budget made available by the executive committee. The director of value management and alliances describes how this is handled:

> *Instead of using a 'cheese slicer' and, for example, forcing all business units to cut 30% out of the project portfolio, a process of informal discussions is initiated between the BDOs to determine how the portfolio can best be optimized. As long as this process works, this approach is preferred instead of escalating to the next management level.*

This consensus-building process generally works well, and as a result, the business/IT board receives an overview of the major programs and only needs to endorse the outcome of the portfolio management process. The director of value management and alliances concludes: "Through a good portfolio management process, we strive for seamless decision making."

Once the portfolio of programs is optimized, the business investment committee (for projects above €500,000) or executive committee (for projects above €5,000,000) still has to release the funding before design, construction, user acceptance testing (UAT) and implementation can start. This might appear as a duplicated decision structure, but it acts as a final check, and it also gives the final authority and decision power back to the business executives. The VP of BDO for passenger operations explains:

> *In the end, the business executives decide. This approach helped in getting them engaged in the portfolio management process because they get their control back, although until now they have never 'used' it. Another important aspect in this context is that we try is to make the time between the business idea and approval on the investment committee as short as possible, as this period is perceived as IT being slow.*

Referring back to Val IT, the goals of the Val IT portfolio and investment management domains are, respectively, to ensure that optimal value is secured by the enterprise across its investment portfolio and to ensure that individual investments contribute to optimal value. The above description illustrates the adoption of some of the processes that Val IT proposes in these areas. The way the business drivers are defined for a business unit and how this leads to a prioritized list of programs in line with the available budget, clearly illustrates PM1, PM2 and PM4 (**figure 3.21**). The business case development process is also displayed in the previous section, covering IM1 through IM6.

| Figure 3.21—Illustration of processes at KLM | |
|---|---|
| **Val IT 2 Management Processes** | **Illustrated at KLM** |
| **Portfolio Management** | |
| PM1 Establish strategic direction and target investment mix. | X |
| PM2 Determine the availability and sources of funds. | X |
| PM3 Manage the availability of human resources. | |
| PM4 Evaluate and select programmes to fund. | X |
| PM5 Monitor and report on investment portfolio performance. | |
| PM6 Optimise investment portfolio performance. | |
| **Investment Management** | |
| IM1 Develop and evaluate the initial programme concept business case. | X |
| IM2 Understand the candidate programme and implementation options. | X |
| IM3 Develop the programme plan. | X |
| IM4 Develop full life-cycle costs and benefits. | X |
| IM5 Develop the detailed candidate programme business case. | X |
| IM6 Launch and manage the programme. | X |
| IM7 Update operational IT portfolios. | |
| IM8 Update the business case. | |
| IM9 Monitor and report on the programme. | |
| IM10 Retire the programme. | |

### Reported Benefits, Lessons Learned and Future Challenges

During the onsite interviews, the following benefits, lessons learned and future challenges were reported. In terms of benefits, the implementation and ongoing assurance of GEIT has restored trust between business and IT and resulted in an increased alignment of investment to strategic goals. The communication and discussions on portfolio management have also improved management awareness and understanding and supported the transformation from cost toward a value culture. Also, more tangible benefits were reported, including lowered IT continuity cost per business production unit, and the freeing up of funds for innovation (**figure 3.22**).

In the course of the journey so far, a number of lessons have been learned. These lessons include the importance of senior management commitment and business engagement; managing change; providing adequate and appropriate support resources; and taking a pragmatic, practical and evolutionary approach.

KLM still has challenges ahead in further maturing the GEIT and value management. These challenges include a better process for measuring and managing the benefits realization, a continuous alignment of required business and IT resources and consolidation of the whole investment portfolio at the group level.[29]

More on accountability in general can be found in section 1.9.

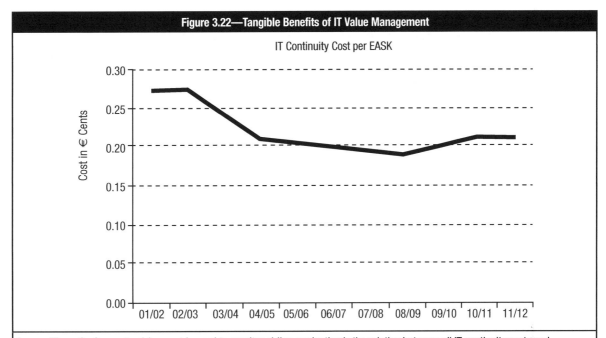

**Figure 3.22—Tangible Benefits of IT Value Management**

**Lower IT continuity cost**—A key metric used to monitor airline production is the relation between all IT continuity costs and Equivalent Available Seat Kilometers (EASK), which represents the total number of seats and cargo capacity multiplied by the total number of kilometers flown by the airline fleet. The graph shows that although many business investments involving IT, such as e-Tickets, more web-based sales and web-based check-in, resulted in a year-on-year increase in the total IT budget, the unit cost of providing IT services (IT continuity cost) per airline production unit decreased by more than 20 percent. (The slight upward curve for the next 3 years is due to a temporary decrease of production in response to the world economic crisis.) This substitution of labor by IT also resulted in lower business cost per unit, because IT is cheaper than labor.

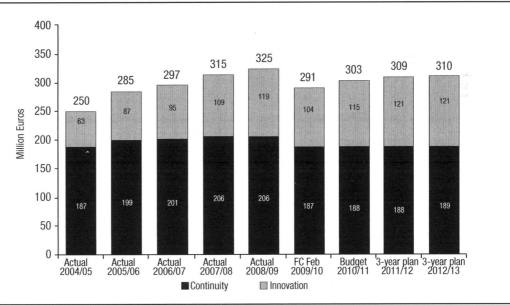

**Increased innovation capacity**—In addition to direct cost savings, the innovation capacity has increased as lower, or at least stable, IT continuity costs contributed to freeing up financials for IT-based innovation. Again here, the CIO office develops metrics to demonstrate this outcome, of which one example is shown above. This bar chart shows a relatively stable IT continuity budget, enabling the increase of the total IT budget to go almost entirely to new innovation, which has increased from 25 percent in 2004-2005 to 39 percent in 2010-2011.

## 3.11 VALUE DELIVERY FRAMEWORKS

The lack of value governance is the main cause of waste in IT investments. Surveys have consistently revealed that 20 to 70 percent of large-scale investments in IT-enabled business change are wasted, challenged or fail to bring a return to the enterprise. Some references that provide validation of this are:
- A 2004 IBM survey of *Fortune* 1000 CIOs found that, on average, CIOs believe that 40 percent of all IT spending brought no return to their enterprise.[30]
- A 2011 study conducted by The Standish Group found that only 37 percent of all IT projects succeeded while the remainder (63 percent) were either challenged or failed.[31]

ISACA's global surveys over the past decade[32] generally support these findings as well as a study by Barua et al.,[33] which states that "spending on IT continues to show long-term growth throughout the economy, reflecting an apparent belief in the economic benefits of IT. However, we also see organizations struggle in practice to demonstrate such benefits." This phenomenon is often called the "IT black hole": Large sums go in, but no returns come out.[34]

As one survey measuring costs and value shows, it is obvious that ineffective or immature value governance practices are responsible for the poor showing that, in many enterprises, less than eight percent of the IT budget is actually spent on initiatives that create value for the enterprise.[35]

Value delivery frameworks, such as Val IT, can help organizations to find the appropriate guidance in implementing their value governance practices and processes.

### Val IT Principles

The Val IT framework[36] consists of a set of guiding principles and a number of processes conforming to these principles, which are further defined as a suite of key management practices. A summary of these principles is as follows:
- IT-enabled investments will:
  - Be managed as a portfolio of investments
  - Include the full scope of activities required to achieve business value
  - Be managed through their full economic life cycle
- Value delivery practices will:
  - Recognize that there are different categories of investments that will be evaluated and managed differently
  - Define and monitor key metrics and respond quickly to any changes or deviations
  - Engage all stakeholders and assign appropriate accountability for the delivery of capabilities and the realization of business benefits
  - Be continually monitored, evaluated and improved

A brief explanation of each of the above principles follows:
- **IT-enabled investments will be managed as a portfolio of investments**—Optimizing investments requires the ability to evaluate and compare investments, objectively select those with the highest potential to create value, and manage all the investments to maximize value.
- **IT-enabled investments will include the full scope of activities required to achieve business value**—Realizing value from IT-enabled investments requires more than delivering IT solutions and services. It also requires changes to some or all of the following: the nature of the business itself; business processes, skills and competencies; and organization, all of which must be included in the business case for the investment.
- **IT-enabled investments will be managed through their full economic life cycle**—Business cases must be kept current from the initiation of an investment until any resulting service is retired. This principle recognizes that there will always be some degree of uncertainty and that variability over time in costs, risk, benefits, strategy and organizational and external changes must be taken into account in determining whether funding should be continued, increased, decreased or stopped.

- **Value delivery practices will recognize that there are different categories of investments that will be evaluated and managed differently**—Such categories might be based on management discretion, magnitude of costs, types of risk, importance of benefits (e.g., achievement of regulatory compliance), types and extent of business change.
- **Value delivery practices will define and monitor key metrics and respond quickly to any changes or deviations**—Metrics must be established and regularly monitored for the performance of: the overall portfolio; individual investments, including intermediate (or lead) metrics and end (or lag) metrics; IT services; IT assets; and other resources resulting from an investment, to ensure that value is created and continues to be created throughout the investment life cycle.
- **Value delivery practices will engage all stakeholders and assign appropriate accountability for the delivery of capabilities and the realization of business benefits**—Both the IT function and the other parts of the business must be engaged and accountable—the IT function for IT capabilities and the business for the business capabilities required to realize value.
- **Value delivery practices will be continually monitored, evaluated and improved**—As enterprises gain experience with Val IT practices, learning can be applied so that the selection of investments and the management of them improve each year.

## Val IT Domains

To fulfill the Val IT value management goal of enabling the enterprise to realize optimal value at an affordable cost with an acceptable level of risk from IT-enabled investments, the Val IT principles are applied within three domains:
- Value governance
- Portfolio management
- Investment management

In pursuing governance maturity, a process perspective has typically proven to be effective.

> **Note:** Processes are a collection of interacting activities undertaken in accordance with management practices. Processes take input from one or more sources (including other processes), manipulate the input, utilize resources according to the policies, and produce output (including output to other processes). Processes should have clear business reasons for existing, accountable owners, clear roles and responsibilities around the execution of each process, and the means to undertake and measure performance.

The three domain areas as defined by the Val IT framework each comprise a number of processes, and within each process are various management practices that enable the respective processes (**figure 3.16**). The relationship between Val IT principles, processes and practices is shown in **figure 3.23**.

**Figure 3.23—Relationship Among Val IT Principles, Processes and Practices**

**Val IT supports the enterprise goal of**
creating optimal value from IT-enabled investments at an affordable cost, with an acceptable level of risk
  **and is guided by**
    a set of principles applied in value management processes
    **that are enabled by**
      key management practices
      **and are measured by**
        performance against goals and metrics

Source: ISACA, *Enterprise Value: Governance of IT Investments, The Val IT™ Framework 2.0*, USA, 2008, figure 5

**Figure 3.24** shows the interrelationships and process flows of the various process blocks within and between each of the domain areas as defined in the Val IT framework. In interpreting the previous figures, it should be noted that, although by necessity the domains and processes are presented in a sequence, it does not imply that each follows from its predecessor. While there is some logic to the sequence, many of the processes and the key management practices within them will and should be followed both in parallel and iteratively. Depending on the nature, scope, size and impact of an investment, certain processes may be repeated a number of times with a stage-gate review after each iteration.

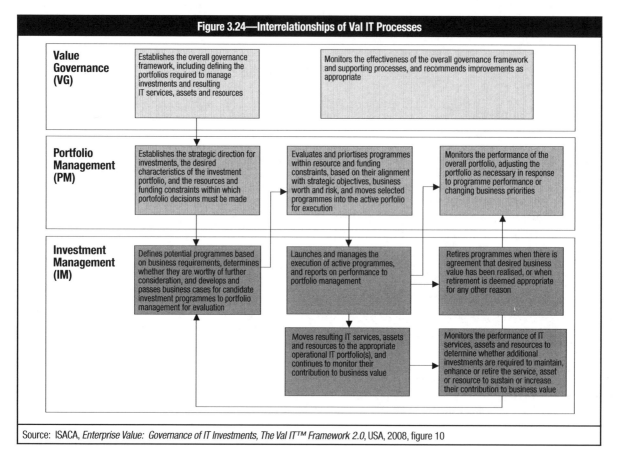

## Val IT Terminology

Realizing value from business change requires effective communication—a critical requirement that is difficult to achieve without the widespread acceptance of a consistent terminology. In many cases, various parts of an enterprise tend to adopt different meanings for key words and, in some situations, fail entirely to ascribe any meaning to other important terms and concepts. For consistency, a number of important terms related to value are defined in the Val IT framework. While an enterprise may choose to use different terms, or embrace different meanings, it is important for clarity to understand how the terms are defined and used by the framework:

- **Project**—A structured set of activities concerned with delivering a defined capability (that is necessary, but not sufficient, to achieve a required business outcome) to the enterprise. It is based on an agreed-on schedule and budget.
- **Program**—A structured grouping of interdependent projects that are both necessary and sufficient to achieve a desired business outcome and create value. These projects could involve, but are not limited to, changes in the nature of the business, business processes, the work performed by people and the competencies required to carry out the work, enabling technology and organizational structure. The investment program is the primary unit of investment in the Val IT framework.
- **Portfolio**—Groupings of objects of interest (investment programs, IT services, IT projects, other IT assets or resources) managed and monitored to optimize business value. The investment portfolio is of primary interest in the Val IT framework.

What has been missing for many years is ready access to a structured approach—a comprehensive, proven, practice-based structured governance framework that can provide boards and executive management teams with practical guidance in making IT investment decisions and using IT to create enterprise value. This is now provided by Val IT.

## Val IT Processes and Key Management Practices

The Val IT framework prescribes processes and key management practices for effective value governance, as shown in **figure 3.25**:

- **Establish informed and committed leadership.**—Done with a leadership forum and an effective CIO reporting line commensurate with the importance of IT to the enterprise. It is necessary to develop a sound understanding of the key elements of governance and clear insights into the enterprise strategy for IT, and to ensure alignment and integration of business and IT. The supporting management practices for this are:
  - Develop an understanding of the significance of IT and the role of governance.—All executives should have a sound understanding of strategic IT issues, such as dependence on IT, the other business functions and the executives regarding the actual and potential significance of IT for the enterprise's strategy.
  - Establish effective reporting lines.—Allows the CIO to engage the enterprise leadership as the advocate of the significance of IT for the enterprise
  - Establish a leadership forum.—To help the leadership understand and regularly discuss the opportunities that could arise from business change enabled by current, new or emerging technologies, and to understand their responsibilities in optimizing the value created from those opportunities
  - Define value for the enterprise.—So that there is a clear and shared understanding of what constitutes value for the enterprise and to ensure that it is communicated throughout the enterprise
  - Ensure alignment and integration of business and IT strategies with key business goals.—So that business and IT strategies are integrated—clearly linking enterprise, business and IT goals—and broadly communicated
- **Define and implement processes.**—Involves defining a governance framework for IT value management, including the supporting processes. This is done by assessing the quality and coverage of current processes to define the requirements of future processes so they provide necessary control and oversight, and enable active linkage among strategy, portfolios, programs and projects. Also included is the establishment of necessary organizational structure and implementation of the processes with the associated roles, responsibilities and accountabilities. The supporting management practices for this are:
  - Define the value governance framework.—Establishes an appropriate governance framework that is consistent with the overall enterprise governance environment and generally accepted governance and control principles
  - Assess the quality and coverage of current processes.—Done against the governance framework
  - Identify and prioritize process requirements.—Processes should include planning and prioritizing current and future work within the overall financial planning and human (business and IT) and other resource planning, securing and allocating funds and resources consistent with the priorities, stage-gating of investment programs, monitoring and communicating performance, taking appropriate remedial action, and realizing benefits so optimal value is secured from the investment portfolio and from all IT services, assets and resources.
  - Define and document the processes.—To include goals and metrics. It is required to define, implement and consistently follow processes that provide for clear and active linkage among the enterprise strategy; the portfolios of investment programs; the IT services, assets and resources that execute the strategy; the individual investment programs; and the business and IT projects that make up the programs.
    · Establish, implement and communicate roles, responsibilities and accountabilities.—Done for all personnel in the enterprise in relation to the portfolios of business investment programs, individual investment programs, and IT services, assets and resources, to allow sufficient authority to exercise the roles and responsibilities assigned
    · Establish organizational structures.—Typically involves establishment of appropriate boards, committees and support structures including, but not limited to, one or more investment and services boards (ISBs), an IT strategy committee, an IT planning or steering committee, and an IT architecture board

- **Define portfolio characteristics.**—Involves defining the different types of portfolios. It is necessary to define the categories within the portfolios, including their relative weight. It also requires the development and communication of how these categories will be evaluated in a comparable and transparent manner, and the definition of requirements for stage-gates and other reviews for each category. The supporting management practices for this are:
  – Define portfolio types.—Recognizes that the enterprise has a number of different types of portfolios about which decisions need to be made. Each type of portfolio should be managed and funded according to what it contains. (Val IT is primarily concerned with the portfolio of IT-enabled investments, while COBIT is primarily concerned with portfolios of IT projects, services, assets and resources.)
  – Define categories (within portfolios).—The governance processes must recognize that each type of portfolio might need to be further subdivided into categories, according to the characteristics of its contents, with each category needing different levels of evaluation, approaches to decision making and funding.
  – Develop and communicate evaluation criteria (for each category).—For each category within a portfolio, evaluation criteria must be in place to support fair, transparent, repeatable and comparable evaluation.
  – Assign weightings to criteria.—For each category within a portfolio, the criteria should be weighted to allow an overall relative score to be derived for evaluation purposes.
  – Define requirements for stage-gates and other reviews.—Done for each portfolio category
- **Align and integrate value management with enterprise financial planning.**—Involves reviewing the current enterprise budgeting practices and identifying, and subsequently implementing, the changes necessary for putting in place optimal value management financial planning practices to facilitate business case preparation, investment decision making and ongoing investment management. The supporting management practices are:
  – Review current enterprise budgeting practices.—An examination is carried out of the practices used to set budgets, including their subdivisions, allocations for programs (investments) and business operations (costs), the periods over which they are set, frequency of reporting and review, levels of sign-off and cross-charging provisions.
  – Determine value management financial planning practice requirements.—Consideration is given to the implications for the enterprise of differentiating investments from costs, funding investments out of alignment with budgeting periods, budgets being held by program business sponsors and operating controls based on future value creation rather than year-to-date spending, etc.
  – Identify changes required.—A comparison is done of the financial planning practices needed for value management with current budgeting practices and identification is made of changes needed.
  – Implement optimal financial planning practices for value management.—Practices are established for financial planning with respect to IT-enabled investments so as to facilitate business case preparation, investment decision making, investment management and the creation of optimal value.
- **Establish effective governance monitoring.**—Identifies the key goals and metrics of the value management processes to be monitored and the approaches, methods, techniques and processes for capturing and reporting the measurement information. It establishes how deviations or problems will be identified and how monitoring and reporting on results of remedial actions will be conducted. The supporting management practices are:
  – Identify key metrics.—Defines a balanced set of performance objectives, metrics, targets and benchmarks
  – Define information capture processes and approaches.—Processes should be established to collect relevant, timely, complete, credible and accurate data to report on progress against targets.
  – Define reporting methods and techniques.—Relevant portfolio, program and IT (technological and functional) performance should be reported to the board and executive management in a timely and accurate manner.
  – Identify and monitor performance improvement actions.—Upon review of reports, appropriate management action should be initiated and controlled.
- **Continuously improve value management practices.**—Involves reviewing lessons learned from value management. There should be planning, initiation and monitoring of the necessary changes to improve value governance, portfolio management and investment management processes. The supporting management practice is:
  – Implement lessons learned.—Lessons learned from value management should be regularly reviewed and necessary changes should be planned, implemented and monitored to improve value governance, portfolio management and investment management processes and practices.

### Figure 3.25—Value Governance Processes and Practices Value Governance (VG)

**Value Governance (VG)**

**VG1 Establish informed and committed leadership:**
- VG1.1 Develop an understanding of the significance of IT and role of governance.
- VG1.2 Establish effective reporting lines.
- VG1.3 Establish a leadership forum.
- VG1.4 Define value for the enterprise.
- VG1.5 Ensure alignment and integration of business and IT strategies with key business goals.

**VG2 Define and implement processes:**
- VG2.1 Define the value governance framework.
- VG2.2 Assess the quality and coverage of current processes.
- VG2.3 Identify and prioritise process requirements.
- VG2.4 Define and document the processes.
- VG2.5 Establish, implement and communicate roles, responsibilities and accountabilities.
- VG2.6 Establish organisational structures.

**VG3 Define portfolio characteristics:**
- VG3.1 Define portfolio types.
- VG3.2 Define categories (within portfolios).
- VG3.3 Develop and communicate evaluation criteria (for each category).
- VG3.4 Assign weightings to criteria.
- VG3.5 Define requirements for stage-gates and other reviews (for each category).

**VG4 Align and integrate value management with enterprise financial planning:**
- VG4.1 Review current enterprise budgeting practices.
- VG4.2 Determine value management financial planning practice requirements.
- VG4.3 Identify changes required.
- VG4.4 Implement optimal financial planning practices for value management.

**VG5 Establish effective governance monitoring:**
- VG5.1 Identify key metrics.
- VG5.2 Define information capture processes and approaches.
- VG5.3 Define reporting methods and techniques.
- VG5.4 Identify and monitor performance improvement actions.

**VG6 Continuously improve value management practices:**
- VG6.1 Implement lessons learned.

Source: ISACA, *Enterprise Value: Governance of IT Investments, The Val IT™ Framework 2.0*, USA, 2008, page 3 of laminate

For each Val IT process, the Val IT management guidelines include inputs and outputs, activity descriptions, with RACI (responsible, accountable, consulted and informed) charts, and goals and metrics at different levels. A high-level summary of the management guidelines for each domain is shown in **figure 3.26**.

### Figure 3.26—High-level Management Guidelines for Val IT Domains and Processes

| Domain | Domain Goal | Inputs | Outputs | Process Metrics | Domain Metric |
|---|---|---|---|---|---|
| Value Governance (VG) | To ensure that value management practices are embedded in the enterprise, enabling it to secure optimal value from its IT-enabled investments throughout their full economic life cycle | • Business strategy<br>• Enterprise governance and control framework<br>• Enterprise investment approach | • Leadership commitment<br>• Value governance requirements with roles, responsibilities and accountabilities<br>• Portfolio characteristics and investment categories | • Level of leadership agreement on value governance principles<br>• Level of leadership engagement<br>• Degree of implementation and compliance with value management processes | Maturity of value management processes |
| Portfolio Management (PM) | To ensure that an enterprise secures optimal value across its portfolio of IT-enabled investments | • Business strategy<br>• Portfolio characteristics and investment categories<br>• Available budget and resources<br>• Detailed business cases | • Approved investment programmes<br>• Overall investment portfolio view<br>• Portfolio performance reports | • Level of satisfaction with IT's contribution to business value<br>• Percentage of IT expenditures that have direct traceability to business strategy<br>• Percentage increase in portfolio value over time | Percentage of forecast optimal value, that is secured across the enterprise's portfolio of IT-enabled investments |

| Figure 3.26—High-level Management Guidelines for Val IT Domains and Processes *(cont.)* ||||||
| Domain | Domain Goal | Inputs | Outputs | Process Metrics | Domain Metric |
| --- | --- | --- | --- | --- | --- |
| **Investment Management (IM)** | To ensure that the enterprise's IT-enabled investments contribute to optimal value | • Business strategy<br>• Detailed business requirements<br>• Portfolio characteristics and mix<br>• Available resources | • Detailed business case, including full life-cycle costs and benefits<br>• Programme plan including budget and resources<br>• Programme performance reports<br>• Updated IT operational portfolios | • Number of new ideas per investment category, and percentage that are developed into detailed business cases<br>• Completeness and compliance of business cases (initial and updated)<br>• Percentage of expectedvalue realised | Contribution of individual IT-enabled investments to optimal value |
| Source: ISACA, *Enterprise Value: Governance of IT Investments, The Val IT™ Framework 2.0*, USA, 2008, figure 11 ||||||

## 3.12 BUSINESS CASE DEVELOPMENT AND EVALUATION TECHNIQUES

Most enterprises see a business case as a necessary evil or a bureaucratic hurdle to get over to obtain required financial and other resources. The focus is on the technology project, and the costs of the technology, with only a cursory discussion of benefits or changes that the business might need to make to create or sustain value from use of the technology. Business cases are also all too often treated as "one-off" documents that are rarely looked at again once the required resources have been obtained—other than, possibly, at a postimplementation review.

This approach to business cases can cause challenges down the road. A well-developed and intelligently used business case for a business change program is actually one of the most valuable tools available to management—the quality of the business case and the processes involved in its creation and use throughout the economic life cycle of an investment has an enormous impact on creating and sustaining value. It describes a proposed journey from initial ideas to realizing expected outcomes for beneficiaries (i.e., those whose money is being invested and for whom the return should be secured) and other affected stakeholders.[37]

The preparation of the business case supports the principles of Val IT directly (e.g., one of the Val IT principles is "IT-enabled investments will be managed through their full economic life cycle"). When the business case is kept updated through the economic life cycle of the investment, it ensures that the
cost-benefits of IT-enabled investments are tracked through the life cycle through the business case.

### How Business Cases Relate to Value Management

Val IT consists of three domains: Value Governance (VG), Portfolio Management (PM) and Investment Management (IM), which include 22 processes, as illustrated in **figure 3.27**.

More information on Val IT can be found in section 3.11 Value Delivery Frameworks.

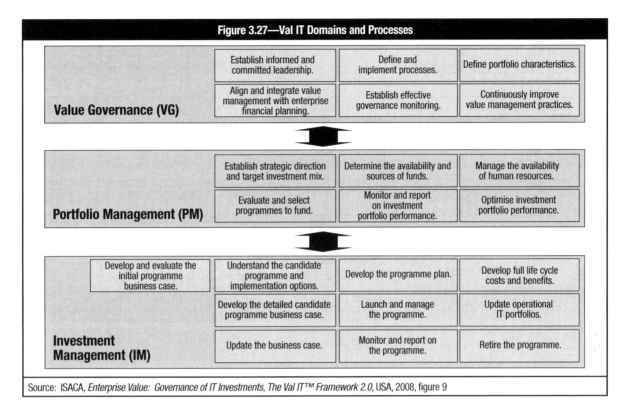

Figure 3.27—Val IT Domains and Processes

Source: ISACA, *Enterprise Value: Governance of IT Investments, The Val IT™ Framework 2.0*, USA, 2008, figure 9

Investment Management (IM) addresses the full life cycle of developing and maintaining business cases, which is fully addressed in the Val IT IM domain. More specifically, the following investment management processes are involved in developing and maintaining business cases (**figure 3.28**):
- IM1 Develop and evaluate initial program concept business case.
- IM2 Understand the candidate program and implementation options.
- IM3 Develop the program plan.
- IM4 Develop full life-cycle costs and benefits.
- IM5 Develop the detailed candidate program business case.
- IM8 Update the business case.

It should be noted that these processes are presented here in a sequential and structured way. In practice, many of these activities can be organized in parallel or in a way suitable and workable to the organization's structure and culture.

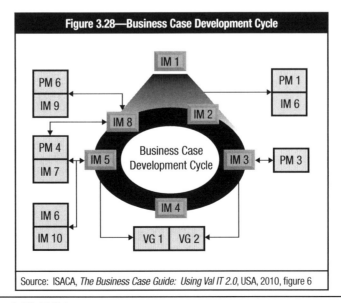

Figure 3.28—Business Case Development Cycle

Source: ISACA, *The Business Case Guide: Using Val IT 2.0*, USA, 2010, figure 6

## Business Case Development

The first five processes in investment management are about the emergence of new investment opportunities in the organization (IM1) and the development of detailed business cases (IM5) for the approved opportunities, including analyses of alternative courses of actions (IM2), definition of a detailed program plan (IM3) and full CBA (IM4). Based on this analysis, full business cases are delivered (IM5), including all the aspects delivered in the previous steps (IM1 to IM4).

## Business Case Maintenance

Once investment program are selected, the business case should be kept up to date throughout the entire life cycle of the investment (IM8).This should be done in preparation for stage-gate reviews or whenever there is any material change that affects the projected costs and/or benefits of the program, including when assumptions or risk changes due to changes to business strategy, or the way the enterprise functions or is organized, or due to the external environment.

As shown in **figure 3.28**, the central business case development processes do not operate in a vacuum. The business case development cycle is highly dependent on the input and outcomes of other value management processes such as PM1 *Establish strategic direction and target investment mix* and provides input to processes such as PM4 *Evaluate and select programs to fund*. Also, inputs and outputs to the COBIT- and Risk IT-related processes should be taken into account.

These are not shown in **figure 3.28**, but the relationships between Val IT, Risk IT and COBIT are extensively addressed in the Val IT 2.0 and Risk IT frameworks.

## Business Case Customization

This section aims to provide a complete and comprehensive approach to the development of a business case for investments in IT or IT-enabled change. While the approach could, and indeed should, be followed for all or most of such investments, there will be differences in emphasis, rigor of analysis and level of detail in different scenarios. These differences will be largely driven by one of (or a combination of) the characteristics shown in **figure 3.29**.

| Figure 3.29—Characteristics Affecting the Business Case Approach | |
|---|---|
| **Characteristic** | **Example** |
| Culture of the enterprise | • Geographic location<br>• Industry<br>• Leadership style |
| Nature of the enterprise | • Private sector<br>• Public/not-for-profit sector |
| Size of the enterprise | • Large<br>• Small to medium enterprise |
| Nature of the investment | • Degree of freedom<br>  – Discretionary<br>  – Nondiscretionary<br>• Type<br>  – Transactional<br>  – Informational<br>  – Transformational<br>  – Infrastructure |

Each of these characteristics can help define what approach and level of rigor may be appropriate for business case development.

The culture of the enterprise is both very specific to geographic location, industry and leadership style, and very broad in terms of impact. There will be different tolerances in terms of the rigor of analysis to support decision making, the level of detail of documentation and the subsequent analysis of performance. While the objective of improving business case processes may be to change the culture, the challenge should not be underestimated. Realistic and achievable objectives should be set, supported by a pragmatic and practical approach.

The nature of the enterprise (private/public) will influence the emphasis and considerations regarding value within the business case. While private-sector enterprises will typically focus more on financial benefits and outcomes, public and not-for-profit enterprises tend to be oriented more towards nonfinancial and service delivery outcomes. These nonfinancial, and often more intangible, aspects within public-sector investments heavily increase their complexity, requiring a stronger rigor of analysis and documentation.

It can be expected that small and medium-sized enterprises will likely need less rigor in documentation because communication and obtaining commitment tends to be less complex in smaller enterprises, and fewer resources are available to develop and maintain highly detailed business cases. The strong linkage to long-term strategic outcomes in large enterprises demands a high rigor of analysis, to ensure that clear and achievable benefits realizations plans are developed in support of achieving those long-term goals.

The nature of the investment impacts the degree of freedom in decision making. For nondiscretionary investments, less rigor is needed for documentation because these investments are done for compliance reasons. Of course, consideration could be given to identifying additional benefits, if this can be done without introducing excessive costs. For discretionary investments, the calculation of the relative value is important to enable a value-optimizing selection; therefore, they need a high rigor of analysis.

Within the group of discretionary investments, different types of investments exist:[38]
- **Transactional investments**—These typically focus on reducing costs and improving efficiency (automation), often require fewer business change programs and, therefore, need less rigor for documentation.
- **Informational investments**—These typically focus on building applications providing information for managing and controlling the enterprise (e.g., decision support systems).
- **Transformational investments**—A clear definition of the desired business outcomes and all of the business initiatives required to achieve this outcome is needed, which implies a high need for rigor of analysis and documentation.
- **Infrastructure investments**—In most cases, it is very difficult to demonstrate the direct business benefits; therefore, the business case of this type of investment typically focuses on costs and risk, and needs a high level of analysis and documentation to demonstrate how the investment will contribute to creating or sustaining value in existing or planned services and to evaluate the risk of doing nothing.

**Figure 3.30** discusses characteristics 2, 3 and 4 against a number of dimensions of the business case, including value emphasis (benefits, costs and risk), value considerations (financial, nonfinancial, strategic alignment and cost), rigor of analysis (high, medium or low) and level of detail of documentation (high, medium or low). The purpose of this discussion is to illustrate the need to tailor the business case development approach according to the specific situation and context. Therefore, the following information is to be regarded as a starting point (and not a strict model) to position the enterprise specific situation.

### Figure 3.30—Scenarios for Applying the Characteristics

| Characteristic | Value Emphasis | Value Considerations | Rigour of Analysis (H, M, L) | Rigour of Documentation (H, M, L) |
|---|---|---|---|---|
| **Nature of the enterprise** | | | | |
| Private sector | Benefits | Financial and non-financial outcomes | M | M |
| Public and not-for-profit sector | Risks | Many outcomes will be non-financial policy and service delivery outcomes. | H - M | H - M |
| **Size of the enterprise** | | | | |
| Large | Benefits | Strong linkage to strategic outcomes | H | M |
| Small to medium | Benefits | Strong linkage to key stakeholder outcomes | M | L |
| **Nature of the investment** | | | | |
| Discretionary | Benefits | Relative value is an important issue in terms of selection. | H | M |
| Non-discretionary | Costs | Consideration could/should be given to identifying additional benefits, if this can be done without introducing excessive costs or risks. | M | L |
| **Investment type** | | | | |
| Transactional | Costs | The focus is on reducing costs and improving organisational efficiency in terms of speed and quality of the information produced. | M | L |
| Informational | Benefits | The key question is how the information will be used to create or sustain value. | M | M |
| Transformational | Benefits | The important thing is to get clarity of the desired outcomes and of all the business initiatives (changes to: the business model, processes, people competencies, the reward system and organisational structures, etc.) required to achieve the outcomes. | H | H |
| Infrastructure | Costs/ Risks | The key questions are: • How will the investment contribute to creating or sustaining value in existing or planned services? • What is the risk of doing nothing? | H | H |

Source: ISACA, *The Business Case Guide: Using Val IT 2.0*, USA, 2010, figure 7

## ENDNOTES

[1] Weill, Peter; Jeanne Ross; *IT Savvy: What Top Executives Must Know to Go from Pain to Gain*, Harvard Business Press, USA, 2009
[2] IT Governance Institute, *IT Governance Global Status Report—2006*, ISACA, USA, 2006 *(www.isaca.org)*
[3] IT Governance Institute, *IT Governance Global Status Report—2008*, ISACA, USA, 2008 *(www.isaca.org)*
[4] Butler Group; *Measuring IT Costs and Value: Maximising the Effectiveness of IT Investment*, UK, 2005, www.butlergroup.com. Quoted in: ISACA, *IT Governance Domains, Practices and Competencies: Optimizing Value Creation From IT Investments*, USA, 2005
[5] ISACA, *Enterprise Value: Governance of IT Investments, The Val IT™ Framework 2.0*, USA, 2008, www.isaca.org/valit
[6] Ward, John, "Delivering Value From Information Systems and Technology Investments: Learning From Success," Forum (monthly newsletter of Cranfield School of Management, UK), August, 2006
[7] *Op cit*, ISACA, *Enterprise Value: Governance of IT Investments*, The Val IT™ Framework 2.0
[8] ISACA, *COBIT 5: Enabling Processes*, USA, 2012
[9] *Op cit* ISACA, *Enterprise Value: Governance of IT Investments, The Val IT™ Framework 2.0*
[10] Office of Government Commerce (OGC), *ITIL® V3*, UK, 2007
[11] *Ibid*.
[12] Highlighted as a best practice at the US National Defense University, Information Resources Management College and the Federal Enterprise Architecture Certification (FEAC) Institute. The Office of Management and Budget (OMB) and Federal Enterprise Architecture Project Management Office (FEAPMO) have cited this as an example of how to effectively develop business-driven segment architectures.
[13] ISACA, *Board Briefing on IT Governance, 2nd Edition*, USA, 2003

[14] De Haes, Steven; Wim Van Grembergen; "IT Governance Structures, Processes and Relational Mechanisms: Achieving IT/Business Alignment in a Major Belgian Financial Group," in proceedings of the Hawaii International Conference on System Sciences (HICSS), 2005

[15] Van Grembergen, Wim; Ronald Saull; Steven De Haes; "Linking the IT Balanced Scorecard to the Business Objectives at a Major Canadian Financial Group," *Journal of Information Technology Cases and Applications*, 2003

[16] ISACA UK Office of Government Commerce (OGC); *Aligning COBIT® 4.1, ITIL V3 and ISO/IEC 27002 for Business Benefit: A Management Briefing from ITGI and OGC*, USA, 2008

[17] UK Office of Government Commerce (OGC); *An Introduction to ITIL*, UK, 2004

[18] De Haes, Steven; Dirk Gemke; John Thorp; Wim van Grembergen; "KLM's Enterprise Governance of IT Journey: From Managing IT Cost to Managing Business Value," *MISQ Executive*, 2011, ISSN 1540-1960-10:3 (2011), p. 109-120

[19] ISACA, COBIT 5, USA, 2012

[20] Weill, Peter; Sinan Aral; *IT Savvy Pays Off: How Top Performers Match IT Portfolios and Organizational Practices*, USA, 2005

[21] De Haes, Steven; Wim Van Grembergen; "An Exploratory Study Into the Design of an IT Governance Minimum Baseline Through Delphi Research," *Communications of the Association of Information Systems*, vol. 22, 2008

[22] *Op cit* Ward

[23] Gartner, *The Four Levels of Cost Optimization*, Barbara Gomolski, Kurt Potter, Mark Raskino, January 20, 2009, Gartner Foundational, May 20, 2014

[24] *Ibid.*

[25] *Ibid.*

[26] Thorp, John; *The Information Paradox: Realizing the Business Benefits of Information Technology*, McGraw-Hill Ryerson, USA, 2003; De Haes, Steven; Wim Van Grembergen; "An Exploratory Study Into IT Governance Implementations and Its Impact on Business/IT Alignment," *Information Systems Management*, vol. 26, no. 2, 2009

[27] *Op cit* Weill, Ross

[28] De Haes, Steven; Dirk Gemke; John Thorp; Wim Van Grembergen; "Analyzing IT Value Management at KLM Through the Lens of Val IT," *ISACA Journal*, vol. 5, 2011

[29] *Op cit* De Haes, Van Grembergen, 2008; *Op cit* ISACA, 2008; *Op cit* Thorp; Van Grembergen, Wim; Steven De Haes; *Enterprise Governance of IT: Achieving Strategic Alignment and Value*, Springer, USA, 2009

[30] Watters, Doug, "IBM Strategy and Change Survey of *Fortune* 1000 CIOs," presented at SHARE, New York, 17 August 2004

[31] The Standish Group, blog.standishgroup.com

[32] *Op cit* ISACA, 2006; *Op cit* ISACA, IT Governance Institute, *IT Governance Global Status Report—2008*

[33] Barua, Anitesh; Laurence Brooks; Kirstin Gillon; Robert Hodgkinson; Rajiv Kohli; Sean Worthington; Bob Zukis; "Creating, Capturing and Measuring Value From IT Investments: Could We Do Better?" *Communications of the Association for Information Systems*, Vol. 27, 2010.

[34] De Haes, Steven; Wim Van Grembergen; *Enterprise Governance of Information Technology: Achieving Alignment and Value, Featuring COBIT 5, Second Edition*, Springer, USA, 2015

[35] *Op cit* Butler Group

[36] *Op cit* ISACA, *Enterprise Value: Governance of IT Investments, The Val IT™ Framework 2.0*

[37] ISACA, *The Business Case Guide: Using Val IT™ 2.0*, USA, 2010, www.isaca.org/valit

[38] *Ibid.*

Page intentionally left blank

# Chapter 4: Risk Optimization

## Section One: Overview

Domain Definition .................................................................................................................................................... 140
Domain Objectives .................................................................................................................................................. 140
Learning Objectives ................................................................................................................................................ 140
CGEIT Exam Reference .......................................................................................................................................... 140
Task and Knowledge Statements ............................................................................................................................. 140
Self-assessment Questions ...................................................................................................................................... 145
Answers to Self-assessment Questions ................................................................................................................... 146
Suggested Resources for Further Study .................................................................................................................. 147

## Section Two: Content

4.1 The Application of Risk Management at the Strategic, Portfolio,
 Program, Project and Operations Levels ........................................................................................................ 148
4.2 Risk Management Frameworks and Standards ............................................................................................... 150
4.3 The Relationship of the Risk Management Approach to Legal and Regulatory Compliance ........................ 158
4.4 Methods to Align IT and Enterprise Risk Management .................................................................................. 160
4.5 The Relationship of the Risk Management Approach to Business Resiliency ............................................... 161
4.6 Risk, Threats, Vulnerabilities and Opportunities Inherent in the Use of IT ................................................... 165
4.7 Types of Business Risk, Exposures and Threats That Can Be Addressed Using IT Resources .................... 172
4.8 Risk Appetite and Risk Tolerance .................................................................................................................... 174
4.9 Quantitative and Qualitative Risk Assessment Methods ................................................................................. 177
4.10 Risk Mitigation Strategies Related to IT in the Enterprise .............................................................................. 179
4.11 Methods to Monitor Effectiveness of Mitigation Strategies and/or Controls ................................................. 181
4.12 Stakeholder Analysis and Communication Techniques .................................................................................. 184
4.13 Methods to Establish Key Risk Indicators ....................................................................................................... 188
4.14 Methods to Manage and Report the Status of Identified Risk ........................................................................ 190
Endnotes .................................................................................................................................................................. 193

# Section One: Overview

## DOMAIN DEFINITION
Ensure that an IT risk management framework exists to identify, analyze, mitigate, manage, monitor and communicate IT-related business risk, and that the framework for IT risk management is in alignment with the enterprise risk management (ERM) framework.

## DOMAIN OBJECTIVES
The objective of this domain is to ensure that appropriate frameworks exist and are aligned with relevant standards to identify, evaluate, analyze, mitigate, manage, monitor and communicate on IT-related business risk as an integral part of an enterprise's governance environment.

The premise of this domain is that the universal need to demonstrate good enterprise governance to shareholders and customers is the driver for increased risk management activities in many organizations. Regulators are specifically concerned about operational and systemic risk, within which technology risk and information security issues are prominent.

## LEARNING OBJECTIVES
The purpose of having this domain is for professionals and executives involved in governance to be able to manage the IT-related risk of doing business in an interconnected digital world, which includes a dependence on entities beyond the direct control of the enterprise.

## CGEIT EXAM REFERENCE
This domain represents 24 percent of the CGEIT exam (approximately 36 questions).

## TASK AND KNOWLEDGE STATEMENTS

### TASKS
There are seven tasks within this domain that a CGEIT candidate must know how to perform. These relate to identifying, evaluating, analyzing, mitigating, managing, monitoring and communicating IT-related business risk.

T4.1 Ensure that comprehensive IT risk management processes are established to identify, analyze, mitigate, manage, monitor and communicate IT risk.
T4.2 Ensure that legal and regulatory compliance requirements are addressed through IT risk management.
T4.3 Ensure that IT risk management is aligned with the enterprise risk management (ERM) framework.
T4.4 Ensure appropriate senior level management sponsorship for IT risk management.
T4.5 Ensure that IT risk management policies, procedures and standards are developed and communicated.
T4.6 Ensure the identification of key risk indicators (KRIs).
T4.7 Ensure timely reporting and proper escalation of risk events and responses to appropriate levels of management.

### KNOWLEDGE STATEMENTS
The CGEIT candidate must have a good understanding of each of the 14 areas delineated by the following knowledge statements. These statements are the basis for the exam. Each statement is defined and its relevance and applicability to this job practice are briefly described as follows:

Knowledge of:

**KS4.1 the application of risk management at the strategic, portfolio, program, project and operations levels**
Business risk is affected by the business environment (management style or culture; risk appetite; and by industry sector factors, such as competition, reputation and national and international regulations) and therefore specific IT risk can be similarly affected. Thus, it is important to consider IT risk within the wider business context at strategic, portfolio, program, project and operational levels.

**KS4.2 risk management frameworks and standards**

A number of risk management frameworks and standards have been published, which should be leveraged because they provide definition and guidance for formulation and implementation of the enterprise risk management (ERM) framework. Once a risk management framework is in place, a common approach can be used across the business, bringing together disparate risk disciplines and functions into a consolidated and consistent approach.

**KS4.3 the relationship of the risk management approach to legal and regulatory compliance**

ERM ensures that management has a process in place to both set objectives and align the objectives with the enterprise's mission or vision, consistent with the enterprise's risk appetite. These enterprise objectives include strategic, operational and reporting issues, but should also consider compliancy requirements relating to the enterprise's applicable laws and regulations.

**KS4.4 methods to align IT and ERM**

Management of business risk is an essential component of the responsible administration of any enterprise. Almost every business decision requires the executive or manager to balance risk and reward. Therefore, it is important to align/integrate IT risk management approaches into the overall ERM approach. In the professional field, frameworks do exist that provide guidance to align IT and ERM.

**KS4.5 the relationship of the risk management approach to business resiliency**

There is a big difference between concept and delivery. Creating useful business resilience strategies and building business resilient IT systems requires substantial investment and deliberate focus. Enterprises should create a comprehensive approach to building resilient IT systems that increase business agility and withstand disruption. This approach traverses the entire organization, focusing on network, applications, communications, and workforce resilience and includes mature and well-controlled IT-related processes that can have a positive influence on reducing the business impact when events happen. The purpose of a well-established business continuity process is to continue critical business operations and maintain availability of information at a level acceptable to the enterprise in the event of a significant disruption.

**KS4.6 risk, threats, vulnerabilities and opportunities inherent in the use of IT**

Information technology (IT) has become pervasive in current dynamic and often turbulent business environments. This major IT dependency implies a huge vulnerability that is inherently present in IT environments. System and network downtime has become far too costly for any organization these days as doing business globally around the clock has become the standard. For example, downtime in the banking and health care sector is a risk factor accompanied by a wide spectrum of external threats, such as errors and omissions, abuse, cybercrime and fraud.

IT not only has the potential to support existing business strategies, but also to shape new strategies. In this mindset, IT becomes not only a success factor for survival and prosperity, but also an opportunity to differentiate and to achieve competitive advantage. Often, these opportunities are driven by new technological evolutions such as BYOD (bring your own device) and cloud computing. It is important to always consider both inherent values and risk of such evolutions.

**KS4.7 types of business risk, exposures and threats that can be addressed using IT resources**

Whether for compliance, effectiveness or efficiency, IT enablement of business has dramatically increased in recent years. As complexities of business evolve, the integral role of IT is extended to that of assisting business in handling risk, which is an inevitable part of business strategies, processes and operations. Therefore, included as part of business enablement, IT has the task of assisting in the management of business risk, including examples such as access controls, process controls and continuous monitoring.

**KS4.8  risk appetite and risk tolerance**
Risk appetite and tolerance are concepts that are frequently used, but where the potential for misunderstanding is high. Some use the concepts interchangeably, others see a clear difference. According to the Committee of Sponsoring Organizations of the Treadway Commission [COSO] ERM definitions (which are equivalent to International Organization for Standardization [ISO] 31000 definitions):
- **Risk appetite**—The broad-based amount of risk a company or other entity is willing to accept in pursuit of its mission (or vision).
- **Risk tolerance**—The acceptable variation relative to the achievement of an objective (and often best measured in the same units as those used to measure the related objective)

**KS4.9  quantitative and qualitative risk assessment methods**
Several methods for risk assessment exist, ranging between high level and mostly qualitative to very quantitative, with hybrid methods in between. The organizations' culture, resources, skills and knowledge, environment, risk appetite, etc., will determine which methodology to use.

The different methods have some common limitations:
- No method is fully objective, and results of risk assessments are always dependent on the persons performing them and their skills and views.
- IT-related risk data (such as loss data and IT risk factors) is often of poor quality—putting structures or models in place can help achieve greater objectivity and provide a common language for discussion in the risk analysis.

**KS4.10  risk mitigation strategies related to IT in the enterprise**
The purpose of defining a risk response is to bring risk in line with the defined risk tolerance for the enterprise after due risk analysis. In other words, a response needs to be defined such that future residual risk (current risk with the risk response defined and implemented) is well within risk tolerance limits (usually depending on budgets available).

**KS4.11  methods to monitor effectiveness of mitigation strategies and/or controls**
There is a strong need to have proven quantitative and qualitative methods that can measure and ascertain the extent to which IT-related activities contribute to or support business success. By having such methods, active monitoring and corrective action can be taken to reduce the risk that IT activities deviate from the expected business contributions they are supposed to be making.

**KS4.12  stakeholder analysis and communication techniques**
The intended audiences for better risk management approaches are extensive. Each of these stakeholder groups, both internal (e.g., board) and external (e.g., regulator) have different expectations which should be considered for the management of IT-related risk.

To reach these stakeholders, communication is a key component of risk management. The benefits of open communication include:
- Executive management understanding the actual exposure to IT risk, enabling definition of appropriate and informed risk responses
- Awareness among all internal stakeholders of the importance of integrating risk and opportunity in their daily duties
- Transparency to external stakeholders on the actual level of risk and risk management processes in use

The consequences of poor communication include:
- A false sense of confidence at the top on the degree of actual exposure related to IT, and lack of a well-understood direction for risk management from the top down
- Over-communication on risk to the external world, especially if risk is at an elevated or barely acceptable level. This may deter potential clients or investors, or generate needless scrutiny from regulators.
- The perception that the enterprise is trying to cover up known risk to stakeholders

**KS4.13 methods to establish key risk indicators (KRIs)**

KRIs can be defined as parameters that show that the enterprise is subject to, or has a high probability of being subject to, a risk that exceeds the defined risk tolerance. KRIs allow management to document and analyze trends, and provide a forward-looking perspective, signaling required actions before the risk actually becomes a loss. In practice, KRIs are used in reporting or in dashboards. They not only warn about and flag possible issues or areas that contain risk, but if selected well they can provide management with a holistic overview of the current risk management situation.

**KS4.14 methods to manage and report the status of identified risk**

Risk monitoring and control are where the "rubber meets the road." Enterprises should establish risk identification, analysis, monitoring and reporting processes such that they are able to handle risk events once they begin occurring

## RELATIONSHIP OF TASK TO KNOWLEDGE STATEMENTS

The task statements are what the CGEIT candidate is expected to know how to do. The knowledge statements delineate what the CGEIT candidate is expected to know to perform the tasks. The task and knowledge statements are approximately mapped in **figure 4.1** insofar as it is possible to do so. Note that although there often is overlap, each task statement will generally map to several knowledge statements.

### Figure 4.1—Task and Knowledge Statements Mapping—Risk Optimization Domain

| Task Statement | Knowledge Statements |
|---|---|
| T4.1 Ensure that comprehensive IT risk management processes are established to identify, analyze, mitigate, manage, monitor and communicate IT risk. | KS4.1 the application of risk management at the strategic, portfolio, program, project and operations levels.<br>KS4.2 risk management frameworks and standards (for example, Risk IT, the Committee of Sponsoring Organizations of the Treadway Commission Enterprise Risk Management—Integrated Framework (2004) [COSO ERM], International Organization for Standardization [ISO] 31000)<br>KS4.5 the relationship of the risk management approach to business resiliency (for example, business continuity planning [BCP] and disaster recovery planning [DRP]<br>KS4.6 risk, threats, vulnerabilities and opportunities inherent in the use of IT<br>KS4.9 quantitative and qualitative risk assessment methods<br>KS4.10 risk mitigation strategies related to IT in the enterprise<br>KS4.11 methods to monitor effectiveness of mitigation strategies and/or controls |
| T4.2 Ensure that legal and regulatory compliance requirements are addressed through IT risk management. | KS4.3 the relationship of the risk management approach to legal and regulatory compliance<br>KS4.10 risk mitigation strategies related to IT in the enterprise |
| T4.3 Ensure that IT risk management is aligned with the enterprise risk management (ERM) framework. | KS4.2 risk management frameworks and standards (for example, Risk IT, the Committee of Sponsoring Organizations of the Treadway Commission Enterprise Risk Management—Integrated Framework (2004) [COSO ERM], International Organization for Standardization [ISO] 31000)<br>KS4.4 methods to align IT and enterprise risk management (ERM)<br>KS4.7 types of business risk, exposures and threats (for example, external environment, internal environment, information security) that can be addressed using IT resources<br>KS4.8 risk appetite and risk tolerance<br>KS4.9 quantitative and qualitative risk assessment methods |
| T4.4 Ensure appropriate senior level management sponsorship for IT risk management. | KS4.12 stakeholder analysis and communication techniques |

| Figure 4.1—Task and Knowledge Statements Mapping—Risk Optimization Domain *(cont.)* ||
|---|---|
| **Task Statement** | **Knowledge Statements** |
| T4.5 Ensure that IT risk management policies, procedures and standards are developed and communicated. | KS4.2 risk management frameworks and standards (for example, Risk IT, the Committee of Sponsoring Organizations of the Treadway Commission Enterprise Risk Management—Integrated Framework (2004) [COSO ERM], International Organization for Standardization [ISO] 31000)<br>KS4.9 quantitative and qualitative risk assessment methods<br>KS4.10 risk mitigation strategies related to IT in the enterprise |
| T4.6 Ensure the identification of key risk indicators (KRIS). | KS4.8 risk appetite and risk tolerance<br>KS4.13 methods to establish key risk indicators (KRIs) |
| T4.7 Ensure timely reporting and proper escalation of risk events and responses to appropriate levels of management. | KS4.14 methods to manage and report the status of identified risk |

## SELF-ASSESSMENT QUESTIONS

CGEIT self-assessment questions support the content in this manual and provide an understanding of the type and structure of questions that have typically appeared on the exam. Questions are written in a multiple-choice format and designed for one best answer. Each question has a stem (question) and four options (answer choices). The stem may be written in the form of a question or an incomplete statement. In some instances, a scenario or a description problem may be included. These questions normally include a description of a situation and require the candidate to answer two or more questions based on the information provided. Many times a question will require the candidate to choose the **MOST** likely or **BEST** answer among the options provided.

In each case, the candidate must read the question carefully, eliminate known incorrect answers and then make the best choice possible. Knowing the format in which questions are asked, and how to study and gain knowledge of what will be tested, will help the candidate correctly answer the questions.

4-1  Which of the following should be implemented at the highest levels of an enterprise?

   A. An enterprise risk register
   B. A risk management board
   C. A risk owner
   D. A risk council


4-2  IT risk associated with the outsourcing of IT services is **BEST** managed through the:

   A. creation of multiple sourcing strategies.
   B. inclusion of controls and service level agreements (SLAs) into contracts.
   C. development of policies and procedures.
   D. performance of due diligence audits.


4-3  The **MOST** direct approach to correcting vulnerabilities and mitigating IT risk is through:

   A. reduction.
   B. retention.
   C. sharing.
   D. avoidance.

## ANSWERS TO SELF-ASSESSMENT QUESTIONS
Correct answers are shown in **bold**.

4-1  A. An enterprise risk register is a management tool that is used within the context of the risk management board.
   B. A risk management board is made up of managers who are responsible for the reporting of the enterprise's risk response to the board.
   C. Although the highest level of management is responsible for risk, the risk owner is tied to a low-level project, program or business unit.
   **D. The IT risk council works under the authority of the enterprise risk committee, which is ultimately accountable for risk management-related activities at the enterprise level.**

4-2  A. Creating multiple sourcing strategies is one of the risk responses and is also important, but this will occur after the fact or an event.
   **B. Mitigating controls and requirements are normally included in contracts and agreements.**
   C. Policies and procedures are important, but are not the best way to manage risk.
   D. Audits and the right to audit are important, but this will occur after the fact or an event.

4-3  **A. Risk reduction is the activity of actively mitigating a risk through the implementation of a control or safeguard.**
   B. Risk retention is used for small risk; when it occurs, it is corrected after the fact.
   C. Risk sharing or transfer is the process of assigning risk to another enterprise, usually through the purchase of an insurance policy or by outsourcing the service
   D. Risk avoidance involves not pursuing a business activity in order to avoid the associated risk.

---

**NOTE:** For more self-assessment questions, you may also want to obtain a copy of the *CGEIT® Review Questions, Answers & Explanations Manual 4th Edition*, which consists of 250 multiple-choice study questions, answers and explanations.

## SUGGESTED RESOURCES FOR FURTHER STUDY

In addition to the resources cited throughout this manual, the following resources are suggested for further study in this domain (publications in **bold** are stocked in the ISACA Bookstore):

Chapman, Robert J.; *Simple Tools and Techniques for Enterprise Risk Management, Second Edition*, John Wiley & Sons Inc., USA, 2012

Committee of Sponsoring Organizations of the Treadway Commission (COSO); *Enterprise Risk Management: Understanding and Communicating Risk Appetite,* USA, 2012

COSO; *Enterprise Risk Management for Cloud Computing*, USA, 2012

Jordan, Ernest; Luke Silcock; *Beating IT Risks*, John Wiley & Sons Inc., USA, 2005

**ISACA, COBIT 5, USA, 2012, www.isaca.org/cobit**

**ISACA, *COBIT 5 for Risk*, USA, 2013**

**ISACA, *CRISC Review Manual 6th Edition*, USA, 2015**

**ISACA, *CSX Cybersecurity Fundamentals Study Guide*, USA, 2014**

**ISACA, *The Risk IT Framework*, USA, 2009**

**ISACA, T*he Risk IT Practitioner Guide*, USA, 2009**

National Institute of Standards and Technology (NIST), *NIST Special Publication 800-39: Managing Information Security Risk*, USA, 2011

Westerman, George; Richard Hunter; *IT Risk: Turning Business Threats Into Competitive Advantage*, Harvard Business School Press, USA, 2007

## Section Two: Content

### 4.1 THE APPLICATION OF RISK MANAGEMENT AT THE STRATEGIC, PORTFOLIO, PROGRAM, PROJECT AND OPERATIONS LEVELS

Business risk is affected by the business environment (management style or culture; risk appetite; and industry sector factors such as competition, reputation, and national and international regulations) and, therefore, specific IT risk can be similarly affected. Thus, it is important to consider IT risk within the wider business context at operational, portfolio, program, project and strategic levels.

While risk management is not entirely new and risk taking is an everyday part of managing an enterprise, the necessary focus and recognition of risk management on the part of business management seems to be lacking when it comes to information risk management or IT risk management.

### Risk and Risk Management

Risk is defined as the possibility of an event occurring that will have an impact on the achievement of objectives, and it is typically measured in terms of likelihood and impact.

The universal need to demonstrate good enterprise governance to shareholders and customers is the driver for increased risk management activities in large enterprises.[1] There are many other types of enterprise risk besides financial risk. Regulators are specifically concerned about operational and systemic risk. These are two areas in which technology risk and information security issues are prominent.

IT risk is a business risk—specifically, the business risk associated with the use, ownership, operation, involvement, influence and adoption of IT with an enterprise. The management of risk is a cornerstone of GEIT, ensuring that the strategic objectives of the business are not jeopardized by IT failures. Risk associated with technology issues is increasingly evident on board agendas because the impact on the business of an IT failure can have devastating consequences. Risk is, however, as much about failing to grasp an opportunity to use IT—for example, to improve competitive advantage or operating efficiency—as it is about doing something badly or incorrectly.

### Risk Hierarchy

IT risk is a component of the overall risk universe of the enterprise. Other types of risk an enterprise faces include:
- Strategic risk
- Environmental risk
- Market risk
- Credit risk
- Operational risk
- Compliance risk

In many enterprises, IT-related risk is considered to be a component of operational risk (e.g., in the financial industry in the Basel III framework). However, even strategic risk can have an IT component to it, especially where IT is the key enabler of new business initiatives. The same applies for credit risk, where poor IT (security) can lead to lower credit ratings.

There are several published risk frameworks and standards, most of which reflect that IT risk is part of a risk hierarchy, with business strategic risk at the highest level. For example, as shown in **figure 4.2**, the UK Office of Government Commerce (OGC), in its Management of Risk (M_o_R®) framework, has defined four levels of risk to help ensure that IT risk is considered by business executives from the strategy level down to the operational level.

Figure 4.2—Four Levels of Risk (UK Office of Government Commerce [OGC])

Source: ISACA, *IT Governance Domain Practices and Competencies: Information Risks: Whose Business Are They?*, USA, 2005, figure 1

The M_o_R framework visualizes these four levels of risk in a pyramid, with appropriate escalations to higher levels for significant risk as seen in **figure 4.3**.

| Figure 4.3—Risk Level Descriptions | |
|---|---|
| **Risk Level** | **Description** |
| Strategic-level risk | Risk to IT achieving its objectives, i.e., commercial, financial, political, environmental, directional, cultural, acquisition, quality, business continuity and growth |
| Program-level risk | Risk involving procurement or acquisition, funding, organizational, projects, security, safety and business continuity |
| Project-level risk | Risk concerning people, technical aspects, cost, schedule, resources, operational support, quality, provider failure and security |
| Operational-level risk | Risk regarding people, technical aspects, cost, schedule, resources, operational support, quality, provider failure, security, infrastructure failure, business continuity and customer relations |

The **strategic level** is where choices are made about risk in relation to innovation and plans for delivering the business strategy. The other three levels are concerned with the actual delivery of the enterprise's strategy.

The focus at the **program** and **project levels** is on medium-term goals to deliver the organization's strategic objectives. Program and project managers juggle things from the top level to the bottom, and they advise on the choices that are made about risk. At any time during the program's life there may be circumstances or situations that could have a detrimental impact on the program. Such circumstances or situations are the risk and issues that the program must manage and resolve. As part of the tool set for dealing with issues at the project level, managers will need a program risk policy or a strategic-level risk policy to give overall guidance and direction on how risk should be managed.

The emphasis at the **operational level** is on short-term goals to ensure ongoing continuity of business services.

The context of risk management varies significantly from the strategic (enterprise) level to the IT operations level. At a minimum, risk should be analyzed; even if no immediate action is taken, the awareness of risk will influence strategic decisions for the better. Often, the most damaging IT risk is not well understood. At the strategic level, it needs to be understood that risk taking is an essential element of business today and success comes to those enterprises that detect, identify and manage risk most effectively. At this level risk management involves responding to the real impact on the business, identifying the issues and making sure that real and important risk is being addressed.

At the **portfolio level**, the enterprise's IT investments are differentiated into various categories, according to the investment categorization schema that has been adopted (see discussion in chapter 3 Benefits Realization). These investment categories need to be treated differently from each other—both in terms of analysis of risk (nature, likelihood, impact, etc.) and benefits or returns.

> **Example:**
>
> In the nondiscretionary category there is little or no room to consider whether to invest. To "keep the lights on," the continual periodic investment required for transaction processing systems (such as enterprise resource planning [ERP]) is not open to question because these systems function as the basic engine for running day-to-day business operations. The risk to the business of any failure of such investments may be highly damaging to business reputation and customer relations. Therefore, analysis of the risk for this category of investment is different from that for the IT investment category of business transformation (e.g., in a merger or acquisition scenario), in which the risk of business impact can also be equally high, but is of a different nature.

At the **program level**, risk management is applied to the three essential elements of effective IT investment management (see discussion in chapter 3 Benefits Realization): the business case, program management and benefits realization.

> **Examples:**
>
> 1. In the business case, risk management is applied to the identification and analysis of risk, which results in the appraisal and optimization of the risk or return for the IT-enabled investment for which the business case is made.
> 2. In executing program management, risk management is seen as managing the risk embedded in the various interlinked activities (e.g., results chain) that have been identified and recognized as part of the program.
> 3. In benefits realization, risk management is applied so that the risk is managed or mitigated to ensure that benefits as promised in the business case are actually realized or are adjusted according to risk analysis along the execution cycle of the program.

At the **project level**, risk management is applied so that the new or enhanced capabilities that have been defined as a result of doing the project are actually delivered. In this case, risk management encompasses a number of aspects in the process of project execution such as risk detection and identification, analysis (including quantification, likelihood and impact analysis), handling strategies and monitoring. Integrating risk management into the project management activities results in better project performance in terms of reducing cost and schedule overruns while delivering to scope.

The focus at the **IT operational level** is on IT delivering services in accordance with expected service management levels. Risk management is seen in terms of the deviation from these service levels and their associated business impact. This process-oriented approach used in the Information Technology Infrastructure Library (ITIL) and COBIT frameworks has gained ground in recent years.[2] Taking a process-oriented approach to IT services will ensure that process owners are tasked with knowing and managing the risk inherent in the processes under their charge. This makes for proactive risk management, in which accountability and remediation for risk is part of standard practice.

## 4.2 RISK MANAGEMENT FRAMEWORKS AND STANDARDS

A number of risk management frameworks and standards have been published and should be leveraged because they provide definition and guidance for formulation and implementation of the ERM framework. Once a risk management framework is in place, a common approach can be used across the business, bringing together disparate risk disciplines and functions into a consolidated and consistent approach.

There are a number of risk management frameworks and standards. They provide guidance for the application of risk management and have various originations. Some of the more popular ones are described in **figure 4.4**.

*Section Two: Content*     *Chapter 4—Risk Optimization*

| Figure 4.4—Risk Management Frameworks and Standards Compared | | | | | | | | | | |
|---|---|---|---|---|---|---|---|---|---|---|
| Principle/Feature | Risk IT | COSO ERM-Integrated Framework, 2004 | ISO/FDIS 31000:2009 | AS/NZS 4360:2004 | ARMS, 2002 | ISO 20000:2005 Parts 1 and 2 | PMBOK | ISO/IEC 27005:2008 | ISO/IEC 27001:2013 | ISO/IEC 27002:2005 |
| **Risk IT Principles** | | | | | | | | | | |
| Always connect to business objectives | ■ | ■ | ■ | ■ | ■ | ▓ | ▓ | ■ | ■ | ■ |
| Align the management of IT-related business risk with overall ERM | ■ | ▓ | ■ | ■ | ■ | □ | □ | ▓ | ▓ | ▓ |
| Balance the costs and benefits of managing risk | ■ | ■ | ■ | ■ | ■ | □ | ▓ | ▓ | ▓ | ▓ |
| Promote fair and open communication of IT risk | ■ | ■ | ■ | ■ | ■ | ▓ | ▓ | ▓ | ▓ | ▓ |
| Establish the right tone from the top while defining and enforcing personal accountability for operating within acceptable and well-defined tolerance levels | ■ | ■ | ■ | ■ | ■ | □ | ▓ | ■ | ■ | ■ |
| Are a continuous process and part of daily activity | ■ | ■ | ■ | ■ | ■ | ▓ | ▓ | ■ | ■ | ■ |
| **Additional Features** | | | | | | | | | | |
| Availability (to the general public) | ■ | ▓ | ▓ | ▓ | ■ | □ | ▓ | ▓ | ▓ | ▓ |
| Comprehensive view on IT (related) risk | ■ | □ | □ | □ | □ | □ | □ | ■ | ■ | ■ |
| Dedicated focus on risk management practices for specific IT areas (project management, service management, security, etc.) | ▓ | □ | □ | □ | □ | ■ | ■ | ■ | ■ | ■ |
| Provide a detailed process model with management guidelines and maturity models | ■ | ▓ | ▓ | ▓ | ▓ | ▓ | ▓ | ▓ | ▓ | ▓ |

Legend:
Black—Principle/feature is fully covered.
Gray—Principle/feature is partially covered.
White—Principle/feature is not covered.

Source: Adapted from ISACA, *The Risk IT Framework*, USA, 2009, figure 42

## Risk IT Framework

ISACA's Risk IT framework is intended to help implement GEIT and enhance IT-related risk management.[3] The Risk IT framework is closely aligned with the COBIT framework, as is ISACA's Val IT[4] framework. Risk IT is comprised of a set of principles, domains and processes, as shown in **figure 4.5**.

Using the Risk IT framework allows enterprises to make appropriate risk-aware decisions. The framework addresses many issues that enterprises face today, notably their need for:
- An accurate view of significant current and near-future IT-related risk throughout the extended enterprise, and the success with which the enterprise is addressing it
- End-to-end guidance on how to manage IT-related risk, beyond both purely technical control measures and security
- Understanding how to capitalize on an investment made in an IT internal control system already in place to manage IT-related risk
- Understanding how effective IT risk management enables business process efficiency, improves quality and reduces waste and costs
- Integrating overall risk and compliance structures within the enterprise when assessing and managing IT risk
- A common framework or language to help communication and understanding among business, IT, risk and audit management
- Promotion of risk responsibility and its acceptance throughout the enterprise
- A complete risk profile to better understand the enterprise's full exposure, so as to better utilize company resources

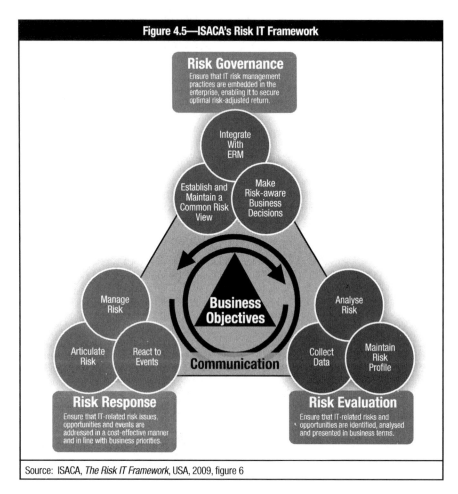

Source: ISACA, *The Risk IT Framework*, USA, 2009, figure 6

The Risk IT framework positions IT risk as a component of the overall risk universe of the enterprise. It takes its place in the risk that an enterprise faces, which includes strategic risk, operational risk, compliance risk, credit risk and market risk. However, even strategic risk can have an IT component, especially when IT is the key enabler of new business initiatives. The same applies for credit risk, especially when poor IT security can lead to lower credit ratings. For this reason, the preferred depiction of IT risk is not within a hierarchic dependency on one of the other risk categories, but as a horizontal risk category applying to all other individual risk categories.

There are three domains in the Risk IT framework: risk governance, risk evaluation and risk response. Each domain is made up of three processes:
- **Risk governance**—The objective of this risk domain is to ensure that IT risk management practices are embedded in the enterprise, enabling it to secure optimal risk-adjusted return. The domain is made up of three processes:
  – Integration with ERM
  – Establishment and maintenance of a common risk view
  – Making risk-aware business decisions
- **Risk evaluation**—The objective of this risk domain is to ensure that IT-related risk and opportunities are identified, analyzed and presented in business terms. The domain is made up of three processes:
  – Analysis of risk
  – Collection of data
  – Maintenance of risk profile
- **Risk response**—The objective of this risk domain is to ensure that IT-related risk issues, opportunities and events are addressed in a cost-effective manner and in line with business priorities. This domain is made up of three processes:
  – Management of risk
  – Articulation of risk
  – Reaction to risk

## COBIT 5 for Risk

*COBIT 5 for Risk* is an information risk view of COBIT 5, which serves as the information risk-specific guidance as it relates to COBIT for ISACA's information risk professionals. The guide is the risk-focused equivalent of the *COBIT 5 for Information Security* publication within the COBIT 5 family of products. **Figure 4.6** shows that *COBIT 5 for Risk* develops two different perspectives in which practical guidance is provided:
- The risk function perspective—How to build and sustain a risk management function leveraging the COBIT 5 enablers
- The risk management perspective—How to mitigate risk using the COBIT 5 enablers

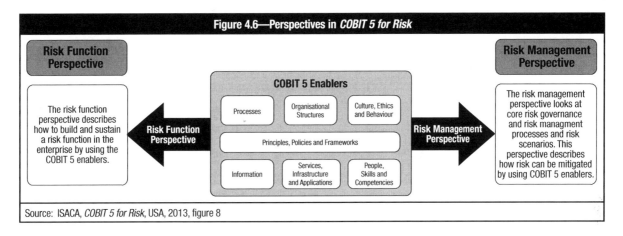

Source: ISACA, *COBIT 5 for Risk*, USA, 2013, figure 8

## COSO ERM Framework

In 2001, COSO initiated a project to develop a framework[5] that would be readily usable by management to evaluate and improve organizations' ERM. The period of the framework's development was marked by a series of high-profile business scandals and failures in which investors, company personnel and other stakeholders suffered tremendous losses. In the aftermath there were calls for enhanced corporate governance and risk management, with new laws, regulations and standards. The need for an ERM framework, providing key principles and concepts, a common language, and clear direction and guidance became even more compelling. According to the COSO ERM framework, ERM encompasses:
- Aligning risk appetite
- Enhancing risk response
- Reducing operational surprises
- Identifying and managing multiple and cross-enterprise risk
- Seizing opportunities
- Improving deployment of capital

The COSO ERM framework outlines objectives and components. There is a direct relationship between objectives, which are what an enterprise strives to achieve, and ERM components, which represent what is needed to achieve them. The relationship is depicted in a three-dimensional matrix in the form of a cube, shown in **figure 4.7**.

The four objectives categories—strategic, operations, reporting and compliance—are represented by the vertical columns, the eight components by horizontal rows, and an enterprise's units by the third dimension. This depiction portrays the ability to focus on the entirety of an organization's ERM, or by objective category, component, enterprise unit or any subset of dimensions.

On May 14, 2013, COSO released an updated version of its Internal Control—Integrated Framework. The 2013 Framework is expected to help organizations design and implement internal control in light of many changes in business and operating environments since the issuance of the original Framework, broaden the application of internal control in addressing operations and reporting objectives, and clarify the requirements for determining what constitutes effective internal control.

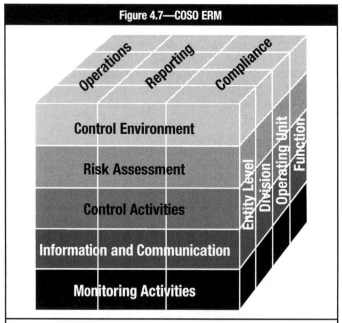

Figure 4.7—COSO ERM

Source: Committee of Sponsoring Organizations of the Treadway Commission, *Enterprise Risk Management—Integrated Framework*, USA, 2013, www.coso.org.
©2013, Committee of Sponsoring Organizations of the Treadway Commission (COSO). Used by permission.

## ISO 31000:2009 Principles and Guidelines on Implementation of Risk Management

The ISO 31000 is a family of standards related to risk management. ISO 31000:2009—*Principles and Guidelines on Implementation of Risk Management* is applicable throughout the life of an enterprise and to a wide range of activities, processes, functions, projects, products, services, assets, operations and decisions. Clause 4 prescribes a number of principles to be adhered to for the effectiveness of an organization's risk management (**figure 4.8**).

| Figure 4.8—ISO 31000 Risk Management Principles ||
|---|---|
| **Risk Management Principle** | **Explanation** |
| Risk management creates value. | Risk management contributes to the demonstrable achievement of objectives and improvement of, for example, human health and safety, legal and regulatory compliance, public acceptance, environmental protection, financial performance, product quality, efficiency in operations, corporate governance and reputation. |
| Risk management is an integral part of organizational processes. | Risk management is part of the responsibilities of management and an integral part of the normal organizational processes of all project and change management processes. Risk management is not a stand-alone activity that is separate from the main activities and processes of the enterprise. |
| Risk management is part of decision making. | Risk management helps decision makers make informed choices. Risk management can help prioritize actions and distinguish among alternative courses of action. Ultimately, risk management can help with decisions regarding whether a risk is unacceptable and whether risk treatment will be adequate and effective. |
| Risk management explicitly addresses uncertainty. | Risk management deals with those aspects of decision making that are uncertain, the nature of that uncertainty and how it can be addressed. |
| Risk management is systematic, structured and timely. | A systematic, timely and structured approach to risk management contributes to efficiency and consistent, comparable and reliable results. |
| Risk management is based on the best available information. | The inputs to the process of managing risk are based on information sources such as experience, feedback, observation, forecasts and expert judgment. However, decision makers should be informed of, and should take into account, any limitations of the data or modeling used or the possibility of divergence among experts. |
| Risk management is tailored. | Risk management is aligned with the enterprise's external and internal context and risk profile. |

| Figure 4.8—ISO 31000 Risk Management Principles *(cont.)* ||
|---|---|
| **Risk Management Principle** | **Explanation** |
| Risk management takes human and cultural factors into account. | The enterprise's risk management recognizes the capabilities, perceptions and intentions of external and internal people that can facilitate or hinder achievement of the organization's objectives. |
| Risk management is transparent and inclusive. | Appropriate and timely involvement of stakeholders and, in particular, decision makers at all levels of the enterprise, ensures that risk management remains relevant and up to date. Involvement also allows stakeholders to be properly represented and to have their views taken into account in determining risk criteria. |
| Risk management is dynamic, iterative and responsive to change. | As internal and external events occur, context and knowledge change, monitoring and review take place, new risk emerges, some risk specifics change, and others disappear. Therefore, an enterprise should ensure that risk management continually senses and responds to change. |
| Risk management facilitates continual improvement and enhancement of the organization. | Enterprises should develop and implement strategies to improve their risk management maturity as well as all other aspects of their organizations. |
| Source: International Organization for Standardization, *ISO 31000:2009: Risk Management—Principles and Guidelines*, Switzerland, 2009 ||

Clause 6 of the standard outlines processes for managing risk. The standard states that the risk management process should be an integral part of management, and embedded in culture and practices, and tailored to the business processes of the enterprise. The risk management process includes five activities:
- Communication and consultation
- Establishing the context
- Risk assessment
- Risk treatment
- Monitoring and review

ISO 31000:2009 has been received as a replacement to the existing standard on risk management, AS/NZS 4360:2004.[6] The objective of the standard is to provide guidance to enable public, private or community enterprises, groups and individuals to achieve the following:
- A more confident and rigorous basis for decision making and planning
- Better identification of opportunities and threats
- Gaining value from uncertainty and variability
- Proactive, rather than reactive, management
- More effective allocation and use of resources
- Improved incident management and reduction in loss and the cost of risk, including commercial insurance premiums
- Improved stakeholder confidence and trust
- Improved compliance with relevant legislation
- Better corporate governance

The standard specifies the elements of the risk management process, but does not propose to enforce uniformity of risk management systems. It is generic and independent of any specific industry or economic sector. The standard states that the design and implementation of the risk management system will be influenced by the varying needs of an enterprise, its particular objectives, its products and services, and the processes and specific practices employed. It also states that the standard should be applied at all stages in the life of an activity, function, project, product or asset, recognizing that the maximum benefit is usually obtained by applying the risk management process from the beginning.

## M_o_R Framework

The OGC published their M_o_R framework for risk management in 2002. The framework is applied enterprisewide to three core elements of a business:
- Strategic—business direction
- Change—turning strategy into action, including program, project and change management
- Operational—day-to-day operation and support of the business

The M_o_R framework is based on the following core concepts:
- **M_o_R Principles**—These are essential for the development of good risk management practice. They are all derived from corporate governance principles; this reiterates that risk management is a subset of an enterprise's internal controls.
- **M_o_R Approach**—The principles need to be adapted and adopted to suit each individual enterprise. Accordingly, an enterprise's approach to the principles needs to be agreed on and defined within risk management policy, process guide and strategies, and supported by the use of risk registers and issue logs.
- **M_o_R Processes**—There are four main process steps (identify, assess, plan and implement), which describe the inputs, outputs and activities involved in ensuring that risk is identified, assessed and controlled.
- **Embedding and Reviewing M_o_R**—Once principles, approach and processes are in place, an enterprise needs to ensure that they are consistently applied across the enterprise, and that their application undergoes continual improvement in order for them to be effective.

In this way, the strategy for managing risk should be led from the top of the enterprise (board-level sponsorship), while embedded in the normal working routines and activities of the enterprise (i.e., also being able to fit into the corporate culture of the business). The framework prescribes a series of activity elements as shown in **figure 4.9**.

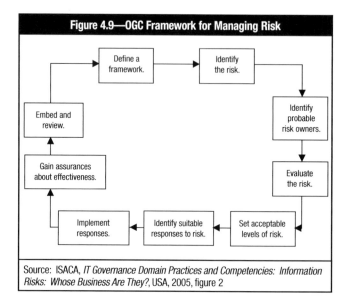

Source: ISACA, *IT Governance Domain Practices and Competencies: Information Risks: Whose Business Are They?*, USA, 2005, figure 2

## OCTAVE

Under the Networked Systems Survivability (NSS) program, the Software Engineering Institute (SEI) of Carnegie Mellon University (CMU) developed a framework for threat and vulnerability assessment, known as OCTAVE (operationally critical threat, asset, and vulnerability evaluation).[7] The objective of the framework is to aid and describe an information security risk evaluation. OCTAVE defines a set of self-directed activities for enterprises to identify and manage their information security risk. OCTAVE examines organizational and technology issues to assemble a comprehensive picture of the information security needs of an enterprise. It contains the following phases:
- Phase 1—Build enterprisewide security requirements.
- Phase 2—Identify infrastructure vulnerabilities.
- Phase 3—Determine security risk management strategy.

Each phase of OCTAVE is designed to produce meaningful results for the enterprise. During phase 1, information assets and their values, threats to those assets, and security requirements are identified using knowledge of the staff from multiple levels within the enterprise, along with standard catalogs of information. For example, known threat profiles and good organizational and technical practices are used to probe staff members for their knowledge of the organization's assets, threats and current protection strategies. This information can then be used to establish the security requirements of the enterprise, which is the goal of the first phase of OCTAVE. Phase 2 builds on the information captured during phase 1 by mapping the information assets of the enterprise to the information infrastructure components (both the physical environment and the networked IT environment) to identify the high-priority infrastructure components. Once this is done, an infrastructure vulnerability evaluation is performed to identify vulnerabilities. As in phase 1, standard catalogs of information are used; for example, standard intrusion scenarios and vulnerability information are used as a basis for the infrastructure vulnerability evaluation. At the conclusion of phase 2 the enterprise has identified the high-priority information infrastructure components, missing policies and practices, and vulnerabilities. Phase 3 builds on the information captured during phases 1 and 2. Risk is identified by analyzing the assets, threats and vulnerabilities identified in OCTAVE's earlier phases in the context of standard intrusion scenarios. The impact and probability of the risk (also called the risk attributes) are estimated and subsequently used to prioritize the risk. The prioritized list of risk specifics is used in conjunction with information from the previous phases to develop a protection strategy for the enterprise and establish a comprehensive plan for managing security risk, which is also the goal of phase 3.

## Other Risk Management Standards and Framework

Other related frameworks important to reference are:
- **The ISO/IEC 27000-series** (also known as the ISMS Family of Standards or ISO27k) comprises information security standards published jointly by ISO and the International Electrotechnical Commission (IEC). The series provides best practice recommendations on information security management, risk and controls within the context of an overall information security management system (ISMS), similar in design to management systems for quality assurance (ISO 9000 series) and environmental protection (ISO 14000 series).
- **ISO/IEC 20000** is the first international standard for IT service management. It was developed in 2005, by ISO/IEC Joint Technical Committee (JTC) 1/SC7 and revised in 2011. It is based on, and intended to, supersede the earlier BS 15000 that was developed by British Standards Institution (BSI) Group. Formally, ISO/IEC 20000-1:2011 (part 1) includes "the design, transition, delivery and improvement of services that fulfill service requirements and provide value for both the customer and the service provider. This part of ISO/IEC 20000 requires an integrated process approach when the service provider plans, establishes, implements, operates, monitors, reviews, maintains and improves a service management system (SMS)." The 2011 version (ISO/IEC 20000-1:2011) comprises guidance on:
  – Service management system general requirements
  – Design and transition of new or changed services
  – Service delivery processes
  – Relationship processes
  – Resolution processes
  – Control processes
- **ITIL M_o_R: Guidance for Practitioners** takes a best practice approach. Every organization must find the right balance between opportunities and threats in managing its risk. M_o_R offers a structured and effective framework for risk management. Its aim is to help organizations achieve their objectives by first identifying risk and then choosing the right response to the threats and opportunities that are created by uncertainty. This publication addresses all organizational activities from strategic, program, project or operational perspectives.
- **NIST Special Publication 800-37: Guide for Applying the Risk Management Framework to Federal Information Systems,** developed by the Joint Task Force Transformation Initiative Working Group, transforms the traditional Certification and Accreditation (C&A) process into the six-step Risk Management Framework (RMF). The RMF provides a disciplined and structured process that integrates information security and risk management activities into the system development life cycle.
- **NIST Special Publication 800-30 Revision 1: Guide for Conducting Risk Assessments,** describes risk assessment in the following manner:[8]

  *Risk assessments are a key part of effective risk management and facilitate decision making at all three tiers in the risk management hierarchy including the organization level, mission/business process level, and information system level.*

  *Because risk management is ongoing, risk assessments are conducted throughout the system development life cycle, from pre-system acquisition (i.e., material solution analysis and technology development), through system acquisition (i.e., engineering/manufacturing development and production/deployment), and on into sustainment (i.e., operations/support).*

- **NIST Special Publication 800-39: Managing Information Security Risk** describes itself as follows:[9]

  *The purpose of Special Publication 800-39 is to provide guidance for an integrated, organizationwide program for managing information security risk to organizational operations (i.e., mission, functions, image, and reputation), organizational assets, individuals, other organizations, and the Nation resulting from the operation and use of federal information systems. This provides a structured, yet flexible approach for managing risk that is intentionally broadbased, with the specific details of assessing, responding to, and monitoring risk on an ongoing basis provided by other supporting NIST security standards and guidelines.*

## 4.3 THE RELATIONSHIP OF THE RISK MANAGEMENT APPROACH TO LEGAL AND REGULATORY COMPLIANCE

ERM ensures that management has a process in place to both set objectives and align the objectives with the enterprise's mission or vision, consistent with the enterprise's risk appetite. These enterprise objectives include strategic, operational and reporting issues, but should also consider compliancy requirements relating to the enterprise's applicable laws and regulations.

The enterprise's business objectives play a defining role in the management of risk. They position the overall risk management and its active and conscious practice within the business (risk awareness), and they shape the ERM framework and treatment of risk in the definition and execution of business strategies.

### Enterprise Goal Categories

Within the context of the established mission or vision, the enterprise's management establishes strategic objectives, selects strategy and establishes related objectives, cascading through the enterprise and aligned with and linked to the strategy. Objectives must exist before management can identify events potentially affecting their achievement. ERM ensures that management has a process in place to both set objectives and align the objectives with the enterprise's mission or vision, consistent with the enterprise's risk appetite. The enterprise's objectives can be viewed in the context of four categories, as shown in **figure 4.10**.

| Figure 4.10—Enterprise Goal Categories | |
|---|---|
| **Enterprise Goal Category** | **Relates to ...** |
| Strategic | High-level goals, aligned with and supporting the enterprise's mission or vision |
| Operational | Effectiveness and efficiency of the enterprise's operations, including performance and profitability goals, which vary based on management's choices about structure and performance |
| Reporting | The effectiveness of the enterprise's reporting, including internal and external reporting and involving financial or nonfinancial information |
| Compliance | The enterprise's compliance with applicable laws and regulations |

This categorization of enterprise objectives allows management and the board to focus on separate aspects of ERM. These distinct, but overlapping, categories (a particular objective can fall under more than one category) address different enterprise needs and may be the direct responsibility of different executives. This categorization also allows distinguishing what can be expected from each category of objectives.

## Objective Setting

Objective setting is a precondition to event identification, risk assessment and risk response. There first must be objectives before management can identify risk to their achievement and take necessary actions to manage the related risk. A distinction is made between strategic objectives and other categories of objectives, collectively called related objectives. These objectives are defined as follows:
- **Strategic objectives**—An enterprise's mission sets out in broad terms what the enterprise aspires to achieve. Whatever term is used, such as mission, vision or purpose, it is important that management (with board oversight) explicitly establishes the enterprise's broad-based reason for being.

From this purpose, management sets its strategic objectives, formulates strategy and establishes related objectives for the enterprise. While an enterprise's mission and strategic objectives are generally stable, its strategy and related objectives are more dynamic and adjusted for changing internal and external conditions. Strategic objectives are high-level goals aligned with and supporting the enterprise's mission or vision. Strategic objectives reflect management's choice as to how the enterprise will seek to create value for its stakeholders. In considering alternative strategies to achieve its strategic objectives, management identifies risk associated with a range of strategy choices and considers their implications. Various event identification and risk assessment techniques can be used in the strategy-setting process. In this way ERM techniques are used in setting strategy and objectives.
- **Related objectives**—Establishing the appropriate objectives that support and align with the selected strategy, relative to all enterprise activities, is critical to success. By focusing first on strategic objectives and strategy, an enterprise is positioned to develop related objectives at operational levels, achievement of which will create and preserve value. Each set of objectives is linked to and integrated with more specific objectives that cascade through the enterprise to subobjectives established for various activities such as sales, production and engineering, and infrastructure functions.

## Critical Success Factors

By setting objectives at the enterprise and activity levels, an enterprise can identify CSFs. These are key issues or actions that must go right if goals are to be attained. CSFs exist for an enterprise, a business unit, a function, a department or an individual. By setting objectives, management can identify measurement criteria for performance, with a focus on CSFs. Where objectives are consistent with prior practice and performance, the linkage among activities is known. However, where new objectives depart from an enterprise's past practices, management must address the linkages or run increased risk. In such cases, the need for business unit objectives or subobjectives that are consistent with the new direction is even more important. Despite the diversity of objectives across enterprises, certain broad categories can be established as follows:
- **Operations objectives**—Relate to the effectiveness and efficiency of the enterprise's operations. They include related subobjectives for operations, directed at enhancing operating effectiveness and efficiency in moving the enterprise toward its ultimate goal. These include performance and profitability goals and safeguarding resources against loss. They vary based on management's choices about structure and performance. Operations objectives need to reflect the particular business, industry and economic environments in which the enterprise functions. The objectives need, for example, to be relevant to competitive pressures for quality, reduced cycle times to bring products to market or changes in technology. Management must ensure that objectives reflect reality and the demands of the marketplace, and are expressed in terms that allow meaningful performance measurements. A clear set of operations objectives, linked to subobjectives, is fundamental to success. Operations objectives provide a focal point for directing allocated resources; if an enterprise's operations objectives are not clear or well conceived, its resources may be misdirected.
- **Reporting objectives**—Pertain to the reliability of reporting. They include internal and external reporting and may involve financial or nonfinancial information. Reliable reporting provides management with accurate and complete information appropriate for its intended purpose. It supports management's decision making and monitoring of the enterprise's activities and performance. Examples of such reports may include results of marketing programs, daily sales flash reports, production quality, and employee and customer satisfaction results. Reliable reporting provides management with reasonable assurance of preparation of reliable reports for external dissemination. Such reporting includes financial statements and footnote disclosures, management's discussion and analysis, and reports filed with regulatory agencies.

- **Compliance objectives**—Pertain to adherence to relevant laws and regulations. They are dependent on external factors, such as environmental regulation, and tend to be similar across all enterprises in some cases and across an industry in others. Enterprises must conduct their activities and often take specific actions in accordance with relevant laws and regulations. These requirements may relate to markets, pricing, taxes, the environment, employee welfare and international trade. Applicable laws and regulations establish minimum standards of behavior, which the enterprise integrates into its compliance objectives. For example, occupational safety and health regulations might cause an enterprise to define its objective as: "Package and label all chemicals in accordance with regulations." In this case, policies and procedures would deal with communication programs, site inspections and training. An enterprise's compliance record can significantly—either positively or negatively—affect its reputation in the community and marketplace.
- **Project and program objectives**—Major strategies are turned into action by means of programs and projects, and ultimately the changes get embedded into the day-to-day business as usual operations. The challenge is to deliver on-time, on-budget projects that meet the customer's requirements. For project-driven enterprises, projects are king.

## 4.4 METHODS TO ALIGN IT AND ENTERPRISE RISK MANAGEMENT

Management of business risk is an essential component of the responsible administration of any enterprise. Almost every business decision requires the executive or manager to balance risk and reward. Therefore, it is important to align/integrate IT risk management approaches into the overall ERM approach. In the professional field, frameworks do exist that provide guidance to align IT and ERM.

To be successful, risk management should function within a risk management framework which provides the foundations and organizational arrangements that will embed it throughout the enterprise, at all levels.[10]

### IT Risk in the Risk Hierarchy

IT risk is a component of the overall risk universe of the enterprise, as shown in **figure 4.11**. Other risk faced by an enterprise includes strategic risk, project and program risk, operational risk, compliance risk, credit risk and market risk. In many enterprises, IT-related risk is considered to be a component of operational risk (e.g., in the financial industry in the Basel III framework). However, even strategic risk can have an IT component, especially when IT is the key enabler of new business initiatives. The same applies for credit risk, when poor IT security can lead to lower credit ratings. For this reason, with the preferred depiction of IT risk is not a hierarchic dependency on one of the other risk categories, but perhaps is better shown in the example (financial industry) given in **figure 4.11**.

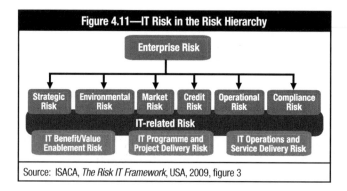

Source: ISACA, *The Risk IT Framework*, USA, 2009, figure 3

IT risk is business risk—specifically, the business risk associated with the use, ownership, operation, involvement, influence and adoption of IT within an enterprise. It consists of IT-related events that could potentially impact the business. It can occur with both uncertain frequency and magnitude, and it creates challenges in meeting strategic goals and objectives. IT risk can be categorized in different ways:
- **IT benefit realization risk**—Associated with (missed) opportunities to use technology to improve efficiency or effectiveness of business processes, or to use technology as an enabler for new business initiatives
- **IT solution delivery/benefit realization risk**—Associated with the contribution of IT to new or improved business solutions, usually in the form of projects and programs
- **IT service delivery risk**—Associated with the performance and availability of IT services, which can bring destruction or reduction of value to the enterprise

Management of business risk is an essential component of the responsible administration of any enterprise. Almost every business decision requires the executive or manager to balance risk and reward.

The all-encompassing use of IT can provide significant benefits to an enterprise, but it also involves risk. Due to the importance of IT to the overall business, IT risk should be treated like other key business risk (such as market risk, credit risk and other operational risk), all of which fall under the highest "umbrella" risk category: failure to achieve strategic objectives. While other risk has long been incorporated into corporate decision-making processes, too many executives tend to relegate IT risk to technical specialists outside the boardroom.

### The Risk IT Framework
The Risk IT framework[11] explains IT risk and enables users to:
- Integrate the management of IT risk into the overall ERM of the organization
- Make well-informed decisions about the extent of the risk, the risk appetite and the risk tolerance of the enterprise
- Understand how to respond to the risk

In brief, this framework allows the enterprise to make appropriate risk-aware decisions.

Practice has shown that the IT function and IT risk are often not well understood by an enterprise's key stakeholders, including board members and executive management; yet these are the very people who should be accountable for risk management within the enterprise. Without a clear understanding of the IT function and IT risk, senior executives have no frame of reference for prioritizing and managing IT risk.

IT risk is **not** purely a technical issue. Although IT subject matter experts are needed to understand and manage aspects of IT risk, business management is the most important stakeholder. Business managers determine what IT needs to do to support their business; they set the targets for IT and, consequently, are accountable for managing the associated risk.

The Risk IT framework fills the gap between generic risk management frameworks such as COSO ERM and ISO 31000 and the British equivalent (ARMS6) and detailed (primarily security-related) IT risk management frameworks. It provides an end-to-end, comprehensive view of all risk related to the use of IT and a similarly thorough treatment of risk management, from the tone and culture at the top to operational issues. In summary, the framework will enable enterprises to understand and manage all significant IT risk types.

The framework provides:
- An end-to-end process framework for successful IT risk management
- A generic list of common potentially adverse IT-related risk scenarios that could impact the realization of business objectives
- Tools and techniques to understand concrete risk to business operations, as opposed to generic checklists of controls or compliance requirements

More information on Risk IT and other frameworks is provided in section 4.2 Risk Management Frameworks and Standards.

## 4.5 THE RELATIONSHIP OF THE RISK MANAGEMENT APPROACH TO BUSINESS RESILIENCY
There is a big difference between concept and delivery. Creating useful business resilience strategies and building business-resilient IT systems requires substantial investment and deliberate focus. Enterprises should create a comprehensive approach to building resilient IT systems that increase business agility and withstand disruption. This approach traverses the entire organization, focusing on network, applications, communications, and workforce resilience and includes mature and well-controlled IT-related processes that can have a positive influence on reducing the business impact when events happen.

The purpose of a well-established business continuity process is to continue critical business operations and maintain availability of information at a level acceptable to the enterprise in the event of a significant disruption.

## Resilience

Resilience is the ability of a system or network to resist failure or to recover quickly from any disruption, usually with minimal recognizable effect.

Enterprises can begin to understand the resilience of their IT systems by measuring uptime. Service providers have long relied on the "five nines" (99.999 percent) concept for planned uptime. But this metric can play a numbers game with availability without considering the true business impact of an outage.

An IT group can justify meeting SLAs through statistical interpretation. For example, a 99.5 percent availability target allows a 50-minute weekly service outage. If an enterprise experiences one business-critical application outage of 100 minutes in a given month, the IT group can say it is exceeding SLAs while the business itself might suffer short-term revenue loss, customer dissatisfaction or possible penalties resulting from regulatory noncompliance and litigation. It is more useful to measure the resilience of IT systems from the end-user perspective. At the service level, metrics for measuring availability to users include:
- **Mean time between failure (MTBF)**—How long a service is operational before it might fail. The maximum MTBF is limited by the MTBF of the least resilient service component.
- **Mean time to repair (MTTR)**—How long it takes to restore a failed service. The minimum MTTR is impacted by the MTTR of the least resilient service component.[12]

More information on availability and availability management can be found in section 5.6 Methods Used to Evaluate and Report on IT Resource Performance.

## The Business Continuity Process

Mature and well-controlled IT-related processes can have a positive influence on reducing the business impact when events happen. An important process in this matter is the business continuity process. In COBIT 5, the business continuity process (DSS04) is described as:[13]

> *Establish and maintain a plan to enable the business and IT to respond to incidents and disruptions in order to continue operation of critical business processes and required IT services and maintain availability of information at a level acceptable to the enterprise.*

The purpose of this process is to "continue critical business operations and maintain availability of information at a level acceptable to the enterprise in the event of a significant disruption."[14] More specifically, this process supports the process and IT-related goals shown in **figure 4.12**.

| Figure 4.12—DSS04 Manage Continuity IT-related Goals and Process Goals and Metrics ||
|---|---|
| **IT-related Goal** | **Related Metrics** |
| 04 Managed IT-related business risk | • Percent of critical business processes, IT services and IT-enabled business programmes covered by risk assessment<br>• Number of significant IT-related incidents that were not identified in risk assessment<br>• Frequency of update of risk profile |
| 07 Delivery of IT services in line with business requirements | • Number of business disruptions due to IT service incidents<br>• Percent of business stakeholders satisfied that IT service delivery meets agreed-on service levels<br>• Percent of users satisfied with the quality of IT service delivery |
| 14 Availability of reliable and useful information for decision making | • Level of business user satisfaction with quality and timeliness (or availability) of management information<br>• Number of business process incidents caused by non-availability of information<br>• Ratio and extent of erroneous business decisions where erroneous or unavailable information was a key factor |
| **Process Goals and Metrics** ||
| **Process Goal** | **Related Metrics** |
| 1. Business-critical information is available to the business in line with minimum required service levels. | • Percent of IT services meeting uptime requirements<br>• Percent of successful and timely restoration from backup or alternate media copies<br>• Percent of backup media transferred and stored securely |
| 2. Sufficient resilience is in place for critical services. | • Number of critical business systems not covered by the plan |
| 3. Service continuity tests have verified the effectiveness of the plan. | • Number of exercises and tests that have achieved recovery objectives<br>• Frequency of tests |
| 4. An up-to-date continuity plan reflects current business requirements. | • Percent of agreed-on improvements to the plan that have been reflected in the plan<br>• Percent of issues identified that have been subsequently addressed in the plan |
| 5. Internal and external parties have been trained in the continuity plan. | • Percent of internal and external stakeholders that have received training<br>• Percent of issues identified that have been subsequently addressed in the training materials |
| Source: ISACA, *COBIT 5: Enabling Processes*, USA, 2012, page 185 ||

The key management practices of this process, according to COBIT 5, are:

- **DSS04.01 Define the business continuity policy, objectives and scope.**
  Define business continuity policy and scope aligned with enterprise and stakeholder objectives.
- **DSS04.02 Maintain a continuity strategy.**
  Evaluate business continuity management options and choose a cost-effective and viable continuity strategy that will ensure enterprise recovery and continuity in the face of a disaster or other major incident or disruption.
- **DSS04.03 Develop and implement a business continuity response.**
  Develop a BCP based on the strategy that documents the procedures and information in readiness for use in an incident to enable the enterprise to continue its critical activities.
- **DSS04.04 Exercise, test and review the BCP.**
  Test the continuity arrangements on a regular basis to exercise the recovery plans against predetermined outcomes and to allow innovative solutions to be developed and help verify over time that the plan will work as anticipated.
- **DSS04.05 Review, maintain and improve the continuity plan.**
  Conduct a management review of the continuity capability at regular intervals to ensure its continued suitability, adequacy and effectiveness. Manage changes to the plan in accordance with the change control process to ensure that the continuity plan is kept up to date and continually reflects actual business requirements.

- **DSS04.06 Conduct continuity plan training.**
  Provide all concerned internal and external parties with regular training sessions regarding the procedures and their roles and responsibilities in case of disruption.
- **DSS04.07 Manage backup arrangements.**
  Maintain availability of business-critical information.
- **DSS04.08 Conduct post-resumption review.**
  Assess the adequacy of the BCP following the successful resumption of business processes and services after a disruption.

### ISO 22301:2012—Societal Security—Business Continuity Management Systems

The ISO standard ISO 22301:2012 on business continuity management systems requirements replaced BS 25999. ISO 22301:2012 specifies requirements to plan, establish, implement, operate, monitor, review, maintain and continually improve a documented management system to protect against, reduce the likelihood of occurrence, prepare for, respond to, and recover from disruptive incidents when they arise. The requirements specified in ISO 22301:2012 are generic and intended to be applicable to all organizations or parts thereof, regardless of their type, size and nature. The extent of application of these requirements depends on the organization's operating environment and complexity.[15]

The standard provides a formal business continuity framework and will help an enterprise develop a BCP that will keep the business running during and following a disruption. It will also minimize the impact so the enterprise can resume normal service quickly, ensuring that key services and products are still delivered.

### Other Business Continuity Standards

Recognition of the value of business continuity has encouraged a dozen or more countries to establish their own standards and practices. Within some countries, such as the United States, United Kingdom, Australia/New Zealand and Singapore, we can observe dramatic growth in additional standards and practices. In addition to the ISO 22301 business continuity standard, several other well-known standards include:
- ANSI/ASIS/BSI BCM.01:2010:  Business Continuity Management Systems:  Requirements with Guidance for Use
- National Fire Protection Association:  NFPA 1600:2010:  Standard on Disaster/Emergency Management and Business Continuity Programs
- ASIS International:  ASIS SPC.1-2009:  Organizational Resilience:  Security, Preparedness, and Continuity Management Systems—Requirements with Guidance for Use
- Australia/New Zealand Standard AS/NZS 5050:  Business continuity—Managing disruption-related risk
- Singapore Standard SS540:  Business Continuity Management
- Canadian Standard:  CSA Z1600:  Emergency Management and Business Continuity Programs
- Government of Japan BCP Guideline
- Japanese Corporate Code—BCP
- ISO 22313:2012:  Societal security—Business continuity management systems—Guidance
- ISO 24762: Information technology—Security techniques—Guidelines for information and communications technology disaster recovery services
- National Association of Stock Dealers:  NASD 3510/3520:  Business Continuity Plans and Emergency Contact Information
- National Institute of Standards and Technology:  NIST SP 800-34:  Contingency Planning Guide for Federal Information Systems
- New York Stock Exchange:  NYSE Rule 446:  Corporate-Wide BCP

**Note:** A comprehensive list of BCM legislations, regulations, standards and good practice is available at *http://www.thebci.org.*

## 4.6 RISK, THREATS, VULNERABILITIES AND OPPORTUNITIES INHERENT IN THE USE OF IT

IT has become pervasive in current dynamic and often turbulent business environments. This major IT dependency implies a huge vulnerability that is inherently present in IT environments. System and network downtime has become far too costly for any organization these days, as doing business globally around the clock has become the standard. For example, downtime in the banking and health care sector is a risk factor accompanied by a wide spectrum of external threats such as errors and omissions, abuse, cybercrime, and fraud.

IT of course not only has the potential to support existing business strategies, but also to shape new strategies. In this mindset, IT becomes not only a success factor for survival and prosperity, but also an opportunity to differentiate and achieve competitive advantage. Often, these opportunities are driven by new technological evolutions such as BYOD (bring your own device) and cloud computing or concepts such as business process reengineering. It is important to always consider both inherent values and risk of such evolutions.

The use of IT obviously carries risk just as its use has potential rewards. The thinking that ignores risk or only considers the most superficial risk is fallacious in today's context because IT is an essential utility that underpins practically every business activity.

### Risk Categories

In assessing IT risk, each of the following risk categories can be considered:[16]
- Inherent risk
- Control risk
- Detection risk
- Residual risk

- **Inherent risk**—Inherent risk refers to the risk associated with an event in the absence of specific controls (i.e., the susceptibility of an area to error that could be material to, individual from or in combination with other errors, assuming that there are no related internal controls).

For example, the inherent risk associated with operating system (OS) security is ordinarily high because changes to, or even disclosure of, data or programs through OS security weaknesses could result in false management information or competitive disadvantage. By contrast, the inherent risk associated with security for a stand-alone personal computer, when a proper analysis demonstrates it is not used for business-critical purposes, is ordinarily low. Inherent risk for most areas of IT is ordinarily high because the potential effect of errors ordinarily spans several business systems and many users. In assessing the inherent risk, there should be consideration for pervasive and detailed IT controls. At the pervasive IT control level, there should be consideration regarding the level appropriate for the investigation area in question:
  - The integrity of IT management and IT management experience and knowledge
  - Changes in IT management
  - Pressures on IT management that may predispose them to conceal or misstate information (e.g., large business-critical project overruns, hacker activity)
  - The nature of the enterprise's business and systems (e.g., plans for electronic commerce, complexity of the systems, lack of integrated systems)
  - Factors affecting the enterprise's industry as a whole (e.g., changes in technology, IS staff availability)
  - The level of third-party influence on the control of the systems being audited (e.g., because of supply chain integration, outsourced IT processes, joint business ventures and direct access by customers)

At the detailed IT control level there should be consideration regarding the level appropriate for the investigation area in question:
  - The complexity of the systems involved
  - The level of manual intervention required
  - The susceptibility to loss or misappropriation of the assets controlled by the system (e.g., inventory, payroll)
  - The likelihood of activity peaks at certain times in the period of investigation
  - Activities outside the day-to-day routine of IT processing (e.g., the use of OS utilities to amend data)
  - The integrity, experience and skills of the management and staff involved in applying the IT controls

- **Control risk**—Control risk is the risk that an error that could occur in an area of investigation and be material to, individual from or in combination with other errors, and will not be prevented or detected and corrected on a timely basis by the internal control system. For example, the control risk associated with manual reviews of computer logs can be high because activities requiring investigation are often easily missed due to the volume of logged information. The control risk associated with computerized data validation procedures is ordinarily low because the processes are applied consistently. The assessment of the control risk should be scored as high, unless relevant internal controls are:
  – Identified
  – Evaluated as effective
  – Tested and proved to be operating appropriately
- **Detection risk**—Detection risk is the risk that the prescribed substantive procedures will not detect an error that could be material to, individual from or in combination with other errors. For example, the detection risk associated with identifying breaches of security in an application system is ordinarily high because logs for the entire period of the investigation are not available at the time when the investigation is made. The detection risk associated with identification of lack of disaster recovery plans (DRPs) is ordinarily low because existence is easily verified. In determining the level of substantive testing required, there should be consideration of:
  – The assessment of inherent risk
  – The conclusion reached on control risk following compliance testing

  The higher the assessment of inherent and control risk, the more evidence should normally be obtained from the performance of substantive audit procedures.
- **Residual risk**—Residual risk refers to the risk associated with an event when the controls in place to reduce the effect or likelihood of that event are taken into account.

## Risk Scenarios

In order to properly assess risk in a qualitative manner, it is necessary to develop risk scenarios that will be used in the risk assessment. Each scenario should be based on an identified risk and each risk should be identified in one or more scenarios. Each scenario is used to document the level of risk associated with the scenario in relation to the business objectives or operations that would be impacted by the risk event.

*COBIT 5 for Risk* describes a risk scenario as a description of a possible event that, when occurring, will have an uncertain impact on the achievement of the enterprise's objectives. The impact can be positive or negative.

The core risk management process requires risk needs to be identified, analyzed and acted on. Well-developed risk scenarios support these activities and make them realistic and relevant to the enterprise.

## Opportunities and Risk

Risk—the potential for events and their consequences—contains both:
- Opportunities for benefit (upside)
- Threats to success (downside)

### Business Process Reengineering

Radical improvements are not possible without increased risk. Business process reengineering (BPR) projects are frequently implemented when IT enablement is seen. BPR projects are known to have a high rate of failure. Risk associated with changing an existing process must be identified. The ISACA publication *IT Audit and Assurance Standards and Guidelines* identifies several risk areas to consider when planning a BPR project. The risk can be broken down into three broad areas:
1. Design risk
2. Implementation risk
3. Operation or rollout risk

1. **Design risk**

    A good design can improve profitability while satisfying customers. Conversely, a design failure will spell doom to any BPR project. It would be reckless to undertake new projects without dedicated resources capable of committing the time and attention necessary to develop a quality solution. Often, this type of detailed planning may consume more money and time than is available from key personnel. Recognition should be given to the risk that may occur in the BPR design.
    - **Sponsorship risk**—C-level management is not supportive of the effort. Insufficient commitment from the top is just as bad as having the wrong person leading the project. Poor communication is also a major problem.
    - **Scope risk**—The BPR project must be related to the vision and the specifications of the strategic plan. Serious problems will arise if the scope is improperly defined. It is a design failure if politically sacred processes and existing jobs are excluded from the scope of change.
    - **Skill risk**—Absence of radical out-of-the-box thinking will create a failure by dismissing new ideas that should have been explored. "Thinking big" is the most effective way to achieve the highest return on investment (ROI). Participants without broad skills will experience serious difficulty because the project vision is beyond their ability to define an effective action plan.
    - **Political risk**—Sabotage is always possible from people fearing a loss of power or resistance to change. Uncontrolled rumors lead to fear and subversion of the concept. People will resist change unless the benefits are well understood and accepted.

2. **Implementation risk**

    Implementation risk represents another source of potential failures that could occur during the BPR project. The most common implementation risk includes the following:
    - **Leadership risk**—C-level executives may fail to provide enough support for the project to be successful. Leadership failures include disputes over ownership and project scope. Management changes during the BPR project may signal wavering needs that may cause the loss of momentum. Strong sponsors will provide money, time and resources while serving as project champions with their political support.
    - **Technical risk**—Complexity may overtake the definition of scope. The required capability may be beyond that of prepackaged software. Custom functions and design may exceed IT's creative capability or available time. Delays in implementation could signal that the complexity of scope was underestimated. If the key issues are not fully identified, disputes will arise about the definitions of deliverables, which leads to scope changes during implementation.
    - **Transition risk**—The loss of key personnel may create a loss of focus during implementation.
    - **Personnel risk**—Personnel may feel burned out because of workload or their perception that the project is not worth the effort. Reward and recognition are necessary during transition to prevent the project from losing momentum.
    - **Scope risk**—Improperly defined project scope will produce excessive costs with schedule overruns (variance from schedule). Poor planning may neglect the human resources (HR) requirements, which will lead team members to feel that the magnitude of effort is overwhelming. The reaction will cause a narrowing of the scope during implementation, which usually leads to a failure of the original BPR objectives.

3. **Operation or rollout risk**

    It is still possible for the BPR project to fail after careful planning. Common failures during production implementation include negative attitudes and technical flaws. These problems manifest in the form of management risk, technical risk and cultural risk:
    - **Management risk**—Strong, respected leadership is required to resolve power struggles over ownership. Communication problems must be cured to prevent resistance and sabotage. Executive sponsors need to provide sufficient training to prevent an unsuccessful implementation.
    - **Technical risk**—Insufficient support is the most obvious cause of failure in a rollout. Inadequate testing leads to operational problems caused by software problems. Data integrity problems represent a root problem capable of escalating into user dissatisfaction. Perceptions of a flawed system will undermine everyone's confidence.
    - **Risk of nonacceptance of change**—Resistance in the organization is a result of failing to achieve user buy-in. Resistance contributes to the erosion of the benefits. Effective training is often successful in solving user problems. Dysfunctional behavior will increase unless the new benefits are well understood and achieved.

### Cybersecurity

Cybersecurity is an increasingly important risk facing enterprises. Cybersecurity plays a significant role in the ever-evolving cyber landscape and digitization of enterprises. New trends in mobility and connectivity present a broader range of challenges as new attacks continue to develop along with emerging technologies. The importance of understanding cybersecurity risk in a digitized world was stressed by the World Economic Forum: "Pervasive digitization, open and interconnected technology environments, and sophisticated attackers, among other drivers, mean that the risk from major cyber events could materially slow the pace of technological innovation over the coming decade. Addressing the problem will require collaboration across all participants in the 'cyber resilience ecosystem.'"[17]

### Cloud Computing

The promise of cloud computing is arguably revolutionizing the IT services world by transforming computing into a ubiquitous utility, leveraging attributes such as increased agility, elasticity, storage capacity and redundancy to manage information assets.

As with any emerging technology and new initiative, cloud computing offers the possibility of high rewards, but it can also bring the potential for high risk. Much of the risk associated with cloud computing is not new and can be found in enterprises today; for example, managing third-party service providers, data management, disaster recovery, etc. Some examples of cloud computing risk for the enterprise that need to be managed include:

- **Selecting a cloud service provider (CSP)**—Companies need to consider history, reputation and internal controls. They need to ensure that their data are available and can be tracked.
- **Cloud service provider auditing**—Companies may have trouble auditing their data-outsourcing provider. It is important to ensure that this is provided for in the contract for services.
- **Responsibility for security**—This is confusing in the cloud environment and many cloud providers believe it is the client's responsibility. Responsibility must be established in any negotiation for services.
- **Third-party access to sensitive data**—Giving a third party access to data creates a risk of compromise to sensitive data.
- **Sharing information assets with other customers**—Public clouds allow high-availability systems to be developed at service levels not achievable in private networks; however, this creates the risk of comingling information assets with other cloud customers, including competitors.
- **Compliance with legislation**—Complying with different legislation in different countries and regions can be a challenge with cloud computing.
- **Disaster recovery preparation**—In the event of a disaster recovery situation, information or data may not immediately be located due to the dynamic nature of cloud computing.

This risk can lead to a number of different threat events. The CSA lists the following as the top cloud computing threats:[18]
1. Data breaches
2. Data loss
3. Account hijacking
4. Insecure application programming interfaces (APIs)
5. Denial-of-service (DoS)
6. Malicious insiders
7. Abuse of cloud services
8. Insufficient due diligence
9. Shared technology issues

To conduct a risk-based assessment of the cloud computing environment, there are generic risk frameworks such as COSO Enterprise Risk Management—Integrated Framework. There are also IT domain-specific risk frameworks, practices and process models, such as ISO 27001 and IT Infrastructure Library (ITIL). Bottom-up guidance specific to cloud computing also exists from various bodies such as the Cloud Security Alliance (CSA), European Network and Information Security Agency (ENISA), the US National Institute of Standards and Technology (NIST) and COBIT 5/ISACA.
- The Cloud Controls Matrix released by CSA is designed to provide security principles to guide cloud vendors and assist prospective cloud clients in assessing overall security risks of a CSP.

- The NIST guidelines on security and privacy in public cloud computing (NIST Special Publication [SP] 800-144), contain the guidelines required to address public cloud security and privacy.
- Risk IT, based on the COBIT framework from ISACA, fills the gap between generic risk management frameworks and domain-specific frameworks based on the premise that IT risk is not purely a technical issue.
- *Controls and Assurance in the Cloud: Using COBIT 5* provides detailed guidance of controls and assurance in the cloud. As an illustration, this document discussed the relationship between the service and deployment models in the cloud and their cumulative risk (**figure 4.13**).

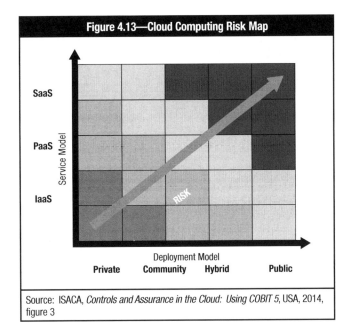

Source: ISACA, *Controls and Assurance in the Cloud: Using COBIT 5*, USA, 2014, figure 3

### Social Media
Initiated as a consumer-oriented technology, social media is increasingly being leveraged as a powerful, low-cost tool for enterprises to drive business objectives, such as enhanced customer interaction, greater brand recognition and more effective employee recruitment. While social media affords enterprises many potential benefits, information risk professionals are concerned about its inherent risk such as data leakage, malware propagation and privacy infringement. Enterprises seeking to integrate social media into their business strategy must adopt a cross-functional, strategic approach that addresses risk, impacts and mitigation steps, along with appropriate governance and assurance measures.

### Big Data
Somewhat related to the rising importance of social media is the increasing attention on big data challenges. According to an ISACA white paper on the impacts and benefits of big data,[19] this is a popularized term for a set of problems and techniques concerning the management and exploitation of very large sets of information. The notion of what a "very large" set of information is can be different for all enterprises. In essence, enterprises are faced with the issue of big data whenever traditional techniques and tools are no longer sufficient to capture, manage and process the data in a reasonable amount of time.

Information ranges from structured to unstructured information, including customer and employee data, metadata, trade secrets, email, video and audio. This can present challenges to the enterprise, which must find a way to govern data in alignment with business requirements, while still allowing for the free flow of information and innovation. Challenges arise when data is:
- Spread across isolated, complex silos
- Duplicated/redundant
- Uncoordinated and lacking standardization

The benefits of big data are plentiful. When correctly managed, big data can provide a more accurate view of the behaviors of consumers in the marketplace, operational efficiency, and potential product development. By using big data, predictions may be more accurate and improvement projects can target exact pain points.

Practically, issues regarding big data can be categorized according to three dimensions:
• Variety of information
• Velocity of information
• Volume of information

These three dimensions, commonly referred to as the three V's of big data, are all growing and becoming even more complex. Information is coming from an ever wider array of channels, sensors and formats. Enterprises seeking to respond to events as they happen must find ways to process data extremely quickly. Information can be altered at a fast pace, which adds to the complexity of the issue.

Building the business case to implement the proper practices to handle big data is challenging. The clear value needs to be demonstrated by the outcomes of big data projects. In the 2014 report, "Generating Value from Big Data Analytics," ISACA proposes that enterprises need to ask the following questions to address key potential challenges before they can, with confidence, realize the gains from big data analytics:
• Does the enterprise have the people, processes and technology in place to build capabilities that will make productive use of data that the enterprise has collected?
• Has the enterprise established roles and responsibilities and identified stakeholders?
• Does the enterprise have (or can it get) data on which to apply advanced analytics?

### Consumerization of IT and Mobile Devices
Mobile devices have had a profound impact on the way business is conducted and on behavior patterns in society. They have greatly increased productivity and flexibility in the workplace, to the extent that individuals are now in a position to work from anywhere at any given time. Likewise, the computing power of smart devices has enabled them to replace desktop personal computers (PCs) and laptops for many business applications.

Manufacturers and service providers alike have created both new devices and new business models such as mobile payments or subscription downloads using a pay-as-you-go model. Simultaneously, consumerization of devices has relegated enterprises, at least in some cases, to followers rather than opinion leaders in terms of which devices are used and how they are used.

The impact of using mobile devices falls into two broad categories:
• The hardware itself has been developed to a level at which computing power and storage are almost equivalent to PC hardware. In accordance with Moore's Law, a typical smartphone represents the equivalent of what used to be a midrange machine a decade ago.
• New mobile services have created new business models that are changing organizational structures and society as a whole.

Consumerization is not limited to devices. New, freely available applications and services provide better user experiences for things like note-taking, video conferencing, email and cloud storage than their respective corporate-approved counterparts. Instead of being provided with company-issued devices and software, employees are using their own solutions that better fit with their lifestyle, user needs and preferences.[20]

**Example: BYOD—Great Opportunity, Great Risk to Manage**[21]

In 1981 the world of data processing would forever change with the introduction of the IBM® PC. PCs were not new, but a PC with business credentials—provided by IBM—was revolutionary.

Many purchased their first PC for use at work. This new device, considered a toy by computing professionals, freed users from the constraints (some would say tyranny) of the data processing (DP) department. By using IBM Lotus® 1-2-3®, end users could create their own tools to enable them to do their jobs, reducing or eliminating dependency on the DP department.

The introduction of terminal emulators a year later enabled PCs to be used as dumb terminals, but with the ability to store data locally. The introduction of local area networks (LANs) completed the first cycle, enabling users to exchange information easily.

By the late 1980s, DP professionals awakened to the potential (or threat) and tried to exert control over end users and their use of technology, largely without success. Spreadsheets had become invaluable and irreplaceable tools in the workplace, and end users were unwilling to give up their newfound freedom.

Thus the age of IT was launched. The switch occurred when end users could create their own unique solutions to business problems. IT became (kicking and screaming in many cases) an enabler to business productivity.

In retrospect, this also heralded the age of information security. Post-PC security was an order of magnitude more difficult, and post-Internet security several more orders of magnitude greater. BYOD simply follows the curve of complexity.

In a remarkably short time the title DP director was replaced with chief information officer (CIO). Suddenly IT was invited into the executive suite and was given a prime seat at the table, all because end users could finally access corporate information, analyzing and manipulating it in ways never thought possible by the DP departments.

BYOD holds the same transformative potential, and just like the PC before it, it cannot be stopped. Small but powerful computers (smartphones) are carried by almost everyone in a business environment, and the trend is for that to increase.

Secretaries and soldiers, executives and custodians, all routinely carry devices vastly more powerful than the original PCs. Moreover, in many cases these people prefer their own devices to those offered by their employers, potentially saving their employers significant amounts of money.

IT can continue to complain about the risk, whine about how difficult these devices make their lives and even impede their further adoption and deployment; however, this behavior hastens the move toward outsourcing to vendors sympathetic to the needs of customers and end users.

It is not a question of if BYOD will be adopted, but when. Smart enterprises are directing their IT departments to find a way to safely enable BYOD. Even the US Department of Defense is adopting BYOD. If they can do it, every enterprise should be able to do it.

**Example: COPE (Corporate Owner, Personally Enabled)**[22]

COPE is a policy and trend alternative to BYOD. It is an IT business strategy through which an organization buys and provides computing resources and devices to be used and managed by employees. It is a hybrid that sits between free-for-all BYOD and traditional company-owned computers that forbade personal use and held zero expectations of privacy for employees.

# 4.7 TYPES OF BUSINESS RISK, EXPOSURES AND THREATS THAT CAN BE ADDRESSED USING IT RESOURCES

Whether for compliance, effectiveness or efficiency, IT enablement of business has dramatically increased in recent years. As complexities of business evolve, the integral role of IT is extended to that of assisting business in handling risk, which is an inevitable part of business strategies, processes and operations. Therefore, included as part of business enablement, IT has the task of assisting in the management of business risk, including examples such as access controls, process controls and continuous monitoring.

## IT Risk Analytics, Monitoring and Reporting

Just as IT can be applied to yield results that were not previously possible in many fields, so too can it prove itself in the field of risk management. More significantly, IT can facilitate the wiring-up, locking down and constant surveillance of the business; and, specifically in the domain of risk management information systems, IT will be relied on for advanced risk analytics, monitoring and reporting.

There are various types of software tools that can be used to perform different types of control monitoring. These tools are organized into the following groups based on the focus of the tool as it relates to monitoring internal control:
- Transaction data
- Conditions
- Changes
- Processing integrity
- Error management

Although automated monitoring tools can be highly effective in a number of situations, they are not without limitations and generally **CANNOT**:
- Determine the propriety of the accounting treatment afforded to individual transactions. This must be determined based on the underlying substance of the transaction itself.
- Address whether an individual transaction was accurately entered into the system. Rather, they can deal only with whether the transactions met internal standards for acceptable transactions (for example, it was valid).
- Determine whether all relevant initial transactions were entered into systems in the proper period. This is typically dependent on human activity.

More information on risk monitoring tools can be found in the CRISC Review Manual.

## Segregation of Duties

Segregation of duties is a key component in maintaining a strong internal control environment because it reduces the risk of fraudulent transactions. When duties for a business process or transaction are segregated, it becomes more difficult for fraudulent activity to occur because it would involve collusion among several employees. There is a wide variety of automated (i.e., IT-based) compliance solutions that address the issue of segregation of duties. Prior to these tools being available, companies typically addressed segregation of duties through a combination of controls:
- Defining transaction authorizations
- Assigning custody of assets
- Granting access to data
- Reviewing or approving authorization forms
- Creating user authorization tables

The automated tools that are typically in use aim to duplicate these efforts as well as provide the enterprise with reporting functionality on segregation of duties violations (i.e., detective controls) and put in place preventive controls. In general, the automated control system contains three elements:
- **Access controls**—Controls that restrict access to the underlying business systems to ensure that only authorized individuals have access
- **Process controls**—Controls that restrict the activities performed by those users
- **Continuous monitoring**—Employs automation to detect, after the fact, system transactions, setup, or data changes that contravene corporate policy

Each of these controls must be viewed with respect to business transactions, system setup and business data because each of these elements must be secure to ensure valid internal controls. For example, each of these may be subject to access controls to ensure that only authorized individuals can view or change them. Similarly, process controls will ensure that only correct actions are taken on each and monitoring controls will track any invalid operations after the fact.

Although enterprises may put many of the typical controls in place, as ERP implementations become more widespread and larger in scope, it becomes overwhelming for departments to keep their manual controls around segregation of duties up to date. As enterprises grow, resources are added and an employee's job functions change to mirror the ongoing changes within an organization. This causes these manual controls to become quickly outdated. By not automating segregation of duties controls, there is, potentially, the issue of these controls becoming a barrier in serving the customer. As manual authorizations are often time consuming and require another step in any business process, this takes time away from serving the customer. These automated compliance solutions aim to provide enterprises with timely and efficient internal controls that do not disrupt their normal business process.

### Risk Management Information System

IT can contribute, directly or indirectly, to the active management of other classes of enterprise risk. A risk management information system (RMIS) can be a very effective tool in monitoring all risk that impacts the enterprise. The danger is that many important classes of risk may be omitted from consideration by the system—many organizations use an RMIS only to monitor physical assets and as a tool principally for the benefit of the insurance function. IT can also assist other risk classes by the specific design characteristics of the IT systems. An accounting system that is easy to validate—a trial balance that can be performed in minutes, for example—assists the auditor to ensure that transactions have not been "lost." Standardized audit trails, recovery routines and data logs will make the detection of fraud, integrity and validity much easier. Similarly, the operation of business risk controls—such as market risk or credit risk for bankers—can be facilitated by their incorporation in IT systems.

### Locked-down Operations

IT can be used to build in business process controls. Applications enforce business rules: mandatory fields required before a record can be saved, lookup fields used to ensure that valid codes are entered, approvals above a certain value routed via work flow for management approval, ATMs not discharging money without a valid account and PIN combination, etc. This is an essential part of controlling normal business operations. It also allows HR to be channeled to do other things as long as the IT systems reliably perform the handling as well as the checking and balancing. For instance, it is much easier to create a letter according to an enterprise's template than to start one from scratch. And in using the template, it is far more likely to achieve a compliant result. IT can maintain a watchful eye on the enterprise information (constant surveillance) and maintain records needed for the provision of evidence in litigation or with which to prosecute.

### Decision Support, Risk Analytics and Reporting

Advanced risk-return decision making requires advanced IT support. It is not feasible to manually calculate the riskiness of today's credit portfolio. Data volumes are huge and the sophisticated models require precise calibration and consistent fine tuning. Quantitative analysis will inevitably turn to IT for the large-scale analysis of risk as IT is typically used when the tasks are challenging and a large number of inputs and mathematical complexity are involved. The objective of all management information systems is to enable faster and better decision making. In the case of risk management information, the decision making regards known and potential risk. The goal of risk management information systems is to achieve compatible and efficient IT systems for capturing, analyzing and, ultimately, reporting risk of all types across the entire enterprise. The consumers of output from the risk management information system are both internal—across all layers of management—and external. Automating risk information management can assist in the embedding of required practices into the enterprise by making "business as usual" risk management activities efficient rather than onerous.

## 4.8 RISK APPETITE AND RISK TOLERANCE

To effectively govern enterprise and IT risk, there must be an:
- Understanding and consensus with respect to the risk appetite and risk tolerance of the enterprise
- Awareness of risk and the need for effective communication about risk throughout the enterprise
- Understanding of the elements of risk culture

### Risk Appetite

Risk appetite is the amount of risk an entity is prepared to accept in pursuit of its mission (or vision). When considering the risk appetite levels for the enterprise, two major factors are important:
- The enterprise's objective capacity to absorb loss, e.g., financial loss, reputation damage
- The (management) culture or predisposition toward risk taking—is it cautious or aggressive? What is the amount of loss the enterprise wants to accept to pursue a return?

Risk appetite can in practice be defined in terms of combinations of frequency and magnitude of risk. Risk appetite can and will be different between enterprises. There is no absolute norm or standard of what constitutes acceptable and unacceptable risk.

Risk appetite can be defined using risk maps. Different bands of risk significance can be defined, indicated by shaded zones on the risk map, as shown in **figure 4.14**.

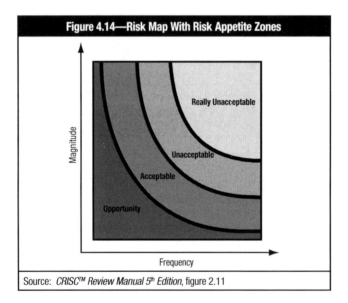

Source: *CRISC™ Review Manual 5th Edition*, figure 2.11

In **figure 4.14**, four bands of significance are defined:
- **Really unacceptable risk**—The enterprise estimates this level of risk is far beyond their normal risk appetite. Any risk found to be in this band might trigger immediate risk response.
- **Unacceptable risk**—Indicates elevated risk (i.e., also above acceptable risk appetite). The enterprise might, as a matter of policy, require mitigation or another adequate response to be defined within certain time boundaries.
- **Acceptable risk**—This indicates normal acceptable levels of risk, usually with no special action required except for maintaining the current controls or other responses.
- **Opportunity**—This indicates very low risk, where (cost) saving opportunities may be found by decreasing the degree of control, or where opportunities for assuming more risk might arise.

The previous risk appetite scheme is an example. Every organization must define their own risk appetite levels and review them on a regular basis. This definition should be in line with the overall risk culture that the organization wants to express (i.e., ranging from very risk adverse to risk taking/opportunity seeking). There is no universal good or wrong, but it needs to be defined, well understood and communicated.

## Risk Tolerance

Risk tolerance is the acceptable level of variation that management is willing to allow for any particular risk as the enterprise pursues its objectives. This is often best measured in the same units as those used to measure the related objective. In other words, risk tolerance is tolerable deviations from the level set by the risk appetite definitions. For example, standards require projects to be completed within the estimated budgets and time, but an overrun of 10 percent budget or 20 percent time are tolerated.

On risk appetite and risk tolerance, the following guidance applies:
- Risk appetite and risk tolerance go hand in hand. Risk appetite is defined at the enterprise level and is reflected in policies set by the executives; at lower (tactical) levels of the organization, or in some entities of the organization, exceptions can be tolerated (or different thresholds defined) as long as the overall exposure at the enterprise level does not exceed the set risk appetite. Any business initiative includes a risk component, so management should have the discretion to exceed risk tolerance levels and pursue some new opportunities. Organizations where policies are "written in stone" could lack the agility and innovation to exploit new business opportunities. Conversely, there are situations in which policies are based on specific legal, regulatory or industry requirements and it is appropriate to have no risk tolerance for failure to comply.
- Risk appetite and tolerance should be defined and approved by senior management and clearly communicated to all stakeholders. A process should be in place to review and approve any exceptions to such standards.
- Risk appetite and tolerance change over time. New technology, new organizational structures, new market conditions, new business strategy and many other factors require the organization to reassess its risk portfolio at regular intervals, and also require the organization to reconfirm their risk appetite at regular intervals, triggering risk policy reviews. In this respect, an enterprise needs to understand that the better risk management it has in place, the more risk that can be taken in pursuit of return.
- The cost of mitigation options can affect risk tolerance. There may be circumstances in which the cost/business impact of risk mitigation options exceeds an enterprise's capabilities/resources, thus forcing higher tolerance for one or more risk conditions. For example, if a regulation says that "sensitive data at rest must be encrypted," yet there is no feasible encryption solution or the cost of implementing a solution is grossly impactful, then the organization may choose to accept the risk associated with regulatory noncompliance—a risk trade-off.

The subject of risk appetite goes to the heart of the relationship between the board, management and (if there is one appointed) chief risk officer (CRO). The board sets a risk appetite, which management subscribes to by installing suitable controls to contain risk. Meanwhile, the CRO or internal auditor will furnish objective reports on the system of internal control. These audit reports will review the extent to which residual risk, after taking account of controls, is acceptable and that, in turn, indicates whether this risk falls in line with the defined risk appetite. This dependency cycle is extremely important and hinges on respective perceptions of risk appetite.

## Process to Determine Risk Appetite

Risk appetite is in many ways a vague concept and something that top management of many enterprises find challenging in terms of what needs to be formalized and how. One way to approach an enterprise's risk appetite is to go through the following five steps:
- **Develop a model**—The model should seek to capture the essential features of risk appetite for each business area in the enterprise in terms of fixed categories of low, medium and high impact or likelihood of risk. Two categories of medium can be used to avoid the practice of placing everything in one medium category. For example, consider in the model enterprise strategic objectives a risk profile that is aligned to the business plans. Determine risk thresholds, and formalize and ratify a risk appetite statement.
- **Define benchmarking factors**—Using this model, define the factors that can be used to benchmark the level of risk that is deemed acceptable to the business.
- **Define levels of risk**—For each of the factors from the model, define what may be viewed as low, medium or high levels of risk tolerance in terms of what can be tolerated and what needs to be much more tightly controlled.
- **Determine risk tolerances**—Go through each part of the business, using models and set scales, to determine where risk tolerances are deemed low, medium or high. For example, service levels for a system uptime require 99.5 percent system availability on a monthly basis; however, isolated cases of 99.2 percent may be tolerated.

- **Provide strong corporate messages**—Messages about levels of risk tolerance to managers in each part of the enterprise. Ensure that they are able to use this information to drive the way risk is assessed and managed. For example, significant areas where there is the potential for financial misreporting may be seen as having a low risk tolerance and, therefore, may be subject to tight risk triggers at both corporate and local levels.

> **Note:** COSO has published thought papers in these areas:
> - *Enterprise Risk Management: Understanding and Communicating Risk Appetite* (January 2012): This thought paper aims to help organizations develop, better articulate, and implement risk appetite. It provides examples of statements of risk appetite and emphasizes the notion that risk appetite should be clearly defined, communicated by management, embraced by the board and continually monitored and updated.
> - *Enterprise Risk Management for Cloud Computing* (June 2012): This thought paper provides guidance on following the principles of the COSO ERM Integrated Framework to assess and mitigate the risk arising from cloud computing.

COBIT 5 also references risk appetite and risk tolerance in process EDM03 *Ensure risk optimisation*. **Figure 4.15** depicts the key governance practices around risk management:
- EDM03.01 Evaluate Risk Management
- EDM03.02 Direct Risk Management
- EDM03.03 Monitor Risk Management

### Figure 4.15—EDM03—Ensure Risk Optimisation

**EDM03 Process Practices, Inputs/Outputs and Activities**

| Governance Practice | Inputs | | Outputs | |
|---|---|---|---|---|
| **EDM03.01 Evaluate risk management.** Continually examine and make judgement on the effect of risk on the current and future use of IT in the enterprise. Consider whether the enterprise's risk appetite is appropriate and that risk to enterprise value related to the use of IT is identified and managed. | **From** | **Description** | **Description** | **To** |
| | APO12.01 | Emerging risk issues and factors | Risk appetite guidance | APO12.03 |
| | | | Approved risk tolerance levels | APO12.03 |
| | Outside COBIT | Enterprise risk management principles | Evaluation of risk management activities | APO12.01 |

**Activities**

1. Determine the level of IT-related risk that the enterprise is willing to take to meet its objectives (risk appetite).
2. Evaluate and approve proposed IT risk tolerance thresholds against the enterprise's acceptable risk and opportunity levels.
3. Determine the extent of alignment of the IT risk strategy to enterprise risk strategy.
4. Proactively evaluate IT risk factors in advance of pending strategic enterprise decisions and ensure that risk-aware enterprise decisions are made.
5. Determine that IT use is subject to appropriate risk assessment and evaluation, as described in relevant international and national standards.
6. Evaluate risk management activities to ensure alignment with the enterprise's capacity for IT-related loss and leadership's tolerance of it.

| Governance Practice | Inputs | | Outputs | |
|---|---|---|---|---|
| **EDM03.02 Direct risk management.** Direct the establishment of risk management practices to provide reasonable assurance that IT risk management practices are appropriate to ensure that the actual IT risk does not exceed the board's risk appetite. | **From** | **Description** | **Description** | **To** |
| | APO12.03 | Aggregated risk profile, including status of risk management actions | Risk management policies | APO12.01 |
| | | | Key objectives to be monitored for risk management | APO12.01 |
| | Outside COBIT | Enterprise risk management (ERM) profiles and mitigation plans | Approved process for measuring risk management | APO12.01 |

### Figure 4.15—EDM03—Ensure Risk Optimisation (cont.)

**EDM03 Process Practices, Inputs/Outputs and Activities**

| Activities |
|---|
| 1. Promote an IT risk-aware culture and empower the enterprise to proactively identify IT risk, opportunity and potential business impacts. |
| 2. Direct the integration of the IT risk strategy and operations with the enterprise strategic risk decisions and operations. |
| 3. Direct the development of risk communication plans (covering all levels of the enterprise) as well as risk action plans. |
| 4. Direct implementation of the appropriate mechanisms to respond quickly to changing risk and report immediately to appropriate levels of management, supported by agreed-on principles of escalation (what to report, when, where and how). |
| 5. Direct that risk, opportunities, issues and concerns may be identified and reported by anyone at any time. Risk should be managed in accordance with published policies and procedures and escalated to the relevant decision makers. |
| 6. Identify key goals and metrics of risk governance and management processes to be monitored, and approve the approaches, methods, techniques and processes for capturing and reporting the measurement information. |

| Governance Practice | Inputs | | Outputs | |
|---|---|---|---|---|
| | From | Description | Description | To |
| **EDM03.03 Monitor risk management.** Monitor the key goals and metrics of the risk management processes and establish how deviations or problems will be identified, tracked and reported for remediation. | APO12.02 | Risk analysis results | Remedial actions to address risk management deviations | APO12.06 |
| | APO12.04 | • Opportunities for acceptance of greater risk<br>• Results of third-party risk assessments<br>• Risk analysis and risk profile reports for stakeholders | Risk management issues for the board | EDM05.01 |

| |
|---|
| 1. Monitor the extent to which the risk profile is managed within the risk appetite thresholds. |
| 2. Monitor key goals and metrics of risk governance and management processes against targets, analyse the cause of any deviations, and initiate remedial actions to address the underlying causes. |
| 3. Enable key stakeholders' review of the enterprise's progress towards identified goals. |
| 4. Report any risk management issues to the board or executive committee. |
| Source: ISACA, *COBIT 5: Enabling Processes*, USA, 2012, pages 40-41 |

## 4.9 QUANTITATIVE AND QUALITATIVE RISK ASSESSMENT METHODS

Several methods for risk assessment exist, ranging between high level and mostly qualitative to very quantitative, with hybrid methods in between. The organizations' culture, resources, skills and knowledge, environment, risk appetite, etc., will determine which methodology to use.

The different methods have some common limitations:
- No method is fully objective, and results of risk assessments are always dependent on the person performing them and his/her skills and views.
- IT risk-related data (such as loss data and IT risk factors) are very often of poor quality; putting in place some structures or models can help to achieve more objectivity and can provide at least a basis for discussion in the risk analysis.

After identifying risk, risk assessment is the first process in the risk management process. Enterprises use risk assessment to determine the:
- Extent of the potential threat
- Potential impact of the threat
- Risk associated with business processes, operations and IT systems throughout their development life cycle and use

The entire risk management process should be managed at multiple levels in the enterprise, including the operational, project and strategic levels and should form part of the risk management practice.

## Qualitative Risk Assessment

Qualitative risk assessment methods are usually applied when only limited or low quality information is available. A qualitative risk assessment approach uses expert opinions to estimate the likelihood and business impact of adverse events. The likelihood and the magnitude of impact are estimated using a scale. These scales can vary depending on the circumstances and different environments.

When qualitative risk assessments are used, the following strengths, limitations and weaknesses apply:
- In situations where there is only limited or low quality information available, qualitative risk analysis methods are usually applied.
- The major disadvantages of using the qualitative approach are: a high level of subjectivity, great variance in human judgment and lack of a standardized approach during the assessment.
- A qualitative risk assessment is usually less complex compared to quantitative analysis, and consequently, it is also less expensive.

## Quantitative Risk Assessment

The essence of quantitative risk assessment is to determine the likelihood and impact of the event based on statistical methods and data and from there calculate risk over a certain period in a certain environment.

When to use quantitative risk assessments and their strengths, limitations and weaknesses follows:
- A quantitative risk assessment is more objective because it is based on formal empirical data.
- Using purely quantitative methods requires good and reliable data on past and comparable events, and obtaining the data is in many cases very difficult unless the organization has already embraced process improvement and follows an approach such as Six Sigma for IT monitoring and productivity improvement.

## Combining Qualitative and Quantitative Methods—Toward Probabilistic Risk Assessment

Both techniques have some advantages and disadvantages; furthermore, neither of the approaches described above seems to meet all the requirements for the management of IT risk to support extensively overall ERM processes. The complex environment requires more flexible methods.

Analysis based on subjective opinions or estimated data might not be sufficient. There is still the question of certainty: How certain can we be about the results of risk assessment? Some specialized methods exist to increase reliability of risk assessments, but these require deep statistical skills:
- **Probabilistic risk assessment**—Involves working like a mathematical model to construct the qualitative risk assessment approach while using the quantitative risk assessment techniques and principles. In a simple way the statistical models are used and missing data to populate these models are collected using qualitative risk assessment methods (e.g., interviews, Delphi method, etc.).
- **Monte Carlo simulation**—A powerful method for combining qualitative and quantitative approaches. It involves working on the basis of a normal deterministic simulation model as described previously, but iteratively evaluating the model using sets of random numbers as inputs. While deterministic models will provide the expected value, Monte Carlo simulation will give the value as a probability distribution based on the quality of the information provided.

## Practical Guidance on Analyzing Risk

The selection for qualitative or quantitative risk analysis depends on many factors:
- User needs—Is there a need for highly accurate data or is a qualitative approach adequate?
- Availability and quality of the data related to IT risk
- Time available for risk analysis
- Level of comfort and expertise of those experts who are giving input

Statistical data may be available in different quantities and quality, ranging on a continuous scale from almost nonexistent to widely available. At the higher end of the scale (i.e., when a wide choice of statistical data is available), a quantitative assessment may be the preferred risk assessment method; at the other end of the scale, with very little, incomplete or poor data, a qualitative assessment will be the only available solution. Hybrid risk assessment methods may be applied to situations in between both extremes described.

Section Two: Content — Chapter 4—Risk Optimization

## 4.10 RISK MITIGATION STRATEGIES RELATED TO IT IN THE ENTERPRISE

The purpose of defining a risk response is to bring risk in line with the defined risk tolerance for the enterprise after due risk analysis. In other words, a response needs to be defined such that future residual risk (current risk with the risk response defined and implemented) is well within risk tolerance limits (usually depending on budgets available).

Risk mitigation is the management of risk through the use of countermeasures and controls. This is not a one-time effort; rather, it is part of the risk management process cycle.

When risk analysis, after weighing risk vs. potential return, has shown that risk is not aligned with the defined risk tolerance levels, a response is required. When the analysis shows risk deviating from the defined tolerance levels, a response needs to be established. This response can be any of the four possible responses explained in the following sections: risk avoidance, risk reduction/mitigation, risk sharing/transfer or risk acceptance.

### Risk Avoidance

Avoidance means exiting the activities or conditions that give rise to risk. Risk avoidance applies when no other risk response is adequate. This is the case when:
- There is no other cost-effective response that can succeed in reducing the likelihood and magnitude below the defined thresholds for risk appetite.
- The risk cannot be shared or transferred.
- The risk is deemed unacceptable by management.

Some IT-related examples of risk avoidance may include relocating a data center away from a region with significant natural hazards or declining to engage in a very large project when the business case shows a notable risk of failure.

### Risk Reduction/Mitigation

Reduction means that action is taken so the risk can be detected, followed by actions to reduce the likelihood or impact of a risk, or both. The most common ways of responding to risk include:
- Strengthening overall IT risk management practices (i.e., implement sufficiently mature IT risk management processes as defined by ISACA's Risk IT framework).
- Introducing a number of control measures intended to reduce the likelihood of an adverse event happening and/or to reduce the business impact of an event, should it happen. This is discussed in the remainder of this section.

### Risk Sharing/Transfer

Sharing means reducing risk likelihood or impact by transferring or otherwise sharing a portion of the risk. Common techniques include insurance and outsourcing. Examples include taking out insurance coverage for IT-related incidents, outsourcing part of the IT activities or sharing IT project risk with the provider through fixed price arrangements or shared investment arrangements. Some IT-related examples of risk sharing or transfer may include taking out insurance or outsourcing.

### Risk Acceptance

Acceptance means that no action is taken relative to a particular risk, and loss is accepted when/if it occurs. This is different from being ignorant of risk; accepting risk assumes that the risk is known, i.e., an informed decision has been made by management to accept it. If an enterprise adopts a risk acceptance stance, it should consider carefully who can accept the risk—even more so with IT risk. IT risk should be accepted only by business management (and business process owners), in collaboration with and supported by IT, and acceptance should be communicated to senior management and the board. If a particular risk is assessed to be extremely rare but very important (catastrophic), and insurance premiums are prohibitive, management can decide to accept it.

ISACA's publication *The Risk IT Practitioner Guide*[23] includes examples of risk response and offers more detailed guidance on how to select and prioritize risk response. The risk response process and prioritization process is depicted in **figure 4.16**.

*Chapter 4—Risk Optimization*  *Section Two: Content*

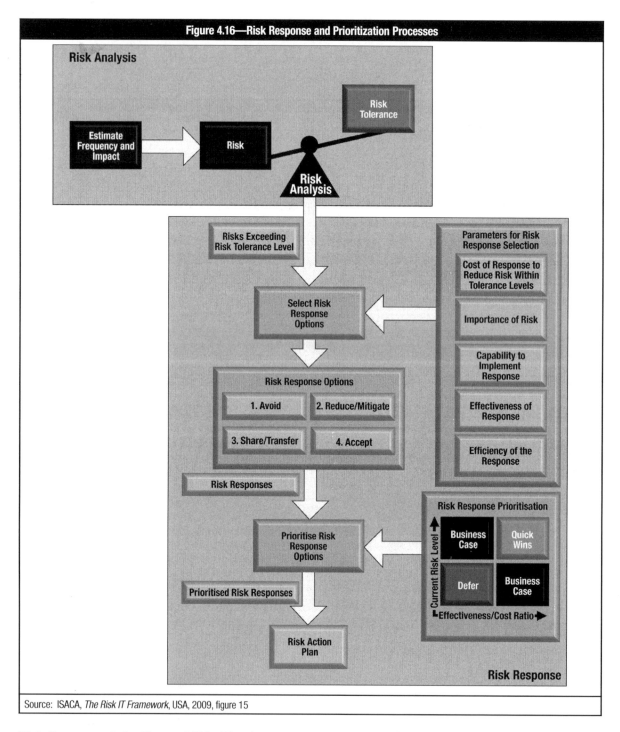

Figure 4.16—Risk Response and Prioritization Processes

Source: ISACA, *The Risk IT Framework*, USA, 2009, figure 15

## Risk Response Selection and Prioritization

The four previous sections listed the available risk response options. What follows is a brief discussion of the selection of an appropriate response (i.e., given the risk at hand, how to respond and how to choose between the available response options). The following parameters need to be taken into account in this process:

- **Cost of the response**—In the case of risk transfer, the cost of the insurance premium; in the case of risk mitigation, the cost (capital expense, salaries, consulting) to implement control measures
- **Importance of the risk addressed by the response**—Its position on the risk map (which reflects combined likelihood and magnitude of impact levels)

- **The enterprise's capability to implement the response**—When the enterprise is mature in its risk management processes, more sophisticated responses can be implemented; when the enterprise is rather immature, some basic responses may be better.
- **Effectiveness of the response**—The extent to which the response will reduce the impact and magnitude of the risk
- **Efficiency of the response**—The relative benefits promised by the response

It is likely that the aggregated required effort for the mitigation responses (i.e., the collection of controls that need to be implemented or strengthened) will exceed available resources. In this case, prioritization is required. Using the same criteria as for risk response selection, risk responses can be placed in a quadrant offering three possible options:
- **Quick wins**—Very efficient and effective responses to high risk
- **Business case to be made**—More expensive or difficult responses to high risk or efficient and effective responses to lower risk, both requiring careful analysis and management decision on investments; the Val IT framework approach may be applied here
- **Deferral**—Costly responses to lower risk

For that reason, the enterprise has to select and prioritize risk responses using the following criteria:
- **Effectiveness of the response**—The extent to which the response will reduce likelihood and the impact of adverse events
- **The enterprise's capability to implement the response**
- **Importance of the risk addressed by the response**—Its position on the risk map (which reflects combined likelihood and impact values)
- **Efficiency of the response**—The relative benefits promised by the response in comparison to:
  – Other IT-related investments—Investing in risk response measures always competes with other IT (or non-IT) investments).
  – Other responses—One response may address several instances of risk while another may not.

### Developing a Risk Action Plan
Risk action planning should be run as a project, with a defined start and end date. The end date is often used to determine the critical path of the project, which refers to those elements of the project that may have a direct impact on whether the end date can be met. A change in the delivery of any project element on the critical path affects the delivery of the entire project; for example, a project that does not receive its equipment from the supplier on time may not be able to meet the scheduled project dates. Critical path elements should be given special consideration for timeliness because delays in these elements increase overall project risk. Through experience and careful evaluation, the risk practitioner can advise the risk owner on the feasibility of project dates, the expected workload associated with the project, the costs of the project and the overall success of the project according to risk management and business goals.[24]

## 4.11 METHODS TO MONITOR EFFECTIVENESS OF MITIGATION STRATEGIES AND/OR CONTROLS

There is a strong need to have proven quantitative and qualitative methods that can measure and ascertain the extent to which IT-related activities contribute to or support business success. By having such methods, active monitoring and corrective action can be taken to reduce the risk that IT activities deviate from the expected business contributions they are supposed to be making.

The controls mandated through risk management must align with security and related policies of the enterprise. The control monitoring function ensures that control requirements are being met, standards are being followed and staff is complying with the policies, practices and procedures of the organization

In this context, three popular methodology frameworks used to measure and determine the success of IT-related controls to business success are:
- Six Sigma
- Service level management
- IT balanced scorecard

Six Sigma is a quantitative approach, IT balanced scorecard (IT BSC) is a qualitative approach and SLM is both quantitative and qualitative. Both Six Sigma and IT BSC are derivatives from original applications in other (non-IT) domains.

Each of these methodology frameworks is briefly described.

## Six Sigma

The Six Sigma approach started at Motorola in the 1980s, and since then it has spread to many other organizations. It provides a common measure of performance across different processes and systems, and it can be used by people to compare, discuss and learn from different operations in different parts of an organization.

### Core Principles

Six Sigma provides guidelines as IT organizations and project teams move through the activities they need to improve their operations and their products. A five-step process guides IT teams through the Six Sigma activities of Define, Measure, Analyze, Improve, Control (DMAIC) as follows:

- **Define**—This step begins a Six Sigma project and produces three output documents. The first document is the project charter. The charter lays out the business case and the problem statement, project scope, project team, project goals and objectives. In addition to the project charter, the second document produced in this step defines and documents the customers that will be served and their needs and expectations. The customers' needs and expectations tell the team what to measure and improve. The third document is a high-level process map that shows the tasks involved in the process and the inputs and outputs of each task. This map shows everyone involved with the project the exact sequence of tasks that are candidates for improvement.
- **Measure**—In this step, the project team creates a data collection plan and then collects data that measure the current state of the process or product targeted for improvement. The data collected reflect customer requirements and show how often the process actually meets customer requirements. The data also show the activity levels of key tasks in the process. After collecting the data, the team calculates the existing sigma measurement for the process. This obvious step of collecting data and documenting the current situation is often overlooked or done poorly because the project team thinks they already know what is wrong and they want to go straight to fixing the problem. Good data collection gets the project off to a start in the right direction.
- **Analyze**—In this step, the project team applies statistical analysis tools to discover and validate root causes of problems. Several of the tools used in this step come from total quality management (TQM). The team uses cause-and-effect diagrams and frequency distribution charts to pinpoint the sources of error in the process being investigated. They use scatter diagrams to test the strength of correlations between one variable and another in the process. They use run charts to track the performance patterns of various tasks and of the process overall. As they pinpoint problems, the team then formulates options for eliminating or reducing these problems and compares the different options. Relevant questions to pose in this step include: How difficult is each option? How much will each cost? What impact will each option have on improving the sigma measure of the process?
- **Improve**—In this step, the team leader works with the project's executive sponsor to select a group of improvement options. They choose the options with the best chance for success and with the greatest impact on the process. With the sponsor's backing, the DMAIC team implements the selected improvements to the process. Best practice calls for the team to implement the improvements one at a time or in small groups of related improvements. After implementing each improvement, the team should collect process performance data and recalculate the sigma measure. The hope is that the sigma measure improves. Recollection and recalculation ensure that either the improvements actually provide valuable results or they are discontinued.
- **Control**—Once a team makes process improvements, it needs to regularly monitor the process to ensure that the improvements stay in place and remain effective. The DMAIC project team defines a set of measurements collected on an ongoing basis to document performance levels of the improved process. In addition, the team creates a response plan that lays out corrective actions if ongoing performance measures indicate that the improvements are beginning to slip. Over the longer term the greatest benefit from the Six Sigma approach is that enterprises reap the very real benefits of process improvements that continue to improve and, thus, deliver more value.

Another Six Sigma methodology is DMADV (define, measure, analyze, design, verify). This methodology is best used when a product or process is not in existence at the organization and needs to be developed.

## Impact

Six Sigma is a defect reduction methodology that transforms enterprises by forcing them to focus on processes. IT is a big user of processes; for example, testing, hardware implementation and software development. For IT, then, that means fewer servers, faster call response times and better project delivery. Quantitatively, Six Sigma provides CIOs with an objective, measurable way to justify technology investments. It also serves as a judgment-free common language between IT and other project stakeholders within the enterprise. Six Sigma gives IT organizations a good tool set that can be used consistently and repeatedly to analyze how IT has infrastructure and processes set up and running.

## Service Level Management

SLM is a key process for every IT service provider organization in that it is responsible for obtaining agreement and documenting service level targets and responsibilities for all activities within IT.[25] If these targets are appropriate and accurately reflect the requirements of the business, then the service delivered by the service providers will align with business requirements and meet the expectations of the customers and users in terms of service quality.

### Core Principles

The goal of the SLM process is to ensure that an agreed-on level of IT service is provided for all current IT services and that future services are delivered to agreed achievable targets. Proactive measures are also taken to seek and implement improvements to the level of service delivered. The purpose of the SLM process is to ensure that all operational services and their performance are measured in a consistent, professional manner throughout the IT organization and that the services and the reports produced meet the needs of the business and customers.

SLM is the name given to the processes of planning, coordinating, drafting, agreeing on, monitoring and reporting on SLAs and the ongoing review of service achievements to ensure that the required and cost-justified service quality is maintained and gradually improved. However, SLM is not only concerned with ensuring that current services and SLAs are managed; it is also involved in ensuring that new requirements are captured and that new or changed services and SLAs are developed to match the business needs and expectations. SLAs provide the basis for managing the relationship between the service provider and the customer, and SLM provides that central point of focus for a group of customers, business units or lines of business. An SLA is a written agreement between an IT service provider and the IT customer(s), defining the key service targets and responsibilities of both parties. The emphasis must be on agreement, and SLAs should not be used as a way of holding one side or the other for ransom. A true partnership should be developed between the IT service provider and the customer so that a mutually beneficial agreement is reached; otherwise, the SLA could quickly fall into disrepute and a blame culture could develop that would prevent any true service quality improvements from taking place. SLM is also responsible for ensuring that all targets and measures agreed on in the SLAs with the business are supported by appropriate underpinning operational level agreements (OLAs) or contracts, with internal support units and external partners and suppliers.

### Impact

SLM provides a consistent interface to the business for all service-related issues. It provides the business with the agreed-on service targets and the required management information to ensure that those targets have been met. Where targets are breached, SLM should provide feedback on the cause of the breach and details of the actions taken to prevent the breach from recurring. Thus, SLM provides a reliable communication channel and a trusted relationship with the appropriate customers and business representatives.

More information on SLM can be found in section 5.9 Service Level Management Concepts.

## IT Balanced Scorecard

BSCs designed specifically to address IT issues surfaced in the mid-1990s and evolved in design, complexity and content over the following decade. Most recently, the content of IT BSCs has become increasingly specific to track individual components of IT management issues. These BSCs cover topics such as GEIT, SLM, ERP, knowledge management and IT audit. Additionally, the increasing publication of management-targeted articles began to emerge during this phase, expanding the literature beyond the previously academic-dominated environment.

### Core Principles

The original BSC concept popularized by Harvard University professors Robert Kaplan and David Norton is based on four fundamental perspectives: financial, customer, internal business process, and learning and growth.[26] By applying a series of specific objectives, measures, targets and initiatives to each perspective, this "balanced" method allows management to plan and evaluate a range of important organizational areas with a single approach. For example, a company using the BSC could track objectives such as increased profitability (financial perspective), decreased customer complaints (customer perspective), improved manufacturing productivity (internal business process perspective) and reduced employee turnover (learning and growth perspective).

### Impact

By assisting in the management of three current drivers, the IT BSC has increased its organizational importance to today's practitioners. These drivers are:
- **Demonstration of IT value**—A recently published study by Accenture[27] found that demonstrating IT value to the business is one of the top five technology challenges for IT executives in Italy, the UK and the United States. The IT BSC assists in managing this issue by providing a straightforward method of reporting on a range of IT metrics, enabling the value of IT to be quantified for business stakeholders.
- **GEIT**—GEIT is considered another important technology challenge, particularly for IT executives in Japan and France (from the same Accenture study). The IT BSC can be specifically customized to address GEIT issues. Using this method, the IT BSC utilizes four perspectives to enable practitioners to manage governance issues: future orientation, operational excellence, stakeholders and corporate contribution of the IT function.
- **Cost cutting and efficiency**—Another key technology challenge for IT executives is cost savings, particularly in Japan and Germany. The IT BSC directly addresses this objective.

More information on the BSC can be found in section 1.10 IT Governance Monitoring Processes/Mechanisms and section 1.11 IT Governance Reporting Processes/Mechanisms.

Automated tools can help in monitoring risk mitigation strategies. More information on tools can be found in section 4.7 Types of Business Risk, Exposures and Threats That Can Be Addressed Using IT Resources.

## 4.12 STAKEHOLDER ANALYSIS AND COMMUNICATION TECHNIQUES

Enterprises have many stakeholders (including shareholders, employees, governments, etc.), and "creating value" means different—and sometimes conflicting—things to each of them. The governance system should consider all stakeholders' value interests when making benefit, risk and resource assessment decisions. For each decision, the following questions can and should be asked: For whom are the benefits? Who bears the risk? What resources are required?

For these stakeholders, communication is a key component of risk management. The benefits of open communication include:
- Executive management understanding the actual exposure to IT risk, enabling definition of appropriate and informed risk responses
- Awareness among all internal stakeholders of the importance of integrating risk and opportunity in their daily duties
- Transparency to external stakeholders on the actual level of risk and risk management processes in use

The consequences of poor communication include:
- A false sense of confidence at the top on the degree of actual exposure related to IT, and lack of a well-understood direction for risk management from the top down
- Over-communication of risk to the external world, especially if risk is at an elevated or barely acceptable level, which may deter potential clients or investors or generate needless scrutiny from regulators
- The perception that the enterprise is trying to cover up known risk to stakeholders

Risk awareness is about acknowledging that risk is an integral part of the business. This does not imply that all risk is avoided or eliminated, but rather that IT risk is understood and known, risk issues are easily identifiable, and the enterprise recognizes and uses the means to manage IT risk.[28]

## Risk Awareness—Risk Culture

A risk-aware culture characteristically offers a setting in which components of risk are discussed openly and acceptable levels of risk are understood and maintained. A risk-aware culture begins at the top with business executives who set direction, communicate risk-aware decision making and reward effective risk management behaviors.

Risk awareness also implies that all levels within an enterprise are aware of how and why to respond to adverse IT events.

A blame culture should be avoided; it is the most effective inhibitor of relevant and efficient communication. In a blame culture, business units tend to point the finger at IT when projects are not delivered on time or do not meet expectations. In doing so, they fail to realize how the business unit's involvement up front affects project success. In extreme cases, the business unit may assign blame for a failure to meet the expectations the unit never clearly communicated. The "blame game" only detracts from effective communication across departments, further fueling delays. Executive leadership must identify and quickly control a blame culture if collaboration is to be fostered throughout the enterprise.

## Risk Communication—What to Communicate

IT risk communication covers a broad array of information flows.[29] Risk IT distinguishes the following major communication streams as shown in **figure 4.17**:
- Policies, procedures, awareness training, continuous reinforcement of principles, etc., are essential communications on the overall strategy the enterprise takes toward IT risk, and it drives all subsequent efforts on risk management.
- Risk management capability and performance information allows monitoring of the state of the "risk management engine" in the enterprise. It is a key performance indicator (KPI) for good risk management, and it has predictive value for how well the enterprise is managing risk and reducing exposure.
- Operational risk management data such as:
  - The risk profile of the enterprise, i.e., the overall portfolio of (identified) risk to which the enterprise is exposed
  - The root cause of loss events
  - Thresholds for risk
  - Options to mitigate (cost and benefits) risk
  - Event/loss data
  - Key risk indicators (KRIs) to support management reporting on risk

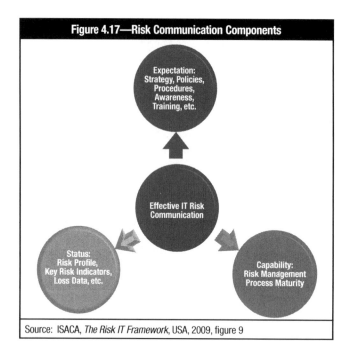

## Effective Risk Communication

To be effective, the information flowing within these three streams of communication should always be:
- Clear
- Concise
- Useful
- Timely
- Designed for the correct target audience
- Available on a need-to-know basis

**Figure 4.18** depicts the key focus areas for risk reporting.

| Figure 4.18—Key Focus Areas for Risk Reporting | |
|---|---|
| **Communication must be:** | **To:** |
| Clear | Enable understanding by all stakeholders. |
| Concise | Focus the reader on the key points.<br><br>Concise information is well structured and complete and avoids peripheral information, jargon and technical terms, except where necessary. |
| Useful | Enable decision making.<br><br>Useful information is relevant and presented at the appropriate level of detail. Usefulness includes consideration of the target audience because information that may be useful to one party may not be useful to another. |
| Timely | Allow action at the appropriate moment to identify and treat the risk.<br><br>For each risk, critical moments exist between its origination and its potential business consequence; a delay in reporting may increase the level of impact.<br><br>**Example:** Communicating a potential problem too late to undertake corrective or preventive action serves no useful purpose. |

| Figure 4.18—Key Focus Areas for Risk Reporting *(cont.)* ||
|---|---|
| **Communication must be:** | **To:** |
| Aimed at the correct target audience | Enable informed decisions. Information must be communicated at the right level of aggregation and adapted for the audience.<br><br>Aggregation must not hide root causes of risk.<br><br>**Example**: A security officer may need technical data on intrusions and viruses to deploy solutions. An IT steering committee may not need this level of detail, but it does need aggregated information to decide on policy changes or additional budgets to treat the same risk. |
| Available on a need-to-know basis | Ensure that information related to IT risk is known and communicated to only those parties with a genuine need.<br><br>A risk register with all documented risk is not public information and should be properly protected against internal and external parties with no need for it. |

## Risk Communication—Stakeholders

**Figure 4.19** provides an overview of the most important communication channels for effective and efficient risk management. It is a summary and does not represent all communication flows among all risk management processes (that information is included in the detailed process descriptions of the process framework). The table's intent is to provide a one-page overview of the main communication flows on IT risk that should exist in one form or another in any enterprise.

The table[30] does not include the source and destination of the information; they can be found in the detailed process model. This table is focused on the most important information each stakeholder needs to process.

| Figure 4.19—Risk Communication Flows |||
|---|---|---|
| **Input** | **Stakeholders** | **Output** |
| • Executive summary IT risk reports<br>• Current IT risk exposure/profile<br>• KRIs | Executive management and board | • Enterprise appetite for IT risk<br>• Key performance objectives<br>• IT risk RACI charts<br>• IT-related policies, expressing management's IT risk tolerance<br>• Risk awareness expectations<br>• Risk culture<br>• Risk analysis request |
| • IT risk management scope and plan<br>• IT risk register<br>• IT risk analysis results<br>• Executive summary IT risk reports<br>• Integrated/aggregated IT risk report<br>• KRIs<br>• Risk analysis request | Chief risk officer (CRO) and enterprise risk committee | • Enterprise appetite for IT risk<br>• Residual IT risk exposures<br>• IT risk action plan |
| • Enterprise appetite for IT risk<br>• IT risk management scope and plan<br>• Key performance objectives<br>• IT risk RACI charts<br>• IT risk assessment methodology<br>• IT risk register | Chief information officer (CIO) | • Residual IT risk exposures<br>• Operational IT risk information<br>• Business impact of the IT risk and impacted business units<br>• Ongoing changes to risk factors |
| • Key performance objectives | Chief financial officer (CFO) | Financial information with regard to IT and IT programmes/projects (budget, actual, trends, etc.) |

### Figure 4.19—Risk Communication Flows (cont.)

| Input | Stakeholders | Output |
|---|---|---|
| • IT risk management scope<br>• Plans for ongoing business and IT risk communication<br>• Risk culture<br>• Business impact of the IT risk and impacted business units<br>• Ongoing changes to IT risk factors | Business management and business process owners | • Control and compliance monitoring<br>• Risk analysis request |
| • Key performance objectives<br>• IT risk action plan<br>• IT risk assessment methodology<br>• IT risk register<br>• Risk culture | IT management (including security and service management) | • IT risk mitigation strategy and plan, including assignment of responsibility and development of metrics |
| • Key performance objectives<br>• IT risk RACI charts<br>• IT risk action plan<br>• Control and compliance monitoring | Compliance and audit | • Audit findings |
| • Key performance objectives<br>• IT risk action plan<br>• IT risk assessment methodology<br>• IT risk register<br>• Audit findings | Risk control functions | • Residual IT risk exposures<br>• IT risk reports |
| • Risk awareness expectations<br>• Risk culture | Human resources (HR) | • Potential IT risk<br>• Support on risk awareness initiatives |
| • Control and compliance monitoring | External auditors | • Audit findings |
| • Public opinion, legislation<br>• IT risk executive summary report<br>• In general, all communications intended for the board and executive management | Regulators | • Requirements for controls and reporting<br>• Summary findings on risk |
| • Executive summary risk reports | Investors | • Risk tolerance levels for their portfolio of investments |
| • Summary IT risk reports, including residual risk, controls maturity levels and audit findings | Insurers | • Insurance coverage (property, business interruption, directors and officers) |
| • Risk awareness expectations<br>• Risk culture | All employees | • Potential IT risk issues |

Source: ISACA, *The Risk IT Framework*, USA 2009, figure 10

## 4.13 METHODS TO ESTABLISH KEY RISK INDICATORS

One of the key objectives of ERM is to promote risk transparency, both in terms of internal risk reporting and external public disclosure. Establishing a robust risk measurement and reporting system is, therefore, critical to ERM success.

KRIs can be defined as a subset of risk indicators that are highly relevant and possess a high probability of predicting or indicating important risk. KRIs allow management to document and analyze trends, and they provide a forward-looking perspective, signaling required actions before the risk actually becomes a loss. In practice, KRIs are used in reporting or in dashboards. They not only warn about and flag possible issues or areas that contain risk, but if selected well they can provide management with a holistic overview of the current risk management situation.

Risk indicators are metrics capable of showing that the enterprise is subject to, or has a high probability of being subject to, a risk that exceeds the defined risk appetite. They are used to measure levels of risk in comparison to defined risk thresholds and alert the enterprise when a risk level approaches a high or unacceptable level of risk. The purpose of a risk indicator is to set in place tracking and reporting mechanisms that alert staff to a developing or potential risk.

## Risk Indicators

Risk indicators are specific to each enterprise. Their selection depends on a number of parameters such as the complexity of the enterprise, whether it is in a highly regulated market and its strategy focus. A suggested approach for identifying risk indicators takes into account the following aspects:
- Consider the different stakeholders in the enterprise. Risk indicators should not be limited to operational risk data, but should also include the more strategic side of risk. Risk indicators can be identified for all stakeholders, in line with the needs for their level of responsibility.
- Make a balanced selection of risk indicators, utilizing both:
  - Risk exposures (lag indicators, indicating risk after events have happened)
  - Risk management capabilities (lead indicators, indicating what capabilities are in place to prevent events from occurring)
- Ensure that the selected indicator drills down to the root cause of the events.

An enterprise may develop an extensive set of metrics to serve as risk indicators; however, it is not possible or feasible to maintain that full set of metrics as KRIs. A KRI is differentiated as being highly relevant and possessing a high probability of predicting or indicating important risk. Criteria to select KRIs include:
- **Sensitivity**—The indicator must be reliable and representative for risk.
- **Impact**—Indicators for risk with high business impact are more likely to be KRIs.
- **Effort to measure**—For different indicators that are equivalent in sensitivity, the one that is easier to measure is preferred.
- **Reliability**—The indicator must possess a high correlation with the risk and be a good predictor or outcome measure.

The complete set of KRIs must also be relevant; indicators for root causes as well as for business impact must be equally represented.

## Risk Indicators as Communication Instruments

In addition to indicating risk, KRIs are particularly important during the communication on risk. They facilitate a dialogue on risk within the organization, based on clear and measurable facts. This results in a less emotion-based or too-intuitive discussion on where to place priorities for risk management. At the same time, KRIs can be used to improve risk awareness throughout the organization due to the factual nature of these indicators.

The following Risk IT components (see also **figure 4.5**) can serve as KRIs:
- The metrics of the three domains and their processes in the Risk IT process model, which are a combination of:
  - Process indicators—Predictors for risk management capabilities, indicating the successful outcome of the risk management process
  - Outcome measures—Indicating risk exposure, measuring actual incidents and related losses
- The process goals and related process metrics defined in the processes of the Risk IT framework
- The maturity model, which gives an indication of the process maturity of the various risk management processes
- Aggregated risk analysis/status results (risk matrix)

Examples of Risk Indicators
The following table contains an example of some possible KRIs for different stakeholders. Both types of indicators, lead and lag, are used. This table is not complete (nor is it intended to be), but it provides some suggestions for KRIs.

The stakeholders that are considered in **figure 4.20** are:
- CIO—This function requires a IT department view on IT risk, which is limited to its own personnel and resources
- CRO—This function requires a broader view on IT risk from across the business, but can be considered operational
- Chief executive officer (CEO)/board of directors—The top of the organization requires a high-level view of risk

| Figure 4.20—Stakeholder KRI Considerations ||||
| Event Category | CIO | CRO | CEO/Board of Directors |
|---|---|---|---|
| Investments/project decision-related events | • Percent of projects on time, on budget<br>• Number and type of deviations from technology infrastructure plan | • Percent of IT projects reviewed and signed off on by quality assurance (QA) that meet target quality goals and objectives<br>• Percent of projects with benefit defined up front | • Percent of IT investments exceeding or meeting the predefined business benefit<br>• Percent of IT expenditures that have direct traceability to the business strategy |
| Business involvement-related events | Degree of approval of business owners of the IT strategic/tactical plans | Frequency of meetings with enterprise leadership involvement where IT's contribution to value is discussed | Frequency of CIO reporting to or attending executive board meetings at which IT's contribution to enterprise goals is discussed |
| Security | Percent of users who do not comply with password standards | Number and type of suspected and actual access violations | Number of (security) incidents with business impact |
| Involuntary staff act: Destruction | • Number of service levels impacted by operational incidents<br>• Percent of IT staff who complete annual IT training plan | • Number of incidents caused by deficient user and operational documentation and training<br>• Number of business-critical processes relying on IT not covered by IT continuity plan | • Cost of IT noncompliance, including settlements and fines<br>• Number of noncompliance issues reported to the board or causing public comment or embarrassment |

## 4.14 METHODS TO MANAGE AND REPORT THE STATUS OF IDENTIFIED RISK

Risk monitoring and control are where the "rubber meets the road." Enterprises should establish risk identification, analysis, monitoring and reporting processes such that they are able to handle risk events once they begin occurring.

The Massachusetts Institute of Technology (MIT) Center for Information Systems Research (CISR) advocates three core disciplines, which together build an effective ERM capability:[31]
- **Risk governance process**—Complete and effective risk-related policies, combined with a mature, consistent process to identify, assess, prioritize and monitor risk over time
- **Risk-aware culture**—Skilled people who know how to identify and assess threats and implement effective risk mitigation
- **Effective IT foundation**—IT infrastructure and applications that have inherently lower risk because they are well architected and well managed

These elements were identified from interviews conducted with more than 50 IT managers. The researchers contend that if an enterprise is severely lacking in any of the three disciplines, it cannot be effective at IT risk management. For example, no level of governance process or expertise can overcome a complex, overly risky (IT) foundation. Similarly, heavy risk governance cannot be effective without the expertise to identify and reduce risk. However, enterprises need not be world-class in all three disciplines; rather, they can be world-class in one, with lower (but still acceptable) levels in the other two. Moreover, enterprises that have ineffective risk management cannot become effective overnight; they build capability over time by using one discipline very well to help the others grow to an effective level.

ISACA recommends that to enable effective governance, IT risk should always be expressed in a business context rather than in the technical language. The following generic elements for expressing IT risk in any enterprise are recommended, providing a framework for business management to be engaged in the risk management process:
- **Business-specific risk**—For example, operational risk of orders not being received
- **Generic common IT risk**—For example, IT availability risk
- **Specific IT risk**—For example, denial-of-service attack on Internet customer order system

In an ISACA governance report,[32] from the cross-reference of the questions, "How would you rate your organization's maturity level on IT governance?" and "How important is IT risk management to your organization?" it is clear that IT governance maturity and the importance of IT risk management follow a linear evolution. This means that the higher the organization's maturity level, the more important that IT risk management becomes. This is shown in **figure 4.21**.

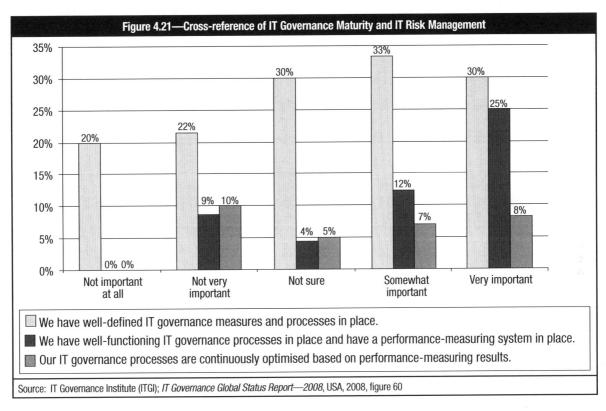

Figure 4.21—Cross-reference of IT Governance Maturity and IT Risk Management

Source: IT Governance Institute (ITGI); *IT Governance Global Status Report—2008*, USA, 2008, figure 60

There are four commonly employed formal risk monitoring activities: the employment of status reports, the use of issue logs, the conduction of evaluations and the use of periodic risk audits.

## Status Reports

The most commonly used mechanism to assess progress on projects and operations is the use of status reports. They are usually issued monthly and they typically follow a prescribed format. For example, they describe budget performance for the past month or identify milestones achieved and missed. A common feature of status reports is that they focus on variances from the plan. For example, a review of cost and schedule status for a project may indicate that it is 10 percent over budget and 12 weeks behind schedule. The combined data may suggest that the project is in trouble. Important questions to ask are: "Will these variances continue to grow?" and "Are there steps we can take to bring the project back on track?" Typically, unfavorable cost and schedule variances indicate that a project is encountering a standard set of problems. Some of these may be related to poor implementation of the project plan such as when inexperienced workers are used on tasks or needed equipment and supplies arrive late. Some are related to excessively optimistic plans such as when salespeople promise clients that the project team will deliver a 10-month job in six months.

These standard problems can be handled in various customary ways. For example, if the cause of overruns is employment of inexperienced workers, more experienced personnel can be put onto the project. Or if a project is based on overly optimistic assumptions, the plan may be re-baselined (i.e., adjusted to capture reality). But cost and schedule variances may also be rooted in nonstandard problems that may indicate the surfacing of new sources of risk. For example, an investigation of persistent schedule slippages for a project may reveal that they are triggered by changes in local government regulations that require that the project's deliverables undergo unanticipated government inspections at frequent intervals. Thus, changes in government regulations leading to increased inspections are a new source of risk that has been discovered by the risk monitoring effort.

## Issue Logs

Issue logs can be used as a tool that assists in risk identification. They are typically filled out on a monthly basis and presented as part of the status report for projects or operations. They are typically divided into two portions: pending issues and resolved issues.
- The pending issues portion lists possible items of concern. Issues are not risk events per se. They are discussion points that need to be addressed because they might ultimately be sources of problems.
- The resolved issues portion of the issue log itemizes previous pending issues that have been taken care of. The date when each issue was resolved should be noted so that management has an idea of how much time is being spent dealing with issues.

Ideally, the pending issues portion of the issue log is quite short and the list of items in the resolved issues portion should be growing longer. This circumstance reflects the fact that as issues arise, they are being handled quickly. A lengthening list of pending issues indicates that the issues are not being handled expeditiously.

The use of issue logs is quite popular and offers two important advantages. First, issue logs provide a systematic way for employees to highlight concerns they have about how things are going in the business. Consequently, the possibility of untoward risk events arising is kept in front of them. Second, issue logs place pressure on employees to handle risk promptly.

## Evaluations

Evaluations are "sanity" checks that are conducted to see whether the fundamental objectives of an undertaking are being achieved. Most enterprises conduct an abundance of evaluations; but they do not call these efforts "evaluation." Examples of evaluation include:
- **Preliminary design reviews and critical design reviews**—Preliminary and detailed design reviews are commonly used approaches to technical evaluation. They are conducted to gain assurance that the design that is being formulated for a product is on target.
- **Pink and red team reviews**—Pink and red team reviews are evaluation efforts carried out during the process of writing important proposals that will lead, everyone hopes, to project funding. The pink team review is held early in the proposal writing effort when a group of colleagues plays the role of the customer and critiques the nascent proposal from a customer perspective. The red team review is held later in the proposal writing effort; as with the pink team review, the proposal is critiqued from the customer's point of view.
- **Walk-throughs**—Customers or technical team members review the merits (or demerits) of a product in great detail. The walk-through concept is familiar to anyone who has purchased a new home. One of the last things a home purchaser does before handing over a check to acquire a property is to walk-through the property carefully, looking for flaws that need to be fixed.
- **Audits**—An audit entails a detailed review of a product or process. Many audits are financial reviews such as when auditors from the tax authorities review a company's books, looking for irregularities. Other audits are reviews of an organization's processes; for example, when a university is undergoing an accreditation review or a factory's operations are being audited to see whether they are in compliance with ISO 9000 standards.
- **Management-by-objectives (MBO) reviews**—MBO reviews are classical management evaluations. Employees and their managers agree that the employees will achieve a well-defined set of objectives by a particular time. When that point is reached, the employee's performance is evaluated to see whether the objectives have been achieved.
- **Performance appraisal reviews**—Performance appraisal reviews are a form of evaluation. They address the effectiveness of employee performance according to a number of defined criteria. Their importance rests on the fact that they help determine whether employees become promoted, win bonuses and gain salary increases or whether they are relegated to some other less valuable role.

Clearly, evaluations serve a risk monitoring function. When evaluations are held, the evaluators are fundamentally looking for signs of nonperformance. However, it should be noted that evaluators are looking for these signs not for the purpose of punishing employees, but for the purpose of identifying problems when they are still small and manageable. In this respect, standard evaluations and risk monitoring have a lot in common.

## Risk Audits

As organizations have grown sensitive to the need to implement good risk management practices, the employment of risk audits has increased dramatically. Risk audits are conscious, systematic attempts to examine an organization's projects, processes and risk management procedures to determine whether things are progressing smoothly or whether problems lurk in the shadows. They are conducted by risk audit teams of highly experienced personnel who are trained on good risk management practices. Risk assessment groups used in major public organizations are examples of risk self-assessments. One important function they play is to review contract terms and statements of work to make sure that they are realistic. If a new initiative is promising more than the enterprise can deliver, the risk assessment group identifies this problem before formal agreements are signed and funds are released, preventing the organization from launching itself down a path that is preordained to result in failure.

## ENDNOTES

[1] IT Governance Institute; *IT Governance Domain Practices and Competencies: Information Risks: Whose Business Are They?*, USA, 2005; ISACA, *Board Briefing on IT Governance, 2nd Edition*, USA, 2003
[2] ISACA, COBIT 4.1, USA, 2007
[3] ISACA, *The Risk IT Framework*, USA, 2009
[4] ISACA, *Enterprise Value: Governance of IT Investments, The Val IT™ Framework 2.0*, USA, 2008
[5] Committee of Sponsoring Organizations of the Treadway Commission (COSO); *COSO Enterprise Risk Management Framework*, USA, 2004, www.coso.org
[6] Standards Australia/Standards New Zealand; *AS/NZS 4360:2004, Australian/New Zealand Standard, Risk Management*, Australia/New Zealand, 2004
[7] Alberts, Christopher; Sandra Behrens; Richard Pethia; William Wilson; *Operationally Critical Threat, Asset and Vulnerability Evaluation^SM (OCTAVE^SM) Framework, Version 1.0*, Carnegie Mellon University, USA, 1999
[8] National Institute of Standards and Technology (NIST), *NIST Special Publication 800-30 Revision 1: Guide for Conducting Risk Assessments*, USA, 2012
[9] NIST: NIST Special Publication 800-39: Managing Information Security Risk, USA, 2011
[10] International Organization for Standardization (ISO); ISO 31000, *Risk Management—Principles and Guidelines on Implementation*, USA, 2009
[11] *Op cit* ISACA, The Risk IT Framework
[12] *Packet® magazine* (Volume 17, No. 1), 2005, Cisco Systems
[13] ISACA, *COBIT® 5: Enabling Processes*, USA, 2012, www.isaca.org/cobit
[14] *Ibid.*
[15] International Organization for Standardization (ISO); *ISO 22301:2012, Societal Security—Business Continuity Management Systems—Requirements*, USA, 2012
[16] ISACA, *IT Audit and Assurance Guideline: G13−Use of Risk Management in Audit Planning*, USA, 2000
[17] World Economic Forum, *Risk and Responsibility in a Hyperconnected World: Pathways to Global Cyber Resilience*, Switzerland, 2014
[18] Cloud Security Alliance, *The Notorious Nine: Cloud Computing Top Threats in 2013*, USA, 2013
[19] ISACA, *Big Data: Impact & Benefits*, USA 2013, www.isaca.org
[20] ISACA, *CSX Cybersecurity Fundamentals Study Guide*, USA, 2014
[21] McIntyre, Jack; "BYOD—Great Opportunity, Great Risks to Manage" *ISACA Now*, 31 July 2012, www.isaca.org
[22] Kaneshige, Tom; "IT Learns to COPE With Mobile Devices," CIO, November 18, 2013, www.cio.com; Sheldon, Robert; "BYOD vs. COPE: Why corporate device ownership could make a comeback," *SearchConsumerization*, July 1, 2013, searchconsumerization.techtarget,com
[23] ISACA, *The Risk IT Practitioner Guide*, USA, 2009
[24] ISACA, *CRISC Review Manual 6th Edition*, USA, 2015
[25] UK Office of Government Commerce (OGC); *ITIL Version 3: Service Design*, UK, 2007
[26] Kaplan, David; Robert Norton; *The Balanced Scorecard: Translating Strategy Into Action*, Harvard Business School Press, USA, 1996
[27] Harris, Jeanne; Jeffrey D. Brooks; *A Matter of IT Value*, Accenture, USA, 2004
[28] *Op cit* ISACA, The Risk IT Practitioner Guide
[29] *Ibid.*
[30] *Op cit* ISACA, The Risk IT Framework
[31] Westerman, George; *Building IT Risk Management Effectiveness*, MIT Center for Information Systems Research (CISR), USA, 2004
[32] IT Governance Institute (ITGI); *IT Governance Global Status Report—2008*, ISACA, USA, 2008

Page intentionally left blank

# Chapter 5: Resource Optimization

## Section One: Overview

Domain Definition ........................................................................................................................................... 196
Domain Objectives ........................................................................................................................................... 196
Learning Objectives ......................................................................................................................................... 196
CGEIT Exam Reference ................................................................................................................................... 196
Task and Knowledge Statements ...................................................................................................................... 196
Self-assessment Questions ................................................................................................................................ 199
Answers to Self-assessment Questions ............................................................................................................. 200
Suggested Resources for Further Study ........................................................................................................... 201

## Section Two: Content

5.1 IT Resource Planning Methods ............................................................................................................... 202
5.2 Human Resource Procurement, Assessment, Training and Development Methodologies ..................... 204
5.3 Processes for Acquiring Application, Information and Infrastructure Resources ................................... 206
5.4 Outsourcing and Offshoring Approaches That May Be Employed to Meet the
Investment Program and Operational Level Agreements and Service Level Agreements ...................... 207
5.5 Methods Used to Record and Monitor IT Resource Utilization and Availability ................................... 210
5.6 Methods Used to Evaluate and Report on IT Resource Performance ..................................................... 212
5.7 Interoperability, Standardization and Economies of Scale ...................................................................... 214
5.8 Data Management and Data Governance Concepts ................................................................................ 215
5.9 Service Level Management Concepts ..................................................................................................... 221
Endnotes ........................................................................................................................................................... 223

# Section One: Overview

## DOMAIN DEFINITION
Ensure the optimization of IT resources, including information, services, infrastructure and applications, and people, to support the achievement of enterprise objectives.

## DOMAIN OBJECTIVES
The objective of this domain is to ensure that IT has sufficient, competent and capable resources to execute current and future strategic objectives and keep up with business demands by optimizing the investment in and use and allocation of IT assets.

The premise of this domain is that the optimal investment and utilization of resources required by IT in its activities and processes assist in the achievement of IT goals, which in turn help in the attainment of enterprise business objectives. With rapidly increasing demand and consequent scarcity of IT talent, there is growing concern at board and senior management levels that neglecting resource management will constrain and risk IT's ability to effectively deliver its services to the business.

## LEARNING OBJECTIVES
The purpose of this domain is for professionals and executives involved in governance to understand, define and execute appropriate resource management practices for the enterprise. This includes defining key resources used for IT processes and activities, defining strategies for their procurement, ensuring their availability and management, and optimizing their use in the enterprise.

## CGEIT EXAM REFERENCE
This domain represents 15 percent of the CGEIT exam (approximately 22 questions).

## TASK AND KNOWLEDGE STATEMENTS

### TASKS
There are seven tasks within this domain that a CGEIT candidate must know how to perform. These relate to having to develop, or assist in the development of, systematic and continuous resource planning, management and evaluation processes.

T5.1  Ensure that processes are in place to identify, acquire and maintain IT resources and capabilities.
T5.2  Evaluate, direct and monitor sourcing strategies to ensure that existing resources are taken into account to optimize IT resource utilization.
T5.3  Ensure the integration of IT resource management into the enterprise's strategic and tactical planning.
T5.4  Ensure the alignment of IT resource management processes with the enterprise's resource management processes.
T5.5  Ensure that a resource gap analysis process is in place so that IT is able to meet strategic objectives of the enterprise.
T5.6  Ensure that policies exist to guide IT resource sourcing strategies that include service level agreements (SLAs) and changes to sourcing strategies.
T5.7  Ensure that policies and processes are in place for the assessment, training and development of staff to address enterprise requirements and personal/professional growth.

**Section One: Overview**  **Chapter 5—Resource Optimization**

## KNOWLEDGE STATEMENTS

The CGEIT candidate must have a good understanding of each of the 9 areas delineated by the following knowledge statements. These statements are the basis for the exam. Each statement is defined and its relevance and applicability to this job practice are briefly described as follows.

Knowledge of:

**KS5.1  IT resource planning methods**
With the evolving nature in which IT interacts with the business, an understanding of current practices, processes, methods and techniques at strategic and tactical levels of business and IT resource planning will be a catalyst to optimize the ways in which IT resources are acquired and used. This knowledge is a prerequisite for strategic sourcing of resources as well as crafting the organizational constructs necessary to achieve the enterprise business objectives.

**KS5.2  human resources (HR) procurement, assessment, training and development methodologies**
With the scarcity of IT talent (due to issues such as attrition) and the growing demands for skills, it is incumbent on management to maintain sufficiency of the requisite IT resources to match requirements. The effective management of human resources, including their optimization, will have a direct impact on the quality, efficiency and effectiveness of IT services provided to the enterprise. Therefore, a clear definition of requirements for this resource area should be provided, addressing HR-related processes such as procurement, assessment and training, and development.

**KS5.3  processes for acquiring application, information and infrastructure resources**
With today's multiplicity of sources for resourcing business and IT services (applications, information and infrastructure), effective resource management requires consideration of the strategic choices to be made in the acquisition of the resources as well as their acquisition processes.

**KS5.4  outsourcing and offshoring approaches that may be employed to meet the investment program and operational level agreements (OLAs) and service level agreements (SLAs)**
The growing trend of outsourcing and offshoring holds both promise and pitfalls for the procurement of strategic resources to deliver IT services. With the maturing of practices, processes used in these areas define the systematic approaches to effectively engage and manage in such environments in line with investment program needs and service and operational level agreements.

**KS5.5  methods used to record and monitor IT resource utilization and availability**
Determining and evaluating business and IT resource demand and utilization is an essential part of the continuous monitoring of the availability of resources to meet planned and unforeseen demands for resources (as in business continuity planning). The adequacy of resources will facilitate continuous availability of IT services, which is a key business requirement.

**KS5.6  methods used to evaluate and report on IT resource performance**
Effective monitoring and reporting of IT resource performance highlights the issues and decisions made to manage IT resources. Without such monitoring and reporting, there would be little or no visibility to management on issues of IT resourcing that may be critical, and for which important management decisions have to be made.

**KS5.7  interoperability, standardization and economies of scale**
In most enterprises, the biggest portion of the IT budget relates to ongoing operations. Effective management of IT operational spending requires effective control of the cost base: the IT assets and the focus where they are needed most. IT assets should be organized so that the required quality of service is provided by the most cost-effective delivery infrastructure. This objective requires the necessary management attention for concepts such as interoperability, standardization and economies of scale.

**KS5.8 data management and data governance concepts**

Information is pervasive throughout any enterprise and includes all information produced and used by the enterprise. Information is required for keeping the enterprise running and well governed, but at the operational level, information is very often the key product of the enterprise itself. In the information life cycle, data management and governance concepts are important in converting data into information, knowledge and value for the enterprise.

**KS5.9 service level management (SLM) concepts**

In today's service driven economy, enterprises are relying more and more on third parties for a variety of IT services. Often, they are not pleased with the service received and are sometimes dependent on third parties whose futures are uncertain. An appropriate SLM process should be in place in the enterprise for obtaining the required service(s). Implementing an SLM process is not an easy and quick task to perform and thus an approach using supportive mechanisms such as COBIT may help in defining or fine-tuning the SLA(s).

## RELATIONSHIP OF TASK TO KNOWLEDGE STATEMENTS

The task statements are what the CGEIT candidate is expected to know how to do. The knowledge statements delineate what the CGEIT candidate is expected to know to perform the tasks. The task and knowledge statements are mapped in **figure 5.1** insofar as it is possible to do so. Note that although there often is overlap, each task statement will generally map to several knowledge statements.

| Figure 5.1—Task and Knowledge Statements Mapping—Resource Optimization Domain | | |
|---|---|---|
| **Task Statement** | | **Knowledge Statements** |
| T5.1 | Ensure that processes are in place to identify, acquire and maintain IT resources and capabilities. | KS5.1 IT resource planning methods<br>KS5.2 human resources (HR) procurement, assessment, training and development methodologies<br>KS5.3 processes for acquiring application, information and infrastructure resources<br>KS5.5 methods used to record and monitor IT resource utilization and availability<br>KS5.6 methods used to evaluate and report on IT resource performance<br>KS5.8 data management and data governance concepts<br>KS5.9 service level management concepts |
| T5.2 | Evaluate, direct and monitor sourcing strategies to ensure that existing resources are taken into account to optimize IT resource utilization. | KS5.3 processes for acquiring application, information and infrastructure resources<br>KS5.4 outsourcing and offshoring approaches that may be employed to meet the investment program and operation level agreements (OLAs) and service level agreements (SLAs)<br>KS5.5 methods used to record and monitor IT resource utilization and availability<br>KS5.6 methods used to evaluate and report on IT resource performance<br>KS5.7 interoperability, standardization and economies of scale |
| T5.3 | Ensure the integration of IT resource management into the enterprise's strategic and tactical planning. | KS5.1 IT resource planning methods<br>KS5.8 data management and data governance concepts |
| T5.4 | Ensure the alignment of IT resource management processes with the enterprise's resource management processes. | KS5.3 processes for acquiring application, information and infrastructure resources |
| T5.5 | Ensure that a resource gap analysis process is in place so that IT is able to meet strategic objectives of the enterprise. | KS5.5 methods used to record and monitor IT resource utilization and availability<br>KS5.6 methods used to evaluate and report on IT resource performance |
| T5.6 | Ensure that policies exist to guide IT resource sourcing strategies that include service level agreements (SLAs) and changes to sourcing strategies. | KS5.4 outsourcing and offshoring approaches that may be employed to meet the investment program and operational level agreements (OLAs) and service level agreements (SLAs)<br>KS5.9 service level management concepts |
| T5.7 | Ensure that policies and processes are in place for the assessment, training and development of staff to address enterprise requirements and personal/professional growth. | KS5.2 human resources (HR) procurement, assessment, training and development methodologies |

## SELF-ASSESSMENT QUESTIONS

CGEIT self-assessment questions support the content in this manual and provide an understanding of the type and structure of questions that have typically appeared on the exam. Questions are written in a multiple-choice format and designed for one best answer. Each question has a stem (question) and four options (answer choices). The stem may be written in the form of a question or an incomplete statement. In some instances, a scenario or a description problem may be included. These questions normally include a description of a situation and require the candidate to answer two or more questions based on the information provided. Many times a question will require the candidate to choose the **MOST** likely or **BEST** answer among the options provided.

In each case, the candidate must read the question carefully, eliminate known incorrect answers and then make the best choice possible. Knowing the format in which questions are asked, and how to study and gain knowledge of what will be tested, will help the candidate correctly answer the questions.

5-1   Which of the following process groups is **MOST** effective at supplying resources to the development of the procurement process?

   A. Acquisition
   B. Service delivery
   C. Demand management
   D. Contract management

5-2   Human resource strategy is typically **BEST** aligned with which of the following objectives?

   A. Having a focus on employee performance
   B. Satisfaction of business needs
   C. Talent retention
   D. Rewarding employees fairly

5-3   Which of the following is the **PRIMARY** objective of business process outsourcing?

   A. Optimizing business processes
   B. Increasing the automation of business processes
   C. Realigning business processes with business strategy
   D. Allowing the enterprise to focus on core competencies

## ANSWERS TO SELF-ASSESSMENT QUESTIONS
Correct answers are shown in **bold**.

5-1 **A. Procurement and acquisition processes are the key process groups for the supply of resources.**
    B. Service delivery does not directly contribute resources to the development of the procurement process.
    C. Demand management does not directly contribute resources to the development of the procurement process. Demand management involves balancing demand to ensure that diverse customer needs are met. Some activities related to demand management are to offer rebates during non-peak business periods.
    D. Contract management does not directly contribute resources to the development of the procurement process. Contract management involves the creation of all contracts entered into, including the master services agreement (MSA) and all schedules and companion agreements for subsidiary entities regarding maintenance, service supply, demand, etc.

5-2   A. Having a focus on employee performance is not necessarily the best sole alignment for HR strategy. Other factors may be as applicable, such as attitude, team spiritedness, etc.
    **B. An effective HR strategy is best aligned with the satisfaction of business needs. The satisfaction of business needs is the ideal foundation for both the formulation and implementation of HR strategy.**
    C. Talent retention is typically only one perspective of HR strategy.
    D. Rewarding employees fairly is typically only one perspective of HR strategy.

5-3   A. Optimizing business processes involves streamlining and improving them, and may not directly apply to all instances of business process outsourcing.
    B. Increasing the automation of business processes makes them more efficient, and may not directly apply to all instances of business process outsourcing.
    C. Realigning business processes with business strategy gains greater effectiveness, and may not directly apply to all instances of business process outsourcing.
    **D. The movement of business processes to be performed by external service providers is the fundamental concept underlying business process outsourcing, allowing the enterprise to focus on core competencies.**

---

**NOTE:** For more self-assessment questions, you may also want to obtain a copy of the *CGEIT® Review Questions, Answers & Explanations Manual 4th Edition*, which consists of 250 multiple-choice study questions, answers and explanations.

## SUGGESTED RESOURCES FOR FURTHER STUDY

In addition to the resources cited throughout this manual, the following resources are suggested for further study in this domain (publications in **bold** are stocked in the ISACA Bookstore):

Carmel, Erran; Paul Tjia; *Offshoring Information Technology: Sourcing and Outsourcing to a Global Workforce*, Cambridge University Press, UK, 2005

Corbett, Michael F.; *The Outsourcing Revolution: Why It Makes Sense and How to Do It Right*, Dearborn Trade Publishing, USA, 2004

**ISACA, *Board Briefing on IT Governance, Second Edition*, USA, 2003**

**ISACA, *COBIT 5*, USA, 2012,** *www.isaca.org/cobit*

**ISACA, *COBIT 5: Enabling Information*, USA, 2013,** *www.isaca.org/cobit*

Thiadens, Theo; *Manage IT! Organizing IT Demand and Supply*, Springer, The Netherlands, 2005

Yoong, Pak; Sid Huff; *Managing IT Professionals in the Internet Age*, Idea Group Inc., Canada, 2007

# Section Two: Content

## 5.1 IT RESOURCE PLANNING METHODS

With the evolving nature in which IT interacts with the business, an understanding of current practices, processes, methods and techniques at strategic and tactical levels of business and IT resource planning will be a catalyst to optimize the ways in which IT resources are acquired and used. This knowledge is a prerequisite for strategic sourcing of resources as well as crafting the organizational constructs necessary to achieve the enterprise business objectives.

It is common in today's business and IT resource planning context to incorporate outsourcing at an early stage. For this to be done effectively, a sourcing strategy needs to be formulated. Entering into deals with service providers on a tactical and piecemeal basis (i.e., without an overarching sourcing strategy) may lock the business/IT planner into a cycle of surprises. No matter how much attention is focused on fine-tuning a contract later, it cannot make up for a badly conceived deal. Multiple service providers interact and overlap and must be considered an ecosystem in which any one change can impact the whole.

### Outsourcing

According to Willcocks and Lacity,[1] two noted authors who have been actively involved in surveying the field of outsourcing and who have published a number of works on the subject, there is a life cycle perspective on managing the sourcing process. Their comprehensive outsourcing process model consists of four phases:

- **Architect phase**—The foundation for outsourcing is laid. This phase consists of the first four building blocks—investigate, target, strategize and design. At the end of this phase, the enterprise knows itself well enough to confidently publicize its needs.
- **Engage phase**—One or more suppliers are selected and the deal is negotiated. This phase consists of the fifth and sixth building blocks—select and negotiate.
- **Operate phase**—The deal is put in place, operationalized and managed through its term. This phase consists of the seventh and eighth building blocks—transition and manage. After this point, if the deal is not working, management rarely has economic or political options other than to continue with the supplier. Outsourcing deals can be prohibitively expensive to renegotiate, terminate and either back-source (bring back in-house) or transfer to another supplier. It is in this phase that the benefits of the previous work that was done (or not done) become visible. The operate phase either proceeds smoothly, as a result of the strategies, processes, documents and relationship management designed in the earlier building blocks, or the phase suffers due to misinterpretations, ambiguities, disagreements and disputes. At this stage such problems can only be corrected through expansive and tedious remedial efforts.
- **Regenerate phase**—Next-generation options are assessed. This phase consists of one building block—refresh. Following this phase, the life cycle begins anew, returning to the architect phase, where the organization prepares for its next-generation deal(s).

### Multisourcing

In IT outsourcing it is quite common for enterprises to use multiple IT service providers, creating a patchwork that, unfortunately, evolves away from a designed pattern. While each individual service provider arrangement needs to be managed according to an overarching sourcing strategy (and through the life cycle, from both contractual and relationship perspectives), engaging with multiple providers creates additional risk that needs to be managed. The rationale behind a multisourcing approach is to engage with the best-in-class for each job (i.e., the most capable and effective at the right price), recognizing that no one provider can be the best-in-class in each service line.

Multisourcing risk arises when the best-in-class benefits do not materialize or become overshadowed by negatives, issues and overheads. These negative synergies have three root causes:
- Poorly shaped clusters of IT services
- Misaligned technology and provider strategies
- Broken end-to-end processes

In each case it is necessary to understand and then tackle the root causes to avoid the major risk in multisourcing.

The key principles of shaping clusters of IT services are to:
- **Give each service provider control over and responsibility for a set of related technologies and platforms**—Most providers excel in service delivery in only a limited range.
- **Group IT activities that relate to "build" or "service development" activities**—This includes activities spanning the cycle from proposing a solution through to managing the implementation of a working technology service.
- **Group IT activities that relate to "run" or "service delivery" activities**—This relates to control over day-to-day operations and service once it has been developed and deployed.

## Good Practices in IT Resource Planning

Enterprises need to be able to manage, govern and allocate resources effectively, addressing concerns such as:
- **Asset management**—All assets utilized by an enterprise must be managed in a governance environment. This environment consists of content, against which to govern (contract schedules, service level agreements [SLAs], policies, etc.) and process (automated work flow supporting all decision making, benchmarking and communication activities). Typically, this includes a repository of equipment detail, location and configuration management. Additionally, this repository is used to provide the linkage among contracts, SLAs, monitoring and performance management, and benchmarking.
- **Contract management**—This covers the definition of all contracts entered into, including the master services agreement (MSA) and all schedules and companion agreements for subsidiary entities regarding maintenance, service supply, demand, etc. It also includes the formal governance processes by which the performance of and change in these contracts remain visible, managed and through which status is known.
- **Relationship management**—This discipline promotes effective communication among parties to the contract and all stakeholders. Enterprises should integrate relationship management with the outsourcing initiative and must consider continuous communications, conflict identification and resolution, effective and creative problem solving and information sharing.
- **SLAs and operational level agreements (OLAs)**—SLAs and OLAs provide the basis against which performance is managed in the governance processes. They must be measurable and comparable over time. Differentiating between SLAs and OLAs provides useful metrics against which to measure the supplier's performance as well as reduce the margin for error in service delivery when driven by key demand patterns.
- **Due diligence**—Due diligence refers to the discovery by both parties of each other's asset base, resources, processes and, most importantly, their capabilities. It is essential to identify and understand the future capabilities required to provide support for early decision making through all major life cycle stages.
- **Baselining and benchmarking**—Baselining involves using the findings from due diligence and expressing them as a normalized set of data from which performance changes can be measured. Benchmarking allows either party to measure its performance and resource requirements against industry norms. In an ISACA global governance survey,[2] 68 percent of the respondents indicated that they use benchmarking to assess the potential cost effectiveness of an outsourcing contract, while 42 percent use postcontract baselining. These activities are key drivers in renegotiating their contracts.
- **Governance processes**—Governance processes are required to identify, manage, audit and disseminate all information related to the outsourcing contract while controlling the relationship between the client organization and service provider. They are used to ensure that all contractual documents, SLAs and OLAs are monitored on an ongoing basis with clear auditability. Typical high-level governance processes include relationship management, service delivery management and contract management. Across these processes there are a number of more detailed processes, including the following:
  – Policy processes acceptance, development and implementation
  – Compliance
  – Dispensation
  – Performance management
  – Business control
  – Change control
  – Environment management
  – Billing analysis and review

- **Governance organization**—It is necessary to define a governance organization or hierarchy that is responsible for tasks such as decision making, ensuring that delivery meets contractual obligations and escalating issues. All of these integrate to form a risk-aware and risk-managed approach to ensure that the activities undertaken by both parties are articulated and transparent. It is necessary that this organization be established and given the correct levels of responsibility, authority, access and visibility within the governance and service demand and supply environments to carry out duties effectively. Effective outsourcing governance must be explicit and have committed executive sponsorship. A three-tier governance structure works effectively if structured in terms of local (day-to-day operational management and issues), regional (divisional/regional or country level) and global tiers.
- **Scope reviews**—Like all organizational activities, governance regimens are subject to lifetime changes. To maintain efficiency of the processes, it is necessary that the governance regimen include a process for revisiting and revising the applicability of each governance process.
- **Roles and responsibilities**—While the actual roles and responsibilities vary in magnitude and complexity with the processes to be outsourced, there are certain key interactive roles on both the client and supplier sides that are crucial to successful implementation and its subsequent governance. Experience has shown that equivalent logical roles should be present at each level in both the client and supplier organizations. These roles are necessary to identify early indications of risk and to ensure that proper management can take place through to resolution.

More information on outsourcing can be found in section 5.4 Outsourcing and Offshoring Approaches That May Be Employed to Meet the Investment Program and Operational Level Agreements and Service Level Agreements.

ISACA's COBIT 5 introduces a "resource governance" process to "Ensure that the resource needs of the enterprise are met in the optimal manner, IT costs are optimized and there is an increased likelihood of benefit realization and readiness for future change."[3] The key governance practices of this process are defined as follows:
- **EDM04.01 Evaluate resource management.**
Continuously examine and make judgement on the current and future need for IT-related resources, options for resourcing (including sourcing strategies), and allocation and management principles to meet the needs of the enterprise in the optimal manner.
- **EDM04.02 Direct resource management.**
Ensure the adoption of resource management principles to enable optimal use of IT resources throughout their full economic life cycle.
- **EDM04.03 Monitor resource management.**
Monitor the key goals and metrics of the resource management processes and establish how deviations or problems will be identified, tracked and reported for remediation.

## 5.2 HUMAN RESOURCE PROCUREMENT, ASSESSMENT, TRAINING AND DEVELOPMENT METHODOLOGIES

With the scarcity of IT talent (due to issues such as attrition) and the growing demands for skills, it is incumbent on management to maintain sufficient IT resources to match requirements. The effective management of human resources (HR), including their optimization, will have a direct impact on the quality, efficiency and effectiveness of IT services provided to the enterprise. Therefore, a clear definition of requirements for this resource area should be provided, addressing HR-related processes such as procurement, assessment and training, and development.

### Human Capital

The human capital of an enterprise consists of the people who work there and on whom the success of the business depends and represents the human factor in the enterprise—the combined intelligence, skills and expertise that gives the enterprise its distinctive character. The human elements of the enterprise, if properly motivated, are those that are capable of learning, changing, innovating and providing creative thrust and ensuring the long-term survival of the enterprise. Human capital can be regarded as the prime asset of an enterprise, and businesses need to invest in that asset to ensure their survival and growth.

## The Objective of Human Resource Management

The purpose of human resource management (HRM) is to ensure that the enterprise is able to achieve success through people.[4] Effective HRM can be the source of organizational capabilities that allow enterprises to learn and capitalize on new opportunities. HRM is typically comprised of the following:
- **HR philosophies**—Describing the overarching values and guiding principles adopted in managing people
- **HR strategies**—Defining the direction in which HRM intends to go
- **HR policies**—Guidelines defining how these values, principles and strategies should be applied and implemented in specific areas of HRM
- **HR processes**—The formal procedures and methods used to put HR strategic plans and policies into effect
- **HR practices**—The informal approaches used in managing people
- **HR programs**—Enable HR strategies, policies and practices to be implemented according to plan

An effective HR strategy is typically aligned with the following objectives:
- It satisfies business needs.
- It is founded on detailed analysis and study, not just on wishful thinking.
- It can be turned into actionable programs that anticipate implementation requirements and problems.
- It is coherent and integrated, being composed of components that fit with and support each other.
- It takes into account the needs of line managers and employees as well as those of the enterprise and its other stakeholders. HR planning should aim to meet the needs of the key stakeholder groups involved in people management in the firm.

## Human Resource Management and IT Personnel

The demand for skilled IT personnel has been consistently high, and IT personnel have historically displayed high turnover rates. Turnover of skilled IT personnel can incur high costs and prove to be disruptive to enterprises. Whenever talented IT personnel leave an organization, costs are incurred to hire and train replacement employees, and there is a cost of losing the employees' (sometimes irreplaceable) knowledge about the enterprise. The hiring costs of skilled IT personnel vary depending on the type of IT job and the specific skills required. Typical estimates range up from a minimum of 25 percent of annual salaries.

Motivating IT professionals to increase productivity and reduce turnover involves a number of factors that IT managers need to manage. The following are some key factors that IT managers should consider in any motivational program to increase productivity and help reduce IT staff turnover:
- Provide strong leadership during periods of rapid and random change.
- Give employees opportunities to correct mistakes; employees should have confidence in their abilities to complete work independently.
- Provide employees with a personal development plan and give them a clearly defined career path.
- Allow people to learn new technologies as they emerge and attend technology conferences.
- Provide the resources people need to do their jobs well.
- Be competitive in terms of salary and benefits; consider annual salary surveys to keep abreast of salary levels.
- Ensure that people perceive that what they do on the job is meaningful work.
- Ask employees what they desire; do not wait for an exit interview.

While economic downturns can greatly lessen the shortage of IT skills in the marketplace, such situations are temporary. IT leadership needs to plan for what will happen when those skills become scarce. This requires a strong human capital management process that balances investment with returns to best protect that resource. One organizational construct that has helped is the creation of a center of excellence (COE) or community of practice (CoP) in the IT organization to develop leaders and new expertise from within, offer employees new opportunities and increase morale.

## 5.3 PROCESSES FOR ACQUIRING APPLICATION, INFORMATION AND INFRASTRUCTURE RESOURCES

With today's multiplicity of sources for resourcing business and IT services (applications, information and infrastructure), effective resource management requires consideration of the strategic choices to be made in the acquisition of the resources as well as their acquisition processes.

The provisioning of IT resources necessary to meet enterprise business requirements is increasingly being viewed from a supply-and-demand perspective. From the demand point of view, the IT supply organization provides IT services and products. These may be supplied by an internal supply organization or by one or more external providers. Together with the supply organization, the demand organization determines what IT services and products are to be supplied. This process is called service level management (SLM). After determination, the services are supplied. This process is known as service process management. Agreements with the demand organization concern the product or IT service to be supplied and the services that support the product and/or the IT service.[5]

### IT Demand

By specifying functional and performance requirements to IT products, IT services and the services supplied with these products or services, the demand organization tries to indicate the desired services. In this way, differences in expected and supplied performances are minimized or prevented. Nevertheless, differences or gaps may result from the following:
- The customer has different expectations of a product and/or service than the supplier.
- The sold product/service is different from the one developed and implemented.
- Employees do not utilize the product and/or service as it was designed and implemented.
- Reports on the product and/or service deviate from reality.
- The customer's expectations with regard to the product and/or service are different from the customer's experiences with the product and/or service.

### IT Supply

The IT supply organization delivers the services and tries to make sure that there are no gaps. It has to ensure that its products and services comply with the demanded functional and performance requirements. To be able to comply with the requirements in a controlled manner, a process-oriented setup of the IT organization is required and, within this organization, an agreed-on architecture for IT capabilities must be kept. In this IT organization, front- and back-office processes are designed. The front-office processes are in direct contact with the customers. The back-office processes operate in a more planned manner. With such a setup, one can quickly and transparently deliver high-quality services that are adjustable within margins and have a high certainty of supply.

### Acquisition and Outsourcing

Important aspects to consider when outsourcing are as follows:
- **Ownership**—Fundamental to outsourcing is accepting that, while service delivery is transferred, accountability remains firmly with the outsourcing organization, which must ensure that the risk is managed and that there is continued delivery of value from the service provider. Transparency and ownership of the decision-making process must reside within the purview of the enterprise. The decision to outsource is a strategic, rather than a procurement, decision.
- **Selection of outsourcing activities**—The enterprise that outsources is effectively reconfiguring its value chain by identifying those activities that are core to its business, retaining them and making noncore activities candidates for outsourcing. Understanding this in the light of governance is key, not only because well-governed enterprises have been shown to increase shareholder value, but more importantly because every enterprise is competing in an increasingly aggressive, global and dynamic market.
- **Balancing flexibility and core competencies**—Establishing and retaining competitive market advantage requires enterprises to respond effectively to competition and changing market conditions. Outsourcing can support this, but only if the enterprise understands which parts of its business truly creates competitive advantage. Disaggregating these parts and giving them to a third party must become a core competency because outsourcing is a strategic mechanism that allows the enterprise to constantly focus its efforts and expertise.

## Information Services Procurement Library

The Information Services Procurement Library (ISPL) can be used in the procurement process. Important considerations are as follows:

- **Focus on IT service procurement**—The acquisition processes that enterprises adopt for IT resources need not be unique; however, there are nuances involved in IT (e.g., service levels) that call for additional attention. ISPL focuses purely on the procurement of information services.
- **Process-based approach**—Within customer and supplier organizations, ISPL can be used at the senior management level to help the enterprise understand the services to be acquired and delivered in a large-scale setting, in the private and public sectors, and to structure their acquisition and delivery.[6] The ISPL framework is based on the premise that a robust process is essential to support the customer of IT services.
- **Holistic and customized IT sourcing**—Understanding the requirements, negotiating and complying with clear agreements, organizing and planning all need to be addressed. The ISPL approach is based on best practices from large-scale IT service contracts in the European private and public sectors. The design of the approach is based on the philosophy that "one set of rules works for all" does not exist. Instead, ISPL offers many methodical building blocks for creating a customized approach. The ISPL approach offers various management instruments for understanding and controlling fast-changing outsourcing arrangements. The basic ISPL acquisition process is shown in **figure 5.2**.

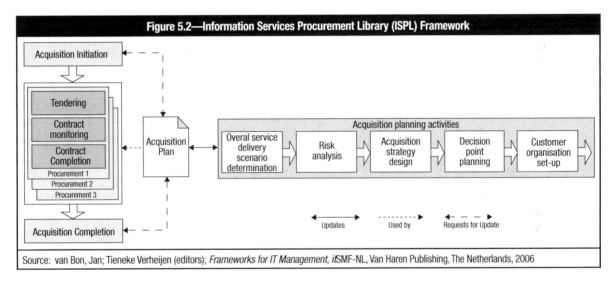

Source: van Bon, Jan; Tieneke Verheijen (editors); *Frameworks for IT Management*, itSMF-NL, Van Haren Publishing, The Netherlands, 2006

More information on SLM can be found in section 5.9 Service Level Management Concepts.

## 5.4 OUTSOURCING AND OFFSHORING APPROACHES THAT MAY BE EMPLOYED TO MEET THE INVESTMENT PROGRAM AND OPERATIONAL LEVEL AGREEMENTS AND SERVICE LEVEL AGREEMENTS

The growing trend of outsourcing and offshoring holds both promise and pitfalls for the procurement of strategic resources to deliver IT services. With the maturing of practices, processes used in these areas define the systematic approaches to effectively engage and manage in such environments in line with investment program needs and SLAs and OLAs.

### Benefits and Risk Considerations for Outsourcing

Surveys consistently show that outsourcing benefits are no longer just about price. The benefits include service quality improvements, scalability, better risk management and freeing internal resources to focus on core, value-adding activities. It is no longer an enterprise's ownership of capabilities that matters, but rather its ability to control and make the most of critical capabilities, whether or not they reside on the enterprise's balance sheet. It is, therefore, necessary to be cognizant of the potential changes to the risk profile of the extended enterprise and its operations when outsourcing.

## Business Process Outsourcing

Outsourcing is a formal agreement with a third party to perform IS or other business functions for an enterprise. Business process outsourcing (BPO) is the form of outsourcing that involves the movement of business processes from inside the enterprise to an external service provider.

With a global telecommunications infrastructure now well established and consistently reliable, BPO initiatives often include shifting work to international providers. Back-office functions such as payroll and benefits administration, customer service, call center and technical support are a few of the processes that enterprises of all sizes have been able to outsource to specialists in those areas. Removing these functions from their internal operations enables enterprises to reduce payroll and other overhead. In an environment where executives search for increased focus on core competencies, BPO offers an opportunity to achieve that and to reach for higher service levels with lower investments. However, with increasing education levels around the world, BPO is no longer confined to routine manufacturing jobs. Today's outsourcing involves complex work that requires extensive preparation and training, including areas such as radiology and IT.

BPO is an interdisciplinary innovation. To be successful, a diverse set of management skills is required. Initiating and implementing a BPO project requires a focus on several human factors, both within the enterprise initiating the project and within the outsourcing vendor. These factors cannot be ignored and need to be handled correctly for the project to succeed. Such a project also requires attention to technology management issues.

In practice, there are three types of BPOs: offshore, onshore and nearshore. They differ in both location and function served. Enterprises typically use any or all of these types, depending on their needs and the BPO initiatives being implemented. In some cases, enterprises use a combination of types to achieve their objectives:

- **Offshore: Larger challenge, greater reward**[7]—Offshore BPO is the most challenging type of this relatively new approach to conducting business, but potentially it is also the most rewarding. It began with movement of factory jobs overseas and has been made both famous and infamous with stories of suddenly prosperous geographic regions mixed with stories of exploitative labor practices. Yet, despite the criticism leveled at some enterprises that outsource processes and functions to international labor markets, the advantages of doing so continue to outweigh the disadvantages. For example, by benefiting from lower wages overseas, US managers can cut overall costs by 25 to 40 percent while building a more secure, more focused workforce in the United States. The complexity of business functions being moved offshore continues to increase. As such, enterprises using the offshore approach have developed a variety of models to ensure continuity. Some have utilized a model known as offshore insourcing, in which the enterprise establishes a wholly-owned subsidiary in the international market and hires local labor. An extension of this is the build-operate-transfer (BOT) model, in which enterprises buy offshore companies specializing in a business process, operate them jointly for a year or so and then transfer the firm to internal control (insource). It is important to bear in mind that there is no one-size-fits-all approach to offshore BPO. With the growing list of enterprises outsourcing at least some business functions to offshore vendors, the range of possible approaches will grow as well. This makes it increasingly likely that the next adopter of offshore BPO will find a model suitable for its needs.
- **Onshore: Outsourcing to same-country-based firms**—It would be an over-simplification to view BPO purely as an international business phenomenon. Many US enterprises outsource back-office functions to US-based firms. A prominent example of this is payroll outsourcing, which is managed by several large US companies. Automatic Data Processing, Inc. (ADP®) is an example of such a firm which provides a range of payroll administration services, time sheets and tax filing and reporting services. There are several reasons why an enterprise will use BPO. The cost savings that result from moving back-office processes to lower-wage environments is the reason most often cited. However, firms can also use BPO to transfer service functions to best-in-class performers to gain a competitive advantage. A firm that outsources customer service functions to a firm that specializes in and provides world-class support in that domain will perform at a higher level in that function than its competitors. Moving to a best-in-class provider may actually increase costs in the short run in the interest of developing a competitive advantage. Under this rationale BPO is a strategic investment that is designed to upgrade service levels at a cost, with the intent of increasing revenues through enhanced competitiveness. What matters most is the acquisition of partners that provide market-shifting capabilities for the firm doing the outsourcing. Many US-based outsourcing firms use the world-class provider strategy to acquire business. Handicapped to the low-cost international rivals, US-based outsourcing firms need to continuously innovate and seek new ways to provide value

to remain in front. They are often favorably considered for services, even if their costs are higher, when strategic advantage is the goal of an enterprise's BPO initiative.

- **Nearshore: Outsourcing to same-continent-based firms**—Nearshore outsourcing is a relatively new phenomenon that refers primarily to the practice of outsourcing on the North American continent. International issues arise when US firms outsource to Mexico, Canada or Central America, but they are likely to be less complex than those that are associated with outsourcing arrangements in, for example, India or China. Nearshore outsourcing allows companies to test the BPO experience without the level of risk involved in going fully offshore. Enterprises that go with a nearshore strategy are often seeking cost savings, but are also occasionally able to find best-in-class providers of the services they need.

## Outsourcing Stakeholders

**Figure 5.3** identifies typical stakeholders in the outsourcing relationship, which itself can take a number of forms. Market-type relationships are categorized as short-term and commodity-style, where there are a number of providers available in the marketplace and switching costs are low. At the other end are the partnership forms of outsourcing arrangements which are typically longer term and require deeper understanding of the client organization. The decision to outsource and subsequently manage that relationship demands effective management to succeed.

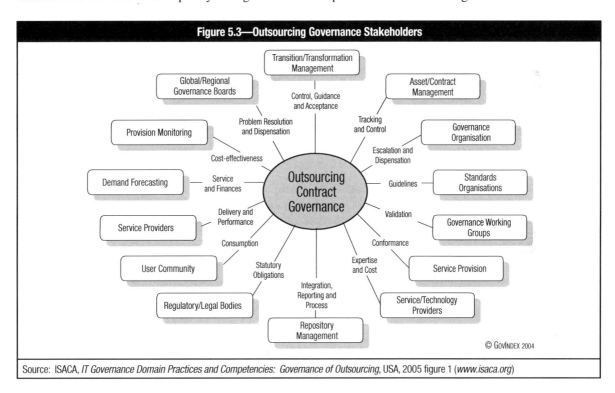

Source: ISACA, *IT Governance Domain Practices and Competencies: Governance of Outsourcing*, USA, 2005 figure 1 (*www.isaca.org*)

Most people who conduct outsourcing contracts include basic control and service execution provisions; however, one of the main objectives of the outsourcing management process, as defined in the outsourcing contract, is to ensure profitability, continuity of service at the appropriate levels and value-add to sustain the commercial viability of both parties. Experience has shown that many enterprises make assumptions about what is included in the outsourcing proposition. Whereas it is neither possible nor cost-effective to define every detail and action contractually, the governance process provides the mechanism to balance risk, service demand, service provision and cost.

## Outsourcing Responsibilities

Following the ISACA definition of governance of enterprise IT, the governance of outsourcing extends both parties' (i.e., client's and supplier's) responsibilities into the following areas:

- Ensuring contractual viability through continuous review, improvement and benefit gain to both parties
- Inclusion of an explicit governance schedule to the contract
- Management of the relationship to ensure that contractual obligations are met through SLAs, OLAs, service credit regimes and gain sharing

- Identification and management of all stakeholders, their relationships and expectations
- Establishment of clear roles and responsibilities for decision making, issue escalation, dispute management, demand management and service delivery
- Allocation of resources, expenditure and service consumption in response to prioritized needs
- Continuous evaluation of performance, cost, user satisfaction and effectiveness
- Ongoing communication across all stakeholders

More information on sourcing in the context of IT resource planning can be found in section 5.1 IT Resource Planning Methods.

## 5.5 METHODS USED TO RECORD AND MONITOR IT RESOURCE UTILIZATION AND AVAILABILITY

Determining and evaluating business and IT resource demand and utilization is an essential part of the continuous monitoring of the availability of resources to meet planned and unforeseen demands for resources (as in business continuity planning). The adequacy of resources will facilitate continuous availability of IT services, which is a key business requirement.

### Demand Management

Demand management is a critical aspect of IT service management.[8] Poorly managed demand is a source of risk for service providers because of uncertainty in demand. Excess capacity generates cost without creating value that provides a basis for cost recovery. The purpose of demand management is to understand and influence customer demand for services and the provision of capacity to meet this demand. At a strategic level this can involve analysis of patterns of business activity and user profiles. At a tactical level, it can involve use of differential charging to encourage customers to use IT services at less busy times. Typically, a service level package (SLP) defines the level of utility and warranty for a service package and is designed to meet the needs of a pattern of business activity (PBA).

### Availability and Capacity Management

The COBIT 5 framework[9] references the need for availability and capacity management. It describes the purpose of availability and capacity management as:

> *Balance current and future needs for availability, performance and capacity with cost-effective service provision. Include assessment of current capabilities, forecasting of future needs based on business requirements, analysis of business impacts, and assessment of risk to plan and implement actions to meet the identified requirements.*

COBIT 5[10] includes the following guidance:

#### *BAI04.01 Assess current availability, performance and capacity and create a baseline.*

Assess availability, performance and capacity of services and resources to ensure that cost-justifiable capacity and performance are available to support business needs and deliver against SLAs. Create availability, performance and capacity baselines for future comparison.

Activities include:

1. Consider the following (current and forecasted) in the assessment of availability, performance and capacity of services and resources: customer requirements, business priorities, business objectives, budget Impact, resource utilizations, IT capabilities and industry trends.
2. Monitor actual performance and capacity usage against defined thresholds, supported where necessary with automated software.
3. Identify and follow up on all incidents caused by inadequate performance or capacity.
4. Regularly evaluate the current levels of performance for all processing levels (business demand, service capacity and resource capacity) by comparing them against trends and SLAs, taking into account changes in the environment.

### BAI04.02 Assess business impact.
Identify important services to the enterprise, map services and resources to business processes, and identify business dependencies. Ensure that the impact of unavailable resources is fully agreed on and accepted by the customer. Ensure that, for vital business functions, the SLA availability requirements can be satisfied.

Activities include:
1. Identify only those solutions or services that are critical in the availability and capacity management process.
2. Map the selected solutions or services to application(s) and infrastructure (IT and facility) on which they depend to enable a focus on critical resources for availability planning.
3. Collect data on availability patterns from logs of past failures and performance monitoring. Use modeling tools that help predict failures based on past usage trends and management expectations of new environment or user conditions.
4. Create scenarios based on the collected data, describing future availability situations to illustrate a variety of potential capacity levels needed to achieve the availability performance objective.
5. Determine the likelihood that the availability performance objective will not be achieved based on the scenarios.
6. Determine the impact of the scenarios on the business performance measures (e.g., revenue, profit, customer services). Engage the business line, functional (especially finance) and regional leaders to understand their evaluation of impact.
7. Ensure that business process owners fully understand and agree to the results of this analysis. From the business owners, obtain a list of unacceptable risk scenarios that require a response to reduce risk to acceptable levels.

### BAI04.03 Plan for new or changed service requirements.
Plan and prioritise availability, performance and capacity implications of changing business needs and service requirements.

Activities include:
1. Review availability and capacity implications of service trend analysis.
2. Identify availability and capacity implications of changing business needs and improvement opportunities. Use modelling techniques to validate availability, performance and capacity plans.
3. Prioritise needed improvements and create cost-justifiable availability and capacity plans
4. Adjust the performance and capacity plans and SLAs based on realistic, new, proposed and/or projected business processes and supporting services, applications and infrastructure changes as well as reviews of actual performance and capacity usage, including workload levels.
5. Ensure that management performs comparisons of actual demand on resources with forecasted supply and demand to evaluate current forecasting techniques and make improvements where possible.

### BAI04.04 Monitor and review availability and capacity.
Monitor, measure, analyse, report and review availability, performance and capacity. Identify deviations from established baselines. Review trend analysis reports identifying any significant issues and variances, initiating actions where necessary, and ensuring that all outstanding issues are followed up.

Activities include:
1. Establish a process for gathering data to provide management with monitoring and reporting information for availability, performance and capacity workload of all information-related resources.
2. Provide regular reporting of the results in an appropriate form for review by IT and business management and communication to enterprise management.
3. Integrate monitoring and reporting activities in the iterative capacity management activities (monitoring, analysis, tuning and implementations).
4. Provide capacity reports to the budgeting processes.

***BAI04.05 Investigate and address availability, performance and capacity issues.***
Address deviations by investigating and resolving identified availability, performance and capacity issues.

Activities include:
1. Obtain guidance from vendor product manuals to ensure an appropriate level of performance availability for peak processing and workloads.
2. Identify performance and capacity gaps based on monitoring current and forecasted performance. Use the known availability, continuity and recovery specifications to classify resources and allow prioritisation.
3. Define corrective actions (e.g., shifting workload, prioritising tasks or adding resources, when performance and capacity issues are identified).
4. Integrate required corrective actions into the appropriate planning and change management processes.
5. Define an escalation procedure for swift resolution in case of emergency capacity and performance problems.

More information on availability management can be found in section 5.6 Methods Used to Evaluate and Report on IT Resource Performance.

## 5.6 METHODS USED TO EVALUATE AND REPORT ON IT RESOURCE PERFORMANCE

Effective monitoring and reporting of IT resource performance highlights the issues and decisions made to manage IT resources. Without such monitoring and reporting there would be little or no visibility to management of IT resourcing issues that may be critical and for which important management decisions have to be made.

Monitoring and reporting on IT resource performance can be found within the configuration management and availability management processes of ITIL. The monitoring and reporting of these process aspects are described as follows.

### Capacity Management Information Systems

The CMIS is typically used as a basis of a successful capacity management process.[11] Information contained within the CMIS is stored and analyzed by all the subprocesses of capacity management because it is a repository that holds a number of different types of data, including business, service, resource or utilization and financial data, from all areas of technology. However, the CMIS is unlikely to be a single database, and data typically exist in several physical locations. Data from all areas of technology, and all components that make up the IT services, can then be combined for analysis and provision of technical and management reporting. Only when all of the information is integrated can end-to-end service reports be produced. The integrity and accuracy of the data within the CMIS need to be carefully managed. If the CMIS is not part of an overall configuration management system (CMS) or service knowledge management system (SKMS), then links between these systems need to be implemented to ensure consistency and accuracy of the information recorded within them. The information in the CMIS is used to form the basis of performance and capacity management reports and views that are to be delivered to customers, IT management and technical personnel. Data are utilized to generate future capacity forecasts and allow capacity management to plan for future capacity requirements. Often, a web interface is provided to the CMIS to provide the different access and views required outside of the capacity management process.

The range of data types typically stored within the CMIS is as follows:
- **Business data**—It is essential to have quality information on the current and future needs of the business. The future business plans of the enterprise need to be considered and the effects on the IT services understood. The business data are used to forecast and validate how changes in business drivers affect the capacity and performance of the IT infrastructure. Business data should include business transactions or measurements, such as the number of accounts, the number of invoices generated or the number of product lines.
- **Service data**—To achieve a service-oriented approach to capacity management, service data should be stored within the CMIS. Typical service data are transaction response times, transaction rates, workload volumes, etc. In general, the SLAs provide the service targets for which the capacity management process needs to record and monitor data. To ensure that the targets in the SLAs are achieved, SLM thresholds should be included so that the monitoring activity can measure against these service thresholds and raise exception warnings and reports before service targets are breached.

- **Component utilization data**—The CMIS also needs to record resource data consisting of utilization, threshold and limit information on all of the technology components supporting the services. Most of the IT components have limitations on the level to which they should be utilized. Beyond this level of utilization, the resource will be overutilized and the performance of the services using the resource will be impaired. For example, the maximum recommended level of utilization on a processor could be 80 percent or the utilization of a shared ethernet local area network (LAN) segment should not exceed 40 percent. Components have various physical limitations beyond which greater connectivity or use is impossible. For example, the maximum number of connections through an application or a network gateway is 100 or a particular type of disk has a physical capacity of 15 GB. The CMIS should, therefore, contain current and past utilization rates and the associated component thresholds for each component and the maximum performance and capacity limits. Over time, this can require vast amounts of data to be accumulated so there need to be good techniques for analyzing, aggregating and archiving these data.
- **Financial data**—The capacity management process requires financial data. For evaluating alternative upgrade options when proposing various scenarios in the capacity plan, the financial cost of the upgrades to the components of the IT infrastructure together with information about the current IT hardware budget must be known and included in the considerations. Most of these data may be available from the financial management for IT services process, but capacity management needs to consider this information when managing the future business requirements.

The data collected from monitoring should be analyzed to identify trends from which the normal utilization and service levels, or baselines, can be established. By regular monitoring and comparison with this baseline, exception conditions in the utilization of individual components or service thresholds can be defined and breaches or near misses in the SLAs can be reported and acted on. The data can be used to predict future resource usage or monitor actual business growth against predicted growth. Analysis of the data may identify issues such as:
- Bottlenecks or hot spots within the infrastructure
- Inappropriate distribution of workload across available resources
- Inappropriate database indexing
- Inefficiencies in the application design
- Unexpected increase in workloads or transaction rates
- Inefficient scheduling or memory usage

## Availability Management

An effective availability management process, consisting of both the reactive and proactive activities, can make a substantial difference and will be recognized as such by the business if the deployment of availability management within an IT organization has a strong emphasis on the needs of the business and customers. The reactive activities of availability management consist of monitoring, measuring, analyzing, reporting and reviewing all aspects of component and service availability. This is to ensure that all agreed-on service targets are measured and achieved. Wherever deviations or breaches are detected they are investigated and remedial action instigated. Most of these activities are conducted within the operations stage of the life cycle and are linked into the monitor and control activities, event and incident management processes. The proactive activities consist of:
- Producing recommendations, plans and documentation on design guidelines
- Setting criteria for new and changed services and continual improvement of service
- Reducing risk in existing services whenever the cost can be justified

The scope of availability management covers the design, implementation, measurement and management of IT service and infrastructure availability. To reinforce the emphasis on the needs of the business and customers there are several guiding principles that underpin the availability management and its focus:
- **Service availability is at the core of customer satisfaction and business success**—There is a direct correlation in most enterprises between the service availability and customer and user satisfaction, where poor service performance is defined as being unavailable.
- **Recognizing that when services fail, it is still possible to achieve business, customer and user satisfaction and recognition**—The way a service provider reacts in a failure situation has a major influence on customer and user perception and expectation.
- **Improving availability**—This can begin only after understanding how the IT services support the operation of the business.

- **Service availability is only as good as the weakest link on the chain**—It can be greatly increased by the elimination of a single point of failure (SPoF) or an unreliable or weak component.
- **Availability is not just a reactive process**—The more proactive the process, the better service availability will be. Availability should not purely react to service and component failure. The more events and failures are predicted, preempted and prevented, the higher the level of service availability.
- **It is cheaper to design the right level of service availability into a service from the start rather than to bolt it on later**—Adding resilience into a service or component is invariably more expensive than designing it in from the start. Once a service gets a bad name for unreliability, it becomes very difficult to change the image. Resilience is also a key consideration of IT service continuity management (IT SCM) and this should be considered at the same time.

Availability management is completed at two interconnected levels:
- **Service availability**—This involves all aspects of service availability and unavailability, and the impact of component availability or the potential impact of component unavailability on service availability.
- **Component availability**—This involves all aspects of component availability and unavailability.

Availability management relies on the monitoring, measuring, analyzing and reporting of the following aspects:
- **Availability**—The ability of a service, component or configuration item (CI) to perform its agreed-on function, when required. It is often measured and reported as a percentage.
- **Reliability**—A measure of how long a service, component or CI can perform its agreed-on function without interruption. The reliability of the service can be improved by increasing the reliability of the service to individual component failure (increasing the component redundancy, e.g., by using load-balancing techniques). It is often measured and reported as mean time between service incidents (MTBSI) or mean time between failures (MTBF).
- **Maintainability**—A measure of how quickly and effectively a service, component or CI can be restored to normal after a failure. It is measured and reported as mean time to restore service (MTRS).
- **Serviceability**—The ability of a third-party supplier to meet the terms of its contract. Often this contract will include agreed-on levels of availability, reliability and/or maintainability for a supporting service or component.

## 5.7 INTEROPERABILITY, STANDARDIZATION AND ECONOMIES OF SCALE

In most enterprises, the biggest portion of the IT budget relates to ongoing operations. Effective management of IT operational spending requires effective control of the cost base: the IT assets and the focus where they are needed most. IT assets should be organized so that the required quality of service is provided by the most cost-effective delivery infrastructure. This objective requires the necessary management attention for concepts such as interoperability, standardization and economies of scale.

### Resource Optimization

A key to successful IT performance is the optimal investment, use and allocation of IT resources (people, applications, technology, facilities, data) in servicing the needs of the enterprise. Most enterprises fail to maximize the efficiency of their IT assets and optimize the costs relating to these assets. Many of the knowledge statements in this chapter contribute to a more optimal resource allocation, often implicitly referring to concepts such as:
- Interoperability
- Standardizations
- Economies of scale

These concepts are now defined more in-depth.

### Economies of Scale

In microeconomics, economies of scale are the cost advantages that an enterprise obtains due to expansion. There are factors that cause a producer's average cost per unit to fall as the scale of output is increased. Economies of scale is a long running concept and refers to reductions in unit cost as the size of a facility and the usage levels of other inputs increase. The simple meaning of economies of scale is doing things efficiently.

### Interoperability

Interoperability is the ability of diverse systems and organizations to work together (interoperate). The term is often used in a technical systems engineering sense or alternatively in a broad sense taking into account social, political and organizational factors that impact system to system performance.

The Institute of Electrical and Electronics Engineers (IEEE) glossary defines interoperability as the ability of two or more systems or components to exchange information and to use the information that has been exchanged.

O'Brien and Marakas define interoperability as being able to accomplish end-user applications using different types of computer systems, operating systems and application software, interconnected by different types of local and wide area networks.[12]

### Standardization

Standardization is the process of developing and implementing. The goals of standardization can be to help with independence of single suppliers (commoditization), compatibility, interoperability, safety, repeatability or quality. In social sciences, including economics, the idea of standardization is close to the solution for a coordination problem, a situation in which all parties can realize mutual gains, but only by making mutually consistent decisions. Standardization is defined as the best technical application of consensual wisdom, inclusive of processes for selection in making appropriate choices for coupled with consistent decisions for maintaining obtained standards. This view includes the case of "spontaneous standardization processes," to produce de facto standards.

## 5.8 DATA MANAGEMENT AND DATA GOVERNANCE CONCEPTS

Information is pervasive throughout any enterprise and includes all information produced and used by the enterprise. Information is required for keeping the enterprise running and well governed, but at the operational level information is very often the key product of the enterprise itself. In the information life cycle, data management and governance concepts are important in converting data into information, knowledge and value for the enterprise.

As discussed in chapter 1, efficient and effective governance and management of enterprise IT require a holistic approach taking into account several interacting components. These components are enablers supporting implementation of a comprehensive governance and management system for enterprise IT. Enablers are broadly defined as anything that can help to achieve the objectives of the enterprise.

The COBIT 5 framework describes seven categories of enablers:[13]
- **Principles, policies and frameworks** are the vehicles to translate the desired strategy into practical guidance for day-to-day management.
- **Processes** describe an organized set of practices and activities to achieve certain objectives and produce a set of outputs in support of achieving overall IT-related goals.
- **Organizational structures** are the key decision-making entities in an enterprise.
- **Culture, ethics and behavior** of individuals and of the enterprise are very often underestimated as a success factor in governance and management activities.
- **Information** is pervasive throughout any organization and includes all information produced and used by the enterprise. Information is required for keeping the organization running and well governed, but at the operational level, information is very often the key product of the enterprise itself.
- **Services, infrastructure and applications** include the infrastructure, technology and applications that provide the enterprise with information technology processing and services.
- **People, skills and competencies** are linked to people and are required for successful completion of all activities and for making correct decisions and taking corrective actions.

## The Information Cycle

The information enabler deals with all information relevant for enterprises, not only automated information. Information can be structured or unstructured, formal or informal.

Information can be considered as being one stage in the "information cycle" of an enterprise. In the information cycle (**figure 5.4**), business processes generate and process data, transforming them into information and knowledge and ultimately generating value for the enterprise. The scope of the information enabler mainly concerns the "information" phase in the information cycle, but the aspects of data and knowledge are also covered in COBIT 5.

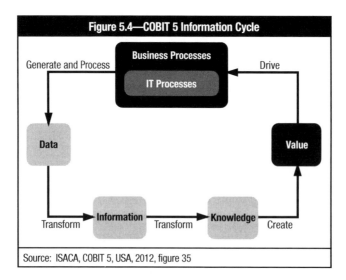

## COBIT 5 Enabler: Information

COBIT 5 has developed a generic model to describe each of the enablers. For the information enabler, this model is visualized in **figure 5.5**. The specifics for the information enabler compared to the generic enabler descriptions are shown in **bold**.

The information model (IM) shows:
- Stakeholders
- Goals
- Life Cycle
- Good Practices

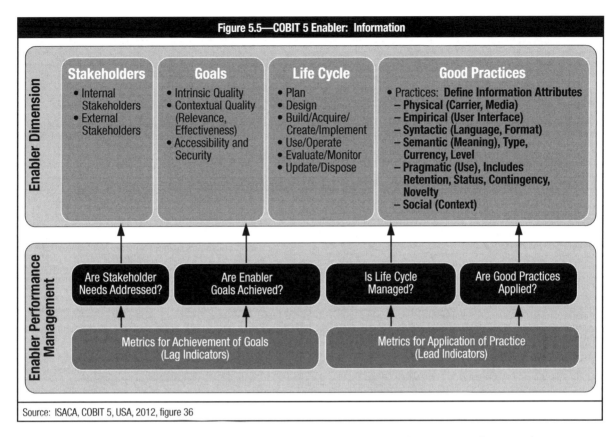

Figure 5.5—COBIT 5 Enabler: Information

Source: ISACA, COBIT 5, USA, 2012, figure 36

- **Stakeholders**—Can be internal or external to the enterprise. The generic model also suggests that, apart from identifying the stakeholders, their stakes need to be identified (i.e., why they care or are interested in the information).

With respect to which information stakeholders exist, different categorizations of roles in dealing with information are possible, ranging from detailed proposals (e.g., suggesting specific data or information roles like architect, owner, steward, trustee, supplier, beneficiary, modeler, quality manager, security manager) to more general proposals, for instance, distinguishing among information producers, information custodians and information consumers:
– Information producer—Responsible for creating the information
– Information custodian—Responsible for storing and maintaining the information
– Information consumer—Responsible for using the information

These categorizations refer to specific activities with regard to the information resource. Activities depend on the life cycle phase of the information. Therefore, to find a categorization of roles that has an appropriate level of granularity for the IM, the information life cycle dimension of the IM can be used. This means that information stakeholder roles can be defined in terms of information life cycle phases (e.g., information planners, information "obtainers," information users). At the same time this means that the information stakeholder dimension is not an independent dimension; different life cycle phases have different stakeholders.

Whereas the relevant roles depend on the information life cycle phase, the stakes can be related to information goals.
- **Goals**—The goals of information are divided in three subdimensions of quality:
– Intrinsic quality—The extent to which data values are in conformance with the actual or true values. It includes:
  · Accuracy—The extent to which information is correct and reliable
  · Objectivity—The extent to which information is unbiased, unprejudiced and impartial
  · Believability—The extent to which information is regarded as true and credible
  · Reputation—The extent to which information is highly regarded in terms of its source or content

– Contextual and representational quality—The extent to which information is applicable to the task of the information user and is presented in an intelligible and clear manner, recognizing that information quality depends on the context of use. It includes:
  - Relevancy—The extent to which information is applicable and helpful for the task at hand
  - Completeness—The extent to which information is not missing and is of sufficient depth and breadth for the task at hand
  - Currency—The extent to which information is sufficiently up to date for the task at hand
  - Appropriate amount of information—The extent to which the volume of information is appropriate for the task at hand
  - Concise representation—The extent to which information is compactly represented
  - Consistent representation—The extent to which information is presented in the same format
  - Interpretability—The extent to which information is in appropriate languages, symbols and units, and the definitions are clear
  - Understandability—The extent to which information is easily comprehended
  - Ease of manipulation—The extent to which information is easy to manipulate and apply to different tasks
– Security/Accessibility quality—The extent to which information is available or obtainable. It includes:
  - Availability—The extent to which information is available when required, or easily and quickly retrievable
  - Restricted Access—The extent to which access to information is restricted appropriately to authorized parties
- **Life cycle**—The full life cycle of information needs to be considered and different approaches may be required for information in different phases of the life cycle. The COBIT 5 Information enabler distinguishes the following phases:
  – Plan—The phase in which the creation and use of the information resource is prepared. Activities in this phase may refer to the identification of objectives, the planning of the information architecture, and the development of standards and definitions (e.g., data definitions, data collection procedures).
  – Design—The phase in which the information origin, flow, usage, storage, archiving and destruction will be considered for being included in the solution
  – Build/Acquire—The phase in which the information resource is acquired. Activities in this phase may refer to the creation of data records, the purchase of data and the loading of external files.
  – Use/Operate—This phase includes:
    - Store—The phase in which information is held electronically or in hard copy (or even just in human memory). Activities in this phase may refer to the storage of information in electronic form (e.g., electronic files, databases, data warehouses) or as hard copy (e.g., paper documents).
    - Share—The phase in which information is made available for use through a distribution method. Activities in this phase may refer to the processes involved in getting the information to places where it can be accessed and used (e.g., distributing documents by email). For electronically held information this life cycle phase may largely overlap with the store phase (e.g., sharing information through database access, file/document servers).
    - Use—The phase in which information is used to accomplish goals. Activities in this phase may refer to all kinds of information usage (e.g., managerial decision making, running automated processes) and also include activities such as information retrieval and converting information from one form to another.
    - Monitor—The phase in which it is ensured that the information resource continues to work properly (i.e., to be valuable). Activities in this phase may refer to keeping information up to date as well as other kinds of information management activities (e.g., enhancing, cleansing, merging, removing duplicate information data in data warehouses).
    - Dispose—The phase in which the information resource is discarded when it is no longer of use. Activities in this phase may refer to information archiving or destroying.
- **Good practice**—The concept of information is understood differently in different disciplines such as economics, communication theory, information science, knowledge management and information systems. Therefore, there is no universally agreed-on definition on what information is. The nature of information can, however, be clarified through defining and describing its properties.

The following scheme is proposed to structure the different properties: It consists of six levels or layers to define and describe properties of information. These six levels present a continuum of attributes, ranging from the physical world of information, where attributes are linked to information technologies and media for information capturing, storing, processing, distribution and presentation, to the social world of information use, sense-making and action.

The following descriptions can be given to the layers and information attributes:
- **Physical world layer**—The world where all phenomena that can be empirically observed take place
  - Information carrier/media—The attribute that identifies the physical carrier of the information (e.g., paper, electric signals, sound waves)
- **Empiric layer**—The empirical observation of the signs used to encode information and their distinction from each other and from background noise
  - Information access channel—The attribute that identifies the access channel of the information (e.g., user interfaces)
- **Syntactic layer**—The rules and principles for constructing sentences in natural or artificial languages; Syntax refers to the form of information.
  - Code/language—The attribute that identifies the representational language/format used for encoding the information. Rules for combining the symbols of the language to form syntactic structures
- **Semantic layer**—The rules and principles for constructing meaning out of syntactic structures. Semantics refers to the meaning of information.
  - Information type—The attribute that identifies the kind of information. Different categorizations are possible (e.g., financial vs. nonfinancial information, internal vs. external origin of the information, forecasted/predicted values vs. observed values, planned values vs. realized values).
  - Information currency—The attribute that identifies the time horizon referred to by the information (i.e., information on the past, the present or the future)
  - Information level—The attribute that identifies the degree of detail of the information (e.g., sales per year, quarter, month)
- **Pragmatic layer**—The rules and structures for constructing larger language structures that fulfill specific purposes in human communication. Pragmatics refers to the use of information.
  - Retention period—The attribute that identifies how long information can be retained before it is destroyed
  - Information status—The attribute that identifies whether the information is operational or historical
  - Novelty—The attribute that identifies whether the information creates new knowledge or confirms existing knowledge (i.e., information vs. confirmation)
  - Contingency—The attribute that identifies the information that is required to precede this information (for it to be considered as information)
- **Social world layer**—The world that is socially constructed through the use of language structures at the pragmatic level of semiotics (e.g., contracts, law, culture).
  - Context—The attribute that identifies the context in which the information makes sense, is used, has value, etc. (e.g., cultural context, subject domain context)

## Further Considerations About Information

Investments in information and related technology are based on business cases which include cost-benefit analysis. Costs and benefits refer not only to tangible, measurable factors, but they also take into account intangible factors such as competitive advantage, customer satisfaction and technology uncertainty. It is only when the information resource is applied or used that an enterprise generates benefits from it, so the value of information is determined solely through its use (internally or by selling it), and information has no intrinsic value. Value can only be generated by putting information into action.

The IM is a new model and is very rich in terms of different components. In order to make it more tangible for the COBIT 5 user, and to make its relevance more clear in the context of the overall COBIT 5 framework, examples of possible uses of the IM are provided in **examples 1**, **2** and **3**.

> **Example 1—Information Model Used for Information Specifications**
>
> When developing a new application, the IM can be used to assist with the specifications of the application and the associated information or data models.
>
> The information attributes of the IM can be used to define specifications for the application and the business processes that will use the information.
>
> For example, the design and specifications of the new system need to specify:
> - Physical layer—Where will information be stored?
> - Empirical layer—How can the information be accessed?
> - Syntactical layer—How will the information be structured and coded?
> - Semantic layer—What sort of information is it? What is the information level?
> - Pragmatic layer—What are the retention requirements? What other information is required for this information to be useful and usable?
>
> Looking at the stakeholder dimension combined with the information life cycle, one can define who will need what type of access to the data during which phase of the information life cycle.
>
> When the application is tested, testers can look at the information quality criteria to develop a comprehensive set of test cases.
>
> Source: ISACA, COBIT 5, USA, 2012, page 84

> **Example 2—Information Model Used To Determine Required Protection**
>
> Security groups within the enterprise can benefit from the attributes dimension of the IM. Indeed, when charged with protection of information, they need to look at:
> - **Physical layer**—How and where is information physically stored?
> - **Empirical layer**—What are the access channels to the information?
> - **Semantic layer**—What type of information is it? Is the information current or relating to the past or to the future?
> - **Pragmatic layer**—What are the retention requirements? Is information historic or operational?
>
> Using these attributes will allow the user to determine the level of protection and the protection mechanisms required.
>
> Looking at another dimension of the IM, security professionals can also consider the information life cycle stages, because information needs to be protected during all phases of the life cycle. Indeed, security starts at the information planning phase, and implies different protection mechanisms for storing, sharing and disposition of information. The IM ensures that information is protected during the full life cycle of the information.
>
> Source: ISACA, COBIT 5, USA, 2012, page 84

> **Example 3—Information Model Used To Determine Ease Of Data Use**
>
> When performing a review of a business process (or an application), the IM can be used to assist with a general review of the information processed and delivered by the process, and of the underlying information systems. The quality criteria can be used to assess the extent to which information is available—whether the information is complete, available on a timely basis, factually correct, relevant, available in the appropriate amount. One can also consider the accessibility criteria—whether the information is accessible when required and adequately protected.
>
> The review can be even further extended to include representation criteria, e.g., the ease with which the information can be understood, interpreted, used and manipulated.
>
> A review that uses the information quality criteria of the IM provides an enterprise with a comprehensive and complete view on the current information quality within a business process.
>
> Source: ISACA, COBIT 5, USA, 2012, page 84

## 5.9 SERVICE LEVEL MANAGEMENT CONCEPTS

In today's service-driven economy, enterprises are relying more and more on third parties for a variety of IT services. Often, they are not pleased with the service received and are sometimes dependent on third parties whose futures are uncertain. An appropriate SLM process should be in place in the enterprise for obtaining the required service(s). Implementing an SLM process is not an easy and quick task to perform, and thus an approach using supportive mechanisms such as COBIT may help in defining or fine-tuning the SLA(s).

### Service Level Management

According to Sturm, Morris and Jander,[14] SLM is "the disciplined, proactive methodology and procedures used to ensure that adequate levels of service are delivered to all IT users in accordance with business priorities and at acceptable cost." The instrument for enforcing SLM is preeminently the SLA. An SLA is an agreement, preferably documented, between a service provider and the customer(s)/user(s) that defines minimum performance targets for a service and how they will be measured.[15]

### Service Level Agreement Types

Generally speaking, there are three basic types of SLAs:
- In-house
- External
- Internal

The differences between those types refer to the parties involved in the definition of the SLA:
- **In-house**—An agreement negotiated between an in-house service provider, such as an IT department, and an in-house client or department, such as marketing, finance or production. This is the most common type of SLA.
- **External**—SLA between an external service provider (third party) and an enterprise. This is the second most common type of SLA.
- **Internal**—SLA used by a service provider to measure the performance of the groups within its own enterprise.[16]

No matter what type of SLA is chosen, it should always be negotiated by an experienced and multidisciplinary team with an equal representation from both the user group and the service provider. Many user group negotiators see the negotiations as a zero-sum challenge, going for maximum service levels at a minimum cost, whereas service provider negotiators seek to get the deal at any cost to gain market share, but with minimum effort and maximum margin. Seeking a balance between these two positions is a vital but very difficult job for a solid SLA and SLM.

An SLA is a necessity between a service provider and service beneficiary because a service can only be called "bad" or "good" if this service is clearly described. Moreover, it formalizes the needs and expectations of the enterprise and serves as such as a kind of "guarantee" for both parties. In this way potential misunderstandings are reduced and a clear view is given on the priorities around the service and the service delivery. SLA are often further translated into more detailed OLAs.

## Service Level Management and COBIT

In COBIT 5 the purpose of process APO09 *Manage service agreements* is to ensure that IT services and service levels meet current and future enterprise needs.[17] This purpose statement can be refined into goals and metrics at different levels (process and IT-related) as visualized in **figure 5.6**.

To achieve the service agreement goals COBIT 5 proposes the following set of management practices:

- **APO09.01 Identify IT services.**
  Analyse business requirements and the way in which IT-enabled services and service levels support business processes. Discuss and agree on potential services and service levels with the business, and compare them with the current service portfolio to identify new or changed services or service level options.
- **APO09.02 Catalogue IT-enabled services.**
  Define and maintain one or more service catalogues for relevant target groups. Publish and maintain live IT-enabled services in the service catalogues.
- **APO09.03 Define and prepare service agreements.**
  Define and prepare service agreements based on the options in the service catalogues. Include internal operational agreements.
- **APO09.04 Monitor and report service levels.**
  Monitor service levels, report on achievements and identify trends. Provide the appropriate management information to aid performance management.
- **APO09.05 Review service agreements and contracts.**
  Conduct periodic reviews of the service agreements and revise when needed.

### Figure 5.6—APO09 Manage Service Agreements

| APO09 Manage Service Agreements | Area: Management<br>Domain: Align, Plan and Organise |
|---|---|
| **Process Description** Align IT-enabled services and service levels with enterprise needs and expectations, including identification, specification, design, publishing, agreement, and monitoring of IT services, service levels and performance indicators. ||
| **Process Purpose Statement** Ensure that IT services and service levels meet current and future enterprise needs. ||
| **The process supports the achievement of a set of primary IT-related goals:** ||
| **IT-related Goal** | **Related Metrics** |
| 07 Delivery of IT services in line with business requirements | • Number of business disruptions due to IT service incidents<br>• Percent of business stakeholders satisfied that IT service delivery meets agreed-on service levels<br>• Percent of users satisfied with the quality of IT service delivery |
| 14 Availability of reliable and useful information for decision making | • Level of business user satisfaction with quality and timeliness (or availability) of management information<br>• Number of business process incidents caused by non-availability of information<br>• Ratio and extent of erroneous business decisions where erroneous or unavailable information was a key factor |
| **Process Goals and Metrics** ||
| **Process Goal** | **Related Metrics** |
| 1. The enterprise can effectively utilise IT services as defined in a catalogue. | Number of business processes with undefined service agreements |
| 2. Service agreements reflect enterprise needs and the capabilities of IT. | • Percent of live IT services covered by service agreements<br>• Percent of customers satisfied that service delivery meets agreed-on levels |
| 3. IT services perform as stipulated in service agreements. | • Number and severity of service breaches<br>• Percent of services being monitored to service levels<br>• Percent of service targets being met |
| Source: ISACA, *COBIT 5: Enabling Processes*, USA, 2012, page 93 ||

## ENDNOTES

[1] Willcocks, Leslie; Mary Lacity; *Global Sourcing of Business and IT Services*, Palgrave Macmillan, UK, 2006
[2] IT Governance Institute, *IT Governance Global Status Report—2008*, USA, 2008
[3] ISACA, COBIT 5, USA, 2012, *www.isaca.org/cobit*
[4] Armstrong, Michael; *Human Resource Management: A Guide to Action*, Kogan Page, UK, 2008
[5] Thiadens, Theo; *Manage IT: Organizing IT Demand and Supply*, Springer, The Netherlands, 2005
[6] Information Services Procurement Library (ISPL), *projekte.fast.de/ISPL*
[7] Duening, Thomas; Rick L. Click; *Essentials of Business Process Outsourcing*, John Wiley & Sons Inc., USA, 2005
[8] Office of Government Commerce (OGC), *ITIL® V3: Service Design*, UK, 2007 and OGC, *ITIL® V3: Service Strategy*, UK, 2007
[9] *Op cit* ISACA, COBIT 5
[10] ISACA, *COBIT 5: Enabling Processes*, USA 2012
[11] *Op cit* Office of Government Commerce (OGC)
[12] O'Brian, James; George Marakas; *Introduction to Information Systems, 13th edition*, McGraw-Hill/Irwin, USA, 2007
[13] *Op cit* ISACA, COBIT 5
[14] Sturm, R.; W. Morris; M. Jander; *Foundations of Service Level Management*, USA, 2000
[15] ISACA, *www.isaca.org/glossary*
[16] *Op cit* Sturm
[17] *Op cit* ISACA, *COBIT 5: Enabling Processes*

Page intentionally left blank

# GENERAL INFORMATION

ISACA is a professional membership association composed of individuals interested in IS audit, assurance, control, security and governance. The CGEIT Certification Working Group is responsible for establishing policies for the CGEIT certification program and developing the exam.

> **Note:** As information regarding the CGEIT examination, requirements and locations and dates may change, please refer to *www.isaca.org/certification* for the most up-to-date information.

## REQUIREMENTS FOR CERTIFICATION

To earn the CGEIT designation, professionals involved with the governance of enterprise IT are required to:
1. Successfully pass the CGEIT exam
2. Submit verified evidence of five (5) or more years of experience managing, serving in an advisory or oversight role, and/or otherwise supporting the governance of the IT-related contribution to an enterprise. There are no substitutions or waivers. This experience is defined specifically by the domains and task statements described in the CGEIT job practice (*www.isaca.org/cgeitjobpractice*). A processing fee must accompany all applications.
3. Adhere to the ISACA Code of Professional Ethics
4. Agree to comply with the CGEIT continuing education policy

A minimum of one (1) year of experience relating to the definition, establishment and management of a framework for the governance of enterprise IT in alignment with the mission, vision and values of the enterprise is required. The type and extent of experience accepted is described in CGEIT domain one (1) (see Framework for the Governance of Enterprise IT at *www.isaca.org/cgeitjobpractice*).

Additional broad experience directly related to any two or more of the remaining CGEIT domains is required. The type and extent of experience accepted is described in CGEIT domains two (2) through five (5) (see the following domains at *www.isaca.org/cgeitjobpractice*). These domains are:
- Strategic Management
- Benefits Realization
- Risk Optimization
- Resource Optimization

Experience must have been gained within the 10-year period preceding the application for certification or within five (5) years from the date of initially passing the exam. Application for certification must be submitted within five (5) years from the passing date of the CGEIT exam. All experience must be verified independently with employers.

Please note that certification application decisions are not final as there is an appeal process for certification application denials. Appeals undertaken by a certification exam taker, certification applicant or by a certified individual are undertaken at the discretion and cost of the exam taker, applicant or individual. Inquiries regarding denials of certification can be sent to *certification@isaca.org*.

*It is important to note that a CGEIT candidate may choose to take the CGEIT exam prior to meeting the experience requirements.*

## DESCRIPTION OF THE EXAM

The CGEIT Certification Working Group oversees the development of the exam and ensures the currency of its content. The exam consists of 150 multiple-choice questions given over a four-hour session that cover the CGEIT job practice areas. The exam covers five GEIT subject areas created from the CGEIT job practice analysis and defines the roles and responsibilities of the professionals performing GEIT work. The job practice was developed and validated using prominent industry leaders, subject matter experts and industry practitioners.

*General Information*

## REGISTRATION FOR THE CGEIT EXAM

The CGEIT exam will be administered twice annually. Please refer to the *ISACA Exam Candidate Information Guide, www.isaca.org/examguide*, for specific dates and registration deadlines, as well as important key information for exam day. Exam registrations can be placed online at *www.isaca.org/examreg*.

## CGEIT PROGRAM ACCREDITATION RENEWED UNDER ISO/IEC 17024:2012

The American National Standards Institute (ANSI) has voted to continue the accreditation for the CISA, CISM, CGEIT and CRISC certifications under ISO/IEC 17024:2012, General Requirements for Bodies Operating Certification Systems of Persons. ANSI, a private, nonprofit organization, accredits other organizations to serve as third-party product, system and personnel certifiers.

ISO/IEC 17024 specifies the requirements to be followed by organizations certifying individual against specific requirements. ANSI describes ISO/IEC 17024 as "expected to play a prominent role in facilitating global standardization of the certification community, increasing mobility among countries, enhancing public safety, and protecting consumers."

ANSI's accreditation:
- Promotes the unique qualifications and expertise ISACA's certifications provide
- Protects the integrity of the certifications and provides legal defensibility
- Enhances consumer and public confidence in the certifications and the people who hold them
- Facilitates mobility across borders or industries

Accreditation by ANSI signifies that ISACA's procedures meet ANSI's essential requirements for openness, balance, consensus and due process. With this accreditation, ISACA anticipates that significant opportunities for CISAs, CISMs, CGEITs and CRISCs will continue to open in the United States and around the world.

## PREPARING FOR THE CGEIT EXAM

The CGEIT exam evaluates a candidate's practical knowledge of the job practice domains listed in this manual and online at *www.isaca.org/cgeitjobpractice*. That is, the exam is designed to test a candidate's knowledge, experience and judgment of the proper or preferred application of IT governance principles, methods and practices. Since the exam covers a broad spectrum of governance for enterprise IT issues, candidates are cautioned not to assume that reading CGEIT study guides and reference publications will fully prepare them for the exam. CGEIT candidates are encouraged to refer to their own experiences when studying for the exam and refer to CGEIT study guides and reference publications for further explanation of concepts or practices with which the candidate is not familiar.

No representation or warranties are made by ISACA in regard to CGEIT exam study guides, other ISACA publications, references or courses assuring candidates' passage of the exam.

## TYPES OF EXAM QUESTIONS

CGEIT exam questions are developed with the intent of measuring and testing practical knowledge and the application of general concepts and standards. All questions are multiple choice and are designed for one best answer.

Every question has a stem (question) and four options (answer choices). The candidate is asked to choose the correct or best answer from the options. The stem may be in the form of a question or incomplete statement. In some instances, a scenario may also be included. These questions normally include a description of a situation and require the candidate to answer two or more questions based on the information provided. The candidate is cautioned to read each question carefully. An exam question may require the candidate to choose the appropriate answer based on a qualifier, such as MOST or BEST. In every case, the candidate is required to ready the question carefully, eliminate known incorrect answers and then make the best choice possible. To gain a better understanding of the types of question that might appear on the exam and how these questions are developed, refer to the Item Writing Guide available at *www.isaca.org/itemwriting*

## ADMINISTRATION OF THE EXAM

ISACA has contracted with an internationally recognized testing agency. This not-for-profit corporation engages in the development and administration of credentialing exams for certification and licensing purposes. It assists ISACA in the construction, administration and scoring of the CGEIT exam.

## SITTING FOR THE EXAM

Candidates are to report to the testing site at the report time indicated on their admission ticket. NO CANDIDATE WILL BE ADMITTED TO THE TEST CENTER ONCE THE CHIEF EXAMINER BEGINS READING THE ORAL INSTRUCTIONS. Candidates who do not attend the scheduled exam date or arrive after the oral instructions have begun will not be allowed to sit for the exam and will forfeit their registration fee. To ensure that candidates arrive in time for the exam, it is recommended that candidates become familiar with the exact location of, and the best travel route to, the exam site prior to the date of the exam. Candidates can use their admission tickets only at the designated test center on the admission ticket.

To be admitted into the test site, candidates must bring a printout of the email OR a printout of the downloaded admission ticket and an acceptable form of photo identification such as a driver's license, passport or government ID. This ID must be a current and original government-issued identification that is not handwritten and that contains both the candidate's name as it appears on the admission ticket and the candidate's photograph. Candidates who do not provide an acceptable form of identification will not be allowed to sit for the exam and will forfeit their registration fee. Candidates are not to write on the admission ticket.

The following conventions should be observed when completing the exam:
- Do not bring study materials (including notes, paper, books or study guides) or scratch paper or notepads into the exam site. For further details regarding what personal belongings can (and cannot) be brought into the test site, please visit www.isaca.org/cgeitbelongings.
- Candidates are not allowed to bring any type of communication, surveillance or recording device (e.g., cellular phone, tablet, smart watch, mobile device, etc.) into the test center. If candidates are viewed with any such device during the exam administration, their exams will be voided and they will be asked to immediately leave the exam site.
- Candidates who leave the testing area without authorization or accompaniment by a test proctor will not be allowed to return to the testing room and will be subject to disqualification.
- Candidates should bring several no. 2 pencils since pencils will not be provided at the exam site.
- As exam venues vary, every attempt will be made to make the climate control comfortable at each exam venue. Candidates may want to dress to their own comfort level.
- Include your exam identification number as it appears on your admission ticket and any other requested information. Failure to do so may result in a delay or errors.
- Read the provided instructions carefully before attempting to answer questions. Skipping over these directions or reading them too quickly could result in missing important information and possibly losing credit points.
- Mark the appropriate area when indicating responses on the answer sheet. When correcting a previously answered question, fully erase a wrong answer before writing in the new one.
- Remember to answer all questions since there is no penalty for wrong answers. Grading is based solely on the number of questions answered correctly. Do not leave any question blank.
- Identify key words or phrases in the question (**MOST, BEST, FIRST**...) before selecting and recording the answer.
- The chief examiner or designate at each test center will read aloud the instructions for entering information on the answer sheet. It is imperative that candidates include their exam identification number as it appears on their admission ticket and any other requested information on their exam answer sheet. Failure to do so may result in a delay or errors.

## BUDGETING TIME

The following are time-management tips for the exam:
- It is recommended that candidates become familiar with the exact location of, and the best travel route to, the exam site prior to the date of the exam.
- Candidates should arrive at the exam testing site at the time indicated on their admission ticket. This will give the candidate time to be seated and get acclimated.
- The exam is administered over a four-hour period. This allows for a little over 1.5 minutes per question. Therefore, it is advisable that candidates pace themselves to complete the entire exam.

**General Information**

---

- Candidates are urged to record their answers on their answer sheet. No additional time will be allowed after the exam time has elapsed to transfer or record answers should candidates mark their answers in the question booklet. The exam will be scored based on the answer sheet recordings only.

## RULES AND PROCEDURES
- Candidates are asked to sign the answer sheet to protect the security of the exam and maintain the validity of the scores.
- Candidates who are discovered engaging in any kind of misconduct—including, but not limited to, giving or receiving help; using notes, papers or other aids; attempting to take the exam for someone else; using any type of communication, surveillance or recording device, including cell phones, during the exam administration; removing test materials, answer sheet or notes from the testing room; or attempting to share test questions or answers or other information contained in the exam (as such are the confidential information of ISACA)—will have their exam voided and be asked to leave the exam site. Candidates who leave the testing area without authorization or accompaniment by a test proctor will not be allowed to return to the testing room and will be subject to disqualification. Candidates who continue to write the exam after the proctor signals the end of the examination time may have their examination voided. Candidates may not access items stored in the personal belongings area until they have completed their exams. The testing agency will report all cases of misconduct to the respective ISACA Committee Working Group for Committee review in order to render any decision necessary. Sharing the confidential test items subsequent to the exam will also be considered misconduct resulting in a voided examinations score.
- Candidates may not take the exam question booklet after completion of the exam.
- Candidates are not permitted to access items stored in the personal belongings area during the exam.
- The *ISACA Exam Candidate Information Guide* includes candidate information about exam registration, dates and deadlines and provides important key candidate details for exam day administration. This publication is available online at *www.isaca.org/examguide*. We encourage exam candidates to review the information in this guide to familiarize themselves with the rules for exam day.

## GRADING THE CGEIT EXAM AND RECEIVING RESULTS
The exam consists of 150 items. Candidate scores are reported as a scaled score. A scaled score is a conversion of a candidate's raw score on an exam to a common scale. ISACA uses and reports scores on a common scale from 200 to 800. A candidate must receive a score of 450 or higher to pass the exam. A score of 450 represents a minimum consistent standard of knowledge as established by ISACA's CGEIT Certification Working Group. A candidate receiving a passing score may then apply for certification if all other requirements are met.

**Passing the exam does not grant the CGEIT designation. To become a CGEIT, each candidate must complete all requirements, including submitting an application for certification (within 5 years of passage of the exam).**

The CGEIT examination contains some questions which are included for research and analysis purposes only. These questions are not separately identified and the candidate's final score will be based only on the common scored questions.

A candidate receiving a score less than 450 is not successful and can retake the exam by registering and paying the appropriate exam fee for any future exam administration. To assist with future study, the result letter each candidate receives includes a score analysis by content area. There are no limits to the number of times a candidate can take the exam.

**Approximately eight weeks after the test date, the official exam results will be mailed to candidates.**
Additionally, with the candidate's consent during the registration process, an e-mail containing the candidates pass/fail status and score will be sent to fully paid candidates. This e-mail notification will only be sent to the address listed in the candidate's profile at the time of the initial release of the results. To ensure the confidentiality of scores, exam results will not be reported by telephone or fax. To prevent the e-mail notification from being sent to the candidate's spam folder, the candidate should add exam@isaca.org to his/her address book, whitelist or safe senders list. Once released, scores will also be available in each ISACA constituent's profile at the My ISACA>myCertifications page of the web site.

In order to become CGEIT-certified, candidates must pass the exam and must complete and submit an application (within 5 years of passage of the exam) for certification (and must receive confirmation from ISACA that the application is approved). The application is available on the ISACA web site at *www.isaca.org/cgeitapp*. Once the application is approved, the applicant will be sent confirmation of the approval. The candidate is not CGEIT-certified, and cannot use the CGEIT designation, until the candidate's application is approved. A processing fee must accompany your CGEIT application for certification.

The score report contains a subscore for each job practice domain. The subscores can be useful in identifying those areas in which the candidate may need further study before retaking the exam. Unsuccessful candidates should note that taking either a simple or weighted average of the subscores does not derive the total scaled score. Candidates receiving a failing score on the exam may request a rescoring of their answer sheet. This procedure ensures that no stray marks, multiple responses or other conditions interfered with computer scoring. Candidates should understand, however, that all scores are subjected to several quality control checks before they are reported; therefore, rescores most likely will not result in a score change. Requests for hand scoring must be made in writing to the certification department within 90 days following the release of the exam results. Requests for a hand score after the deadline date will not be processed. All requests must include a candidate's name, exam identification number and mailing address. A fee must accompany this request.

## CONFIDENTIALITY

By taking an ISACA Exam, the candidate understands and agrees that the Exam (which includes all aspects of the exam, including, without limitation, the test questions, answers, examples and other information presented or contained in the exam) belongs to ISACA and constitutes ISACA's confidential information (collectively, Confidential Information). The candidate agrees to maintain the confidentiality of all of ISACA's Confidential Information at all times and understands that any failure to maintain the confidentiality of ISACA's Confidential Information may result in disciplinary action against the candidate by ISACA or other adverse consequences, including, without limitation, nullification of the candidate's exam, loss of the candidate's credentials, and/or litigation. Specifically, the candidate understands that he/she may not, for example, discuss, publish or share any exam question(s), the candidate's answers to any questions(s) or the exam's format with anyone in any forum or media (i.e., via email, Facebook, LinkedIn or any other form of social media).

Page intentionally left blank

# GLOSSARY

**Note:** Glossary terms are provided for reference within the CGEIT Review Manual. As definitions of terms may evolve due to the changing technological environment, please see www.isaca.org/glossary for the most up-to-date terms and definitions.

## A

**Accountability**—The ability to map a given activity or event back to the responsible party

**Alignment**—A state where the enablers of governance and management of enterprise IT support the goals and strategies of the enterprise

**Architecture**—Description of the fundamental underlying design of the components of the business system, or of one element of the business system (e.g., technology), the relationships among them, and the manner in which they support enterprise objectives

**Asset**—Something of either tangible or intangible value that is worth protecting, including people, information, infrastructure, finances and reputation

**Assurance**—Pursuant to an accountable relationship between two or more parties, an IT audit and assurance professional is engaged to issue a written communication expressing a conclusion about the subject matters for which the accountable party is responsible. Assurance refers to a number of related activities designed to provide the reader or user of the report with a level of assurance or comfort over the subject matter.

> **Scope Note:** Assurance engagements could include support for audited financial statements, reviews of controls, compliance with required standards and practices, and compliance with agreements, licenses, legislation and regulation.

## B

**Balanced scorecard (BSC)**—Developed by Robert S. Kaplan and David P. Norton as a coherent set of performance measures organized into four categories that includes traditional financial measures, but adds customer, internal business process, and learning and growth perspectives

**Benchmarking**—A systematic approach to comparing enterprise performance against peers and competitors in an effort to learn the best ways of conducting business

> **Scope Note:** Examples include benchmarking of quality, logistical efficiency and other metrics.

**Benefit**—In business, an outcome whose nature and value (expressed in various ways) are considered advantageous by an enterprise

**Benefits realization**—One of the objectives of governance. The bringing about of new benefits for the enterprise, the maintenance and extension of existing forms of benefits, and the elimination of those initiatives and assets that are not creating sufficient value

**Budget**—Estimated cost and revenue amounts for a given range of periods and set of books

> **Scope Note:** There can be multiple budget versions for the same set of books.

**Business balanced scorecard**—A tool for managing organizational strategy which uses weighted measures for the areas of financial performance (lag) indicators, internal operations, customer measurements, learning and growth (lead) indicators, combined to rate the enterprise

**Business case**—Documentation of the rationale for making a business investment, used both to support a business decision on whether to proceed with the investment and as an operational tool to support management of the investment through its full economic life cycle

**Business control**—The policies, procedures, practices and organizational structures designed to provide reasonable assurance that the business objectives will be achieved and undesired events will be prevented or detected

**Business continuity plan**—A plan used by an enterprise to respond to disruption of critical business processes. Depends on the contingency plan for restoration of critical systems

**Business dependency assessment**—A process of identifying resources critical to the operation of a business process

**Business process**—An interrelated set of cross-functional activities or events that result in the delivery of a specific product or service to a customer

**Business process reengineering (BPR)**—The thorough analysis and significant redesign of business processes and management systems to establish a better performing structure, more responsive to the customer base and market conditions, while yielding material cost savings

**Business sponsor**—The individual accountable for delivering the benefits and value of an IT-enabled business investment program to the enterprise

# C

**Capability**—An aptitude, competency or resource that an enterprise may possess or require at an enterprise, business function or individual level that has the potential, or is required, to contribute to a business outcome and to create value

**Capability Maturity Model (CMM)**—Contains the essential elements of effective processes for one or more disciplines

It also describes an evolutionary improvement path from *ad hoc*, immature processes to disciplined, mature processes with improved quality and effectiveness.

**Capital expenditure/expense**—An expenditure that is recorded as an asset because it is expected to benefit more than the current period. The asset is then depreciated or amortized over the expected useful life of the asset.

**Change management**—A holistic and proactive approach to managing the transition from a current to a desired organizational state, focusing specifically on the critical human or "soft" elements of change

> **Scope Note:** Includes activities such as culture change (values, beliefs and attitudes), development of reward systems (measures and appropriate incentives), organizational design, stakeholder management, human resources (HR) policies and procedures, executive coaching, change leadership training, team building and communications planning and execution

**Chief executive officer (CEO)**—The highest ranking individual in an enterprise

**Chief financial officer (CFO)**—The individual primarily responsible for managing the financial risk of an enterprise

**Chief information officer (CIO)**—The most senior official of the enterprise who is accountable for IT advocacy, aligning IT and business strategies, and planning, resourcing and managing the delivery of IT services, information and the deployment of associated human resources

> **Scope Note:** In some cases, the CIO role has been expanded to become the chief knowledge officer (CKO) who deals in knowledge, not just information. Also see chief technology officer (CTO).

*Glossary*

**Chief technology officer (CTO)**—The individual who focuses on technical issues in an enterprise

**Scope Note:** Often viewed as synonymous with chief information officer (CIO)

**COBIT 5**—Formerly known as Control Objectives for Information and related Technology (COBIT); now used only as the acronym in its fifth iteration. A complete, internationally accepted framework for governing and managing enterprise information and technology (IT) that supports enterprise executives and management in their definition and achievement of business goals and related IT goals. COBIT describes five principles and seven enablers that support enterprises in the development, implementation, and continuous improvement and monitoring of good IT-related governance and management practices

**Scope Note:** Earlier versions of COBIT focused on control objectives related to IT processes, management and control of IT processes and GEIT Adoption and use of the COBIT framework are supported by guidance from a growing family of supporting products. (See *www.isaca.org/cobit* for more information.)

**Combined Code on Corporate Governance**—The consolidation in 1998 of the "Cadbury," "Greenbury" and "Hampel" Reports

**Scope Note:** Named after the Committee Chairs, these reports were sponsored by the UK Financial Reporting Council, the London Stock Exchange, the Confederation of British Industry, the Institute of Directors, the Consultative Committee of Accountancy Bodies, the National Association of Pension Funds and the Association of British Insurers to address the financial aspects of corporate governance, directors' remuneration and the implementation of the Cadbury and Greenbury recommendations.

**Competencies**—The strengths of an enterprise or what it does well

**Scope Note:** Can refer to the knowledge, skills and abilities of the assurance team or individuals conducting the work.

**Contingency planning**—Process of developing advance arrangements and procedures that enable an enterprise to respond to an event that could occur by chance or unforeseen circumstances

**Continuous improvement**—The goals of continuous improvement (Kaizen) include the elimination of waste, defined as "activities that add cost, but do not add value;" just-in-time (JIT) delivery; production load leveling of amounts and types; standardized work; paced moving lines; and right-sized equipment

**Scope Note:** A closer definition of the Japanese usage of Kaizen is "to take it apart and put it back together in a better way." What is taken apart is usually a process, system, product or service. Kaizen is a daily activity whose purpose goes beyond improvement. It is also a process that, when done correctly, humanizes the workplace, eliminates hard work (both mental and physical) and teaches people how to do rapid experiments using the scientific method and how to learn to see and eliminate waste in business processes.

**Control framework**—A set of fundamental controls that facilitates the discharge of business process owner responsibilities to prevent financial or information loss in an enterprise

**Control Objectives for Enterprise Governance**—A discussion document that sets out an "enterprise governance model" focusing strongly on both the enterprise business goals and the information technology enablers that facilitate good enterprise governance, published by the Information Systems Audit and Control Foundation in 1999

**Control risk**—The risk that a material error exists that would not be prevented or detected on a timely basis by the system of internal controls

**Corporate governance**—The system by which enterprises are directed and controlled. The board of directors is responsible for the governance of their enterprise. It consists of the leadership and organizational structures and processes that ensure the enterprise sustains and extends strategies and objectives.

*Glossary*

**Corporate security officer (CSO)**—Responsible for coordinating the planning, development, implementation, maintenance and monitoring of the information security program

**Critical success factor (CSF)**—The most important issues or actions for management to achieve control over and within its IT processes

**Culture**—A pattern of behaviors, beliefs, assumptions, attitudes and ways of doing things

## D

**Dashboard**—A tool for setting expectations for an enterprise at each level of responsibility and continuous monitoring of the performance against set targets

**Disaster recovery**—Activities and programs designed to return the enterprise to an acceptable condition. The ability to respond to an interruption in services by implementing a disaster recovery plan (DRP) to restore an enterprise's critical business functions

**Disaster recovery plan (DRP)**—A set of human, physical, technical and procedural resources to recover, within a defined time and cost, an activity interrupted by an emergency or disaster

**Due diligence**—The performance of those actions that are generally regarded as prudent, responsible and necessary to conduct a thorough and objective investigation, review and/or analysis

## E

**Enterprise**—A group of individuals working together for a common purpose, typically within the context of an organizational form such as a corporation, public agency, charity or trust

**Enterprise architecture (EA)**—Description of the fundamental underlying design of the components of the business system, or of one element of the business system (e.g., technology), the relationships among them, and the manner in which they support the enterprise's objectives

**Enterprise architecture (EA) for IT**—Description of the fundamental underlying design of the IT components of the business, the relationships among them, and the manner in which they support the enterprise's objectives

**Enterprise governance**—A set of responsibilities and practices exercised by the board and executive management with the goal of providing strategic direction, ensuring that objectives are achieved, ascertaining that risks are managed appropriately and verifying that the enterprise's resources are used responsibly

**Enterprise risk management (ERM)**—The discipline by which an enterprise in any industry assesses, controls, exploits, finances and monitors risk from all sources for the purpose of increasing the enterprise's short- and long-term value to its stakeholders

## G

**Good practice**—A proven activity or process that has been successfully used by multiple enterprises and has been shown to produce reliable results

**Governance**—Ensures that stakeholder needs, conditions and options are evaluated to determine balanced, agreed-on enterprise objectives to be achieved; setting direction through prioritization and decision making; and monitoring performance and compliance against agreed-on direction and objectives

> **Scope Note:** Conditions can include the cost of capital, foreign exchange rates, etc. Options can include shifting manufacturing to other locations, subcontracting portions of the enterprise to third parties, selecting a product mix from many available choices, etc.

**Governance enabler**—Something (tangible or intangible) that assists in the realization of effective governance

**Governance of enterprise IT**—A governance view that ensures that information and related technology support and enable the enterprise strategy and the achievement of enterprise objectives; this also includes the functional governance of IT, i.e., ensuring that IT capabilities are provided efficiently and effectively.

**Governance framework**—A framework is a basic conceptual structure used to solve or address complex issues. An enabler of governance. A set of concepts, assumptions and practices that define how something can be approached or understood, the relationships amongst the entities involved, the roles of those involved, and the boundaries (what is and is not included in the governance system).

## I

**Impact analysis**—A study to prioritize the criticality of information resources for the enterprise based on costs (or consequences) of adverse events. In an impact analysis, threats to assets are identified and potential business losses determined for different time periods. This assessment is used to justify the extent of safeguards that are required and recovery time frames. This analysis is the basis for establishing the recovery strategy.

**Impact assessment**—A review of the possible consequences of a risk

> **Scope Note:** See Impact analysis.

**Information**—An asset that, like other important business assets, is essential to an enterprise's business. It can exist in many forms. It can be printed or written on paper, stored electronically, transmitted by post or by using electronic means, shown on films, or spoken in conversation.

**Information security**—Ensures that within the enterprise, information is protected against disclosure to unauthorized users (confidentiality), improper modification (integrity), and non-access when required (availability)

**Information security governance**—The set of responsibilities and practices exercised by the board and executive management with the goal of providing strategic direction, ensuring that objectives are achieved, ascertaining that risks are managed appropriately and verifying that the enterprise's resources are used responsibly

**Information systems (IS)**—The combination of strategic, managerial and operational activities involved in gathering, processing, storing, distributing and using information and its related technologies

> **Scope Note:** Information systems are distinct from information technology (IT) in that an information system has an IT component that interacts with the process components.

**Information technology (IT)**—The hardware, software, communication and other facilities used to input, store, process, transmit and output data in whatever form

**Internal rate of return (IRR)**—The discount rate that equates an investment cost with its projected earnings

> **Scope Note:** When discounted at the IRR, the present value of the cash outflow will equal the present value of the cash inflow. The IRR and net present value (NPV) are measures of the expected profitability of an investment project.

**Investment and services board (ISB)**—A management structure primarily accountable for managing the enterprise's portfolios of investment programs and existing/current services and, in so doing, managing the level of overall funding to provide the necessary balance between enterprisewide and specific line-of-business needs

**Investment portfolio**—The collection of investments being considered and/or being made

*Glossary*

**IT governance**—The responsibility of executives and the board of directors; consists of the leadership, organizational structures and processes that ensure that the enterprise's IT sustains and extends the enterprise's strategies and objectives

**IT governance framework**—A model that integrates a set of guidelines, policies and methods that represent the organizational approach to IT governance

> **Scope Note:** Per COBIT 4.1, IT governance is the responsibility of the board of directors and executive management. It is an integral part of institutional governance and consists of the leadership and organizational structures and processes that ensure that the enterprise's IT sustains and extends the enterprise's strategy and objectives.

**IT infrastructure**—The set of hardware, software and facilities that integrates an enterprise's IT assets

**IT investment dashboard**—A tool for setting expectations for an enterprise at each level and continuous monitoring of the performance against set targets for expenditures on, and returns from, IT-enabled investment projects in terms of business value

**IT service**—The day-to-day provision to customers of IT infrastructure and applications and support for their use—e.g., service desk, equipment supply and moves, and security authorizations

**IT steering committee**—An executive-management-level committee that assists the executive in the delivery of the IT strategy, oversees day-to-day management of IT service delivery and IT projects, and focuses on implementation aspects

**IT strategic plan**—A long-term plan (i.e., three- to five-year horizon) in which business and IT management cooperatively describe how IT resources will contribute to the enterprise's strategic objectives (goals)

**IT strategy committee**—A committee at the level of the board of directors to ensure that the board is involved in major IT matters and decisions

> **Scope Note:** The committee is primarily accountable for managing the portfolios of IT-enabled investments, IT services and other IT resources. The committee is the owner of the portfolio.

**IT tactical plan**—A medium-term plan (i.e., six- to 18-month horizon) that ranslates the IT strategic plan direction into required initiatives, resource requirements and ways in which resources and benefits will be monitored and managed

# K

**Key goal indicator (KGI)**—A measure that tells management, after the fact, whether an IT process has achieved its business requirements, usually expressed in terms of information criteria

**Key management practice**—Management practices that are required to successfully execute business processes

**Key performance indicator (KPI)**—A measure that determines how well the process is performing in enabling the goal to be reached

> **Scope Note:** A lead indicator of whether a goal will likely be reached, and a good indicator of capabilities, practices and skills. It measures an activity goal, which is an action that the process owner must take to achieve effective process performance.

**Key risk indicator (KRI)**—A subset of risk indicators that are highly relevant and possess a high probability of predicting or indicating important risk

## M

**Management**—Plans, builds, runs and monitors activities in alignment with the direction set by the governance body to achieve the enterprise objectives.

**Maturity**—In business, indicates the degree of reliability or dependency that the business can place on a process achieving the desired goals or objectives

**Maturity model**—See Capability maturity model (CMM)

**Metric**—A quantifiable entity that allows the measurement of the achievement of a process goal

   **Scope Note:** Metrics should be SMART—specific, measurable, attainable, realistic and timely. Complete metric guidance defines the unit used, measurement frequency, ideal target value (if appropriate) and also the procedure to carry out the measurement and the procedure for the interpretation of the assessment.

## N

**Net present value (NPV)**—Calculated by using an after-tax discount rate of an investment and a series of expected incremental cash outflows (the initial investment and operational costs) and cash inflows (cost savings or revenues) that occur at regular periods during the life cycle of the investment

   **Scope Note:** To arrive at a fair NPV calculation, cash inflows accrued by the business up to about five years after project deployment should also be taken into account.

**Net return**—The revenue that a project or business makes after tax and other deductions; often classified as net profit

## O

**Operational risk**—The most important types of operational risk involve breakdowns in internal controls and corporate governance. Such breakdowns can lead to financial losses through error, fraud or failure to perform in a timely manner, or cause the interests of the bank to be compromised in some other way, for example, by its dealers, lending officers or other staff exceeding their authority or conducting business in an unethical or risky manner. Other aspects of operational risk include major failure of information technology systems or events such as security problems or other disasters.

**Outcome measure**—Represents the consequences of actions previously taken; often referred to as a lag indicator

   **Scope Note:** Outcome measures frequently focus on results at the end of a time period and characterize historical performance. They are also referred to as a key goal indicator (KGI) and used to indicate whether goals have been met. These can be measured only after the fact and, therefore, are called "lag indicators."

**Outsourcing**—A formal agreement with a third party to perform IS or other business functions for an enterprise

**Organizational structure**—An enabler of governance and of management. Includes the enterprise and its structures, hierarchies and dependencies.

   **Scope Note:** Example: Steering committee

## P

**Payback period**—The length of time needed to recoup the cost of capital investment

   **Scope Note:** Financial amounts in the payback formula are not discounted. Note that the payback period does not take into account cash flows after the payback period and therefore is not a measure of the profitability of an investment project. The scope of the internal rate of return (IRR), net present value (NPV) and payback period is the useful economic life of the project up to a maximum of five years.

*Glossary*

**Performance**—In IT, the actual implementation or achievement of a process

**Performance driver**—A measure that is considered the "driver" of a lag indicator. It can be measured before the outcome is clear and, therefore, is called a "lead indicator."

   Scope Note: There is an assumed relationship between the two that suggests that improved performance in a leading indicator will drive better performance in the lagging indicator. They are also referred to as key performance indicators (KPI) and are used to indicate whether goals are likely to be met.

**Performance indicators**—A set of metrics designed to measure the extent to which performance objectives are being achieved on an ongoing basis

   Scope Note: Performance indicators can include service level agreements (SLA), critical success factors (CSF), customer satisfaction ratings, internal or external benchmarks, industry best practices and international standards.

**Performance management**—In IT, the ability to manage any type of measurement, including employee, team, process, operational or financial measurements. The term connotes closed-loop control and regular monitoring of the measurement.

**Performance testing**—Comparing the system's performance to other equivalent systems, using "well-defined benchmarks

**Policy**—Overall intention and direction as formally expressed by management

**Portfolio**—A grouping of "objects of interest" (investment programs, IT services, IT projects, other IT assets or resources) managed and monitored to optimize business value. The investment portfolio is of primary interest to Val IT. The IT service, project, asset and other resource portfolios are of primary interest to COBIT.

**Practices**—An enabler of governance and of management. The manner in which processes are performed. Example: committee reporting practices guideline.

**Principle**—An enabler of governance and of management. Comprises the values and fundamental assumptions held by the enterprise, the beliefs that guide and put boundaries around the enterprise's decision making, communication within and outside the enterprise, and stewardship--caring for assets owned by another.

**Process**—Generally, a collection of activities influenced by the enterprise's policies and procedures that takes inputs from a number of sources, (including other processes), manipulates the inputs and produces outputs

   Scope Note: Processes have clear business reasons for existing, accountable owners, clear roles and responsibilities around the execution of the process, and the means to measure performance.

**Process maturity model**—A subjective assessment technique derived from the Software Engineering Institute (SEI) capability maturity model integration (CMMI) concepts and developed as a COBIT management tool. It provides management with a profile of how well developed the IT management processes are.

   Scope Note: It enables management to easily place itself on a scale and appreciate what is required if improved performance is needed. It is used to set targets, raise awareness, capture broad consensus, identify improvements and positively motivate change.

**Program**—A structured grouping of interdependent projects that is both necessary and sufficient to achieve a desired business outcome and create value. These projects could include, but are not limited to, changes in the nature of the business, business processes and the work performed by people as well as the competencies required to carry out the work, the enabling technology and the organizational structure.

**Glossary**

**Project**—A structured set of activities concerned with delivering a defined capability (that is necessary, but not sufficient, to achieve a required business outcome) to the enterprise, based on an agreed-on schedule and budget

**Project portfolio**—The set of projects owned by a company

> **Scope Note:** It usually includes the main guidelines relative to each project, including objectives, costs, time lines and other information specific to the project.

## Q

**Quality assurance (QA)**—A planned and systematic pattern of all actions necessary to provide adequate confidence that an item or product conforms to established technical requirements. (ISO/IEC 24765)

## R

**RACI Chart**—Illustrates who is Responsible, Accountable, Consulted and Informed within an organizational framework

**Reengineering**—A process involving the extraction of components from existing systems and restructuring these components to develop new systems or to enhance the efficiency of existing systems

> **Scope Note:** Existing software systems can be modernized to prolong their functionality. An example is a software code translator that can take an existing hierarchical database system and transpose it to a relational database system. Computer-aided software engineering (CASE) includes a source code reengineering feature.

**Release to production (RTP)**—Term used in software and application development to mean the "final version" of a particular IT product that is released for production

**Reputation risk**—The current and prospective effect on earnings and capital arising from negative public opinion

> **Scope Note:** Reputation risk affects a bank's ability to establish new relationships or services, or continue servicing existing relationships. It may expose the bank to litigation, financial loss or a decline in its customer base. A bank's reputation can be damaged by Internet banking services that are executed poorly or otherwise alienate customers and the public. An Internet bank has a greater reputation risk as compared to a traditional brick-and-mortar bank since it is easier for its customers to leave and go to a different Internet bank and since it cannot discuss any problems in person with the customer.

**Responsible**—In a Responsible, Accountable, Consulted, Informed (RACI) chart, refers to the person who must ensure that activities are completed successfully

**Resource optimization**—One of the governance objectives. Involves effective, efficient and responsible use of all resources—human, financial, equipment, facilities, etc.

**Return on investment (ROI)**—A measure of operating performance and efficiency, computed in its simplest form by dividing net income by the total investment over the period being considered

**Risk**—The combination of the probability of an event and its consequence. (ISO/IEC73)

**Risk analysis**—The initial steps of risk management: analyzing the value of assets to the business, identifying threats to those assets and evaluating how vulnerable each asset is to those threats
Scope Note: It often involves an evaluation of the probable frequency of a particular event as well as the probable impact of that event.

**Risk appetite**—The amount of risk, on a broad level, that an entity is willing to accept in pursuit of its mission

*Glossary*

**Risk assessment**—A process used to identify and evaluate risks and their potential effects

> **Scope Note:** Includes assessing the critical functions necessary for an enterprise to continue business operations, defining the controls in place to reduce organization exposure and evaluating the cost for such controls. Risk analysis often involves an evaluation of the probabilities of a particular event.

**Risk management**—One of the governance objectives. Entails recognizing risk; assessing the impact and likelihood of that risk; and developing strategies, such as avoiding the risk, reducing the negative effect of the risk and/or transferring the risk, to manage it within the context of the enterprise's risk appetite.

**Risk mitigation**—The management of risk through the use of countermeasures and controls

**Risk tolerance**—The acceptable level of variation that management is willing to allow for any particular risk as the enterprise pursues its objectives

**Risk transfer**—The process of assigning risk to another enterprise, usually through the purchase of an insurance policy or by outsourcing the service

**Risk treatment**—The process of selection and implementation of measures to modify risk (ISO/IEC Guide 73:2002)

## S

**Segregation/separation of duties (SoD)**—A basic internal control that prevents or detects errors and irregularities by assigning to separate individuals the responsibility for initiating and recording transactions and for the custody of assets

> **Scope Note:** Segregation/separation of duties is commonly used in large IT organizations so that no single person is in a position to introduce fraudulent or malicious code without detection.

**Service level agreement (SLA)**—An agreement, preferably documented, between a service provider and the customer(s)/user(s) that defines minimum performance targets for a service and how they will be measured

**Service shell**—In IT service and resource management, the term used for services that support an IT product or IT service

**Service pit**—The term used to describe the object of the service shell, the IT product or the IT service

**SMART**—Specific, measurable, attainable, realistic and timely, generally used to describe appropriately set goals

**Stage-gate**—A point in time when a program is reviewed and a decision is made to commit expenditures to the next set of activities on a program or project, to stop the work altogether, or to put a hold on execution of further work

**Stakeholder**—Anyone who has a responsibility for, an expectation from or some other interest in the enterprise.

**Standard**—A mandatory requirement, code of practice or specification approved by a recognized external standards organization, such as the International Organization for Standardization (ISO)

**Strategic planning**—The process of deciding on the enterprise's objectives, on changes in these objectives, and the policies to govern their acquisition and use. Strengths, weaknesses, opportunities and threats (SWOT)—A combination of an organizational audit listing the enterprise's strengths and weaknesses and an environmental scan or analysis of external opportunities and threats

# Glossary

## T

**Threat**—Anything (e.g., object, substance, human) that is capable of acting against an asset in a manner that can result in harm

> **Scope Note:** A potential cause of an unwanted incident (ISO/IEC 13335)

**Threat event**—Any event during which a threat element/actor acts against an asset in a manner that has the potential to directly result in harm

**Transparency**—Refers to an enterprise's openness about its activities and is based on the following concepts:
- How the mechanism functions is clear to those who are affected by or want to challenge governance decisions.
- A common vocabulary has been established.
- Relevant information is readily available.

> **Scope Note:** Transparency and stakeholder trust are directly related; the more transparency in the governance process, the more confidence in the governance.

## V

**Value**—The relative worth or importance of an investment for an enterprise, as perceived by its key stakeholders, expressed as total life cycle benefits net of related costs, adjusted for risk and (in the case of financial value) the time value of money

**Value creation**—The main governance objective of an enterprise, achieved when the three underlying objectives (benefits realization, risk optimization and resource optimization) are all balanced

**Vulnerability**—A weakness in the design, implementation, operation or internal control of a process that could expose the system to adverse threats from threat events

**Vulnerability analysis**—A process of identifying and classifying vulnerabilities

Page intentionally left blank

# Index

A slash (/) indicates that the terms are synonymous within this manual.

"See also" indicates that the terms are related or relevant to one another.

## A

Acceptance/Risk acceptance, 179
Accountability, 30-32, 117-118
Acquisition process, 206, 220
Agility, 65-66
Agility loops, 66
Alignment, 54-55
Applications, 17, 24, 215
Architecture, 23-24, 72-74
AS/NZS 4360:2004, See Australian/New Zealand Standard, (AS/NZS) 4360-2004-Risk Management
Asset management, 203
Australian/New Zealand Standard (AS/NZS) 4360:2004-Risk Management, 155
Availability management, 106, 162, 213-214

## B

Balanced scorecard (BSC), 32, 34-35, 57-58, 69-72, 103, 184
Barriers to the achievement of strategic alignment, 67
    Expression, 67
    Implementation, 67
    Specification, 67
Basel III, 148, 160
Baselining, 203
Basic transaction processing systems, 66
BCG, See Boston Consulting Group
Benchmarking, 203
Benefits, 99, 124-125
Benefits Dependency Network, 78
Benefits realization, 93-94
Big data, 169-170
BiSL, See Business Information Services Library
Boston Consulting Group (BCG), 21
BOT, See Build-operate-transfer
BPO, See Business process outsourcing
BPR, See Business process reengineering
Breakeven analysis, 101
Bring your own device (BYOD), 171
BSC, See Balanced scorecard
Build-operate-transfer (BOT), 208
Business architecture, 72-73
Business case, 80, 113-116, 181
Business continuity plan, 162-164, 210
Business contribution/enterprise contribution, 34-35, 71, 103

Business drivers, 14-15, 29, 123-124
Business governance, 9-10
Business Information Services Library (BiSL), 12
Business process outsourcing (BPO), 208-209
Business process reengineering (BPR), 166-167
Business risk, 148
Business resiliency, 116, 161-162
Business strategy, 55, 131-132
Business unit architecture, 23, 73
BYOD, See Bring your own device

## C

Capability Maturity Model (CMM), 12
Capability Maturity Model Integration (CMMI), 12, 14
Capacity management, 106, 117, 210-213
Capacity Management Information System (CMIS), 212-213
CBA, See Cost-benefit analysis
Change enablement, 28-29
Change management, 39
CI, See Configuration item
Cloud computing, 168-169
CMIS, See Capacity Management Information System
CMM, See Capability Maturity Model
CMMI, See Capability Maturity Model Integration
CMS, See Configuration Management System
COBIT, 11-14, 17-18, 36-38, 56-65, 73-77, 95-100, 111, 153, 162-163, 176-177, 204, 210-212, 215-222
Committee of Sponsoring Organizations of the Treadway Commission (COSO), 153-154
Committee of Sponsoring Organizations of the Treadway Commission Enterprise Risk Management (COSO) Enterprise Management (ERM), 153-154
Communication plan, 37, 39
Communication strategy, 36-37, 39
Compliance objectives, 160
Component availability, 214
Component utilization data, 213
Configuration item (CI), 214
Configuration Management System (CMS), 212
Conformance, 9, 10, 64
Continuous improvement cycle, 4, 43, 118
Continuous service improvement (CSI), 100, 111
Contract management, 199-200, 203
Control risk, 165-166, 233
Corporate contribution, 71
Corporate governance, 9-10, 16, 153-156
COSO, See Committee of Sponsoring Organizations of the Treadway Commission
COSO ERM, See Committee of Sponsoring Organizations of the Treadway Commission (COSO) Enterprise Risk Management (ERM)
Cost optimization, 116

# Index

Cost-benefit analysis (CBA), 100-102, 134
Critical business processes, 66, 162-163
Critical success factor (CSF), 111, 159
Cross-enterprise risk, 153
CSF, See Critical success factor
CSI, See Continuous service improvement
Culture, 17, 134, 185, 215
Current capabilities, 44, 210
Customer orientation, 71

## D

Data, 23-24, 73, 167-170, 173, 208, 212, 221
Data architecture, 23-24, 73
Decision support, risk analytics and reporting, 173
Decision-making entities, 17, 24, 215
Define-Measure-Analyze-Improve-Control (DMAIC), 109, 182
Delivery systems architecture, 23-24, 52, 72-73
Demand management, 199-200, 210
Design risk, 166-167
Detailed candidate program business case, 119, 124, 133
Detection risk, 165-166
DMAIC, See Define-Measure-Analyze-Improve-Control
Due diligence, 203

## E

EA, See Enterprise architecture
Economies of scale, 214
EDM, See Evaluate, direct, monitor
Emerging technologies, 35, 71, 74-76, 103, 115, 129, 168
Enabler, 19, 41, 60, 216-217
Enterprise architecture (EA), 15, 22-24, 72-74
Enterprise financial planning, 119, 122, 130-131, 133
Enterprise goals, 57, 111
Enterprise governance, 9, 10, 131
Enterprise risk management (ERM), 26, 140-144, 150-154, 158-161, 176, 178, 188, 190
ERM, See Enterprise risk management
Evaluate, direct, monitor (EDM), 13-14, 40, 48, 50, 61, 104, 196, 198
Evaluations, 76, 192
Expression barriers, 67

## F

FEAF, See Federal Enterprise Architecture Framework
Federal Enterprise Architecture Framework (FEAF), 23
Financial transparency, 36, 57, 59-60
Forest of frameworks, 11
Framework, 9, 11-12, 118-119, 151-158, 161
Full life-cycle costs and benefits, 113, 119, 124, 132-133, 218
Future orientation, 71

## G

Governance, 9-11, 27-28, 31, 40, 121-122, 191, 203-204
Governance monitoring, 32, 118-119, 122, 130
Governance of key assets, 10
Governance organization, 204
Governance processes, 203

## H

HR philosophies, 205
HR policies, 205
HR practices, 205
HR processes, 205
HR programs, 205
HR strategies, 205
HRM, See Human resource management
Human resource management (HRM), 205

## I

IAASB, See International Auditing and Assurance Standards Board
IFAC, See International Federation of Accountants
Implementation barriers, 67
Implementation risk, 166-167
Information, 17, 24, 168-170, 215-221
Information architecture, 23-24, 73
Information Services Procurement Library (ISPL), 12, 207, 223
Information systems architecture, 23-24, 73
Infrastructure, 12, 77, 83, 105, 134-136, 150, 168
    Technology/IT, 24, 27, 55
    Organizational, 54-55
Infrastructure investments, 77, 135
Inherent risk, 165
Initial program concept business case, 133
Innovation, 53, 61, 74-76, 109
Intangible benefits, 33-34, 102
Internal rate of return (IRR), 33, 77, 98, 101
International Auditing and Assurance Standards Board (IAASB), 40, 46
International Federation of Accountants (IFAC), 9, 45-46
International Organization for Standardization (ISO), 11-12, 100, 104-107, 154-155, 157, 164
Interoperability, 195, 214-215
Investment and services board (ISB), 129
Investment management, 91, 93, 127
Investment portfolio, 98
IRR, See Internal rate of return
ISB, See Investment and services board
ISO 31000, 142-144, 154-155, 161
ISO, See International Organization for Standardization
ISPL, See Information Services Procurement Library
Issue logs, 192

# Index

IT assurance, 15
IT balanced scorecard (IT BSC), 17, 32, 34-35, 58, 70-71, 103, 182, 184
IT BSC, See IT balanced scorecard
IT demand, 15, 121
IT industry good practices, 11, 25, 34, 105-106, 203, 218
IT Infrastructure Library (ITIL), 12, 100, 105-107, 111-112, 117, 150, 157, 168, 212
IT investment/IT-enabled investment, 78-79, 95, 100-102, 126
IT performance management, 32-34
IT performance measurement, 112
IT-related goals, 17-18, 22, 44, 56, 58-60, 64-65, 68, 73, 111, 162, 215, 222
IT-related risk, 14, 25, 40, 140, 142, 148, 151-152, 160-161, 176
IT resources, 15, 67, 124, 141, 143, 202, 204, 206-207, 212, 214
IT risk, 14, 25-26, 148-149, 151-152, 160-161, 165, 169, 174, 176-179, 184-185, 187-191
IT risk management, 148, 151, 152, 160, 161, 176, 179, 187, 188, 190, 191
IT SCM, See IT service continuity management
IT service, 15, 38, 55, 100, 106-107, 157, 160, 163, 183, 202, 206-207, 210, 213-214, 222
IT service continuity management (ITSCM), 107, 214
IT service delivery practices, 106
IT service management, 87, 100, 157, 210
IT strategy, 19, 21-22, 27, 38, 54-56, 67-68, 72-73, 76, 81-83, 96, 109, 120-121, 129
IT strategy committee, 67-68, 129
IT supply, 206
ITIL, See IT Infrastructure Library

## K

Key performance indicator (KPI), 111, 185
Key risk indicator (KRI), 189-190, 236
Knowledge, 12, 39-40, 56, 58, 60-65, 79, 83, 109, 139-140, 143-144
Kotter's Implementation Life Cycle, 29
KPI, See Key performance indicator
KRI, See Key risk indicator

## L

Level of risk/Risk level, 12, 36, 127, 149, 166, 174-175, 184, 188, 209
Lewin/Schein's Change Theory—Unfreeze–Change–Refreeze, 30
Life cycle approach, 43
Life cycle management of IT investments, 93-94
Locked-down operations, 173
Loss data, 142, 177, 185-186

## M

Management, 31-32, 68-69, 72-73, 96-100, 112, 116, 159, 167, 191-192
Management of Risk (M_o_R) Framework, 12, 148-149, 155-157
Maturity models, 87, 89, 102, 118, 151
Mean time between failures (MTBF), 162, 214
Mean time between service incidents (MTBSI), 214
Mean time to restore service (MTRS), 214
Mitigate risk/Risk mitigation, 98, 153, 175, 179-180, 184, 188, 190
M_o_R, See Management of Risk Framework
MTBF, See Mean time between failures
MTBSI, See Mean time between service incidents
MTRS, See Mean time to restore service
Multisourcing, 202

## N

Nearshoring, 209
Net present value (NPV) analysis, 33, 77, 98, 101
NPV, See Net present value analysis

## O

OCTAVE<sup>SM</sup>, See Operationally Critical Threat, Asset and Vulnerability Evaluation<sup>SM</sup>
Offshoring, 117, 195, 201, 204
Onshoring, 208
Operation/Rollout risk, 166-167
Operational capability, 80
Operational excellence, 71
Operational IT portfolios, 119, 124, 128, 133
Operational risk, 148, 185
Operationally Critical Threat, Asset and Vulnerability Evaluation<sup>SM</sup> (OCTAVE<sup>SM</sup>), 156-157, 193
Operations objectives, 159
Organizational structure, 17-18, 24-26, 32, 60, 94, 113, 117, 128-129, 170, 175, 215
Outcome measures, 34, 71, 111, 189
Outsourcing responsibilities, 209
Outsourcing/IT outsourcing, 117, 202

## P

Pain points and trigger events, 14-16, 29, 170
Patterns of business activity (PBA), 210
Payback period, 101
PBA, See Patterns of business activity
PDCA, See Plan-Do-Check-Act
People, 17, 18, 82, 167, 215
Performance drivers, 34, 71
Performance management, 203
Performance measures, 33, 86, 89, 100, 102, 182, 211

Plan-Do-Check-Act (PDCA), 23, 106, 110
Plan, build, run, monitor, 30, 104
Policy, 203
Portfolio categorization, 77, 97, 130
Portfolio characteristics, 131, 132
Portfolio management, 32, 77, 95-100, 118-119, 127-128, 130-133
Portfolio management principles, 95-100
Portfolio types, 110, 130-131
Practices, 11, 34-35, 67-68, 127, 129-132, 176-177, 203-204
Principles, 17, 22, 79, 126-127, 154-155, 183-184, 215
Probabilistic risk assessment, 178
Process, 38, 61-64, 82, 115-116, 131-132, 162-163, 175-177, 222
Process capability model, 104
Process change, 25
Process goal, 60, 65, 111, 163, 189
Process improvement, 28, 91, 104-105, 109, 117, 178, 182
Process metrics, 111
Process purpose, 104, 222
Procurement process, 207
Program management, 15, 78-19, 91, 93-94, 121, 150
Program plan, 94, 113-114, 119, 133-134
Program risk, 149, 160
Project management, 26-27, 40, 44, 82, 108, 149-151

## R

RACI chart, 26, 32
Related objectives, 159
Relationship management, 203
Reporting objectives, 153, 159
Residual risk, 165-166
Resource optimization, 239
Responsibility, 168, 193
Results Chain™ technique, 78
Return on investment (ROI), 32-34, 76, 89, 92, 96-98, 101-102, 167
Risk analysis, 37, 177, 187-188, 239-240
Risk and concerns with cloud computing, 78
Risk appetite, 142, 174-176
Risk assessment, 96, 155, 193
Risk audits, 193
Risk avoidance, 179
Risk communication, 177, 185-188
Risk culture, 187-188
Risk hierarchy, 148, 160
Risk IT framework, 151-152, 161, 179, 189
Risk management frameworks, 140-141, 143-144, 150-151, 157, 160-161, 169
Risk management standards, 157
Risk mitigation, 179, 240

Risk monitoring, 143, 190
Risk reporting, 36, 186-188
Risk response, 152
Risk transfer, 240
ROI, See Return on investment
Roles and responsibilities, 204
Root cause, 15, 110, 182, 185, 187, 189, 202

## S

Scenarios, 136, 166
Scope reviews, 204
Scorecards, 35
SDLC, See System development life cycle
SEI, See Software Engineering Institute
Sensitivity, 189
Service data, 212
Service knowledge management system (SKMS), 212
Service level agreement (SLA), 183, 198, 211, 221, 238, 240
Service level management (SLM), 107, 181-184, 206-207, 212, 221
Service measurement, 102
Single point of failure (SPoF), 214
Six Sigma, 12, 109, 178, 181-183
SKMS, See Service knowledge management system
SLA, See Service level agreement
SLM, See Service level management
SMART, See Specific, measurable, attainable, realistic and timely
Social media, 169
Software Engineering Institute (SEI), 12, 156
SPC, See Statistical process control
Specific, measurable, attainable, realistic and timely (SMART), 112
Specification barriers, 67
SPoF, See Single point of failure
Stage-gates/Stage-gating, 98, 128, 130-131, 134
Stakeholder needs, 32, 56-57
Stakeholders, 25, 36, 187-188, 209, 216-217
Standardization, 106, 155, 193, 195, 215, 240
Standards, 11-12, 14, 22-23, 42, 45-46, 72, 83, 105, 151, 157, 161, 164, 166, 168, 193, 226
Statistical process control (SPC), 109, 164
Strategic fit, 54, 82
Strategic investments, 77
Strategic objectives, 159
Strategic risk, 148
Strengths, weaknesses, opportunities, threats (SWOT) analysis, 19, 20, 75
SWOT analysis, See Strengths, weaknesses, opportunities, threats (SWOT) analysis
System development life cycle (SDLC), 87, 105, 108

## T

Target investment mix, 95-96, 119, 124, 133-134
Technical capability, 80
Technology architecture, See Delivery systems architecture
The Open Group Architecture Framework (TOGAF), 12, 23
TOGAF, See The Open Group Architecture Framework
Total quality management (TQM), 12, 45, 110, 182
TQM, See Total quality management
Training, 82, 195
Transactional investments, 77, 135
Transfer risk/Risk transfer, 180

## V

Value delivery practices, 126-127
Value governance, 118, 127, 131
Value of information, 219
Value management, 78, 100, 118-125, 127, 129-134

Page intentionally left blank

# EVALUATION

ISACA continuously monitors the swift and profound professional, technological and environmental advances affecting the IT governance profession. Recognizing these rapid advances, the *CGEIT® Review Manual* is updated annually.

To assist ISACA in keeping abreast of these advances, please take a moment to evaluate the *CGEIT® Review Manual 7th Edition*. Such feedback is valuable to fully serve the profession and future CGEIT exam registrants.

To complete the evaluation on the web site, please go to *www.isaca.org/studyaidsevaluation*.

Thank you for your support and assistance.

# READY FOR YOUR CGEIT EXAM?
# LET ISACA HELP YOU GET PREPARED.

Successful Certified in the Governance of Enterprise IT® (CGEIT®) exam candidates know the importance of properly preparing for the challenging CGEIT exam. That is why they turn to ISACA's study resources and review courses—for the knowledge and expertise necessary to earn a CGEIT certification.

**CGEIT book resources:**
- CGEIT Review Manual 7th Edition
- CGEIT Review Questions, Answers & Explanations Manual 4th Edition

**CGEIT review course:**
- Chapter-sponsored Review Courses (*www.isaca.org/cgeitreview*)

To learn more about ISACA's certification exam prep materials, visit *www.isaca.org/bookstore*.